The Sciences of Change

Who are you... *really*?

Ian Clarke

Copyright © Deilightful Media Ltd 2024
All Rights Reserved.

ISBN 979-8-89415-260-8
Deilightful DEIQ01 P-0000000-0-2

Legal Statements

This book has been published with all efforts taken to minimise human error and ensure the material factual correctness of all meaningful statements it makes. The author Ian Clarke and originating publisher Deilightful Media have assumed full accountability thereto for remedying anything found to be (and subsequently flagged to us) verifiable negligence, intentional deception, unfair defamation or legally unsubstantiable character assassination.

However all re-publisher(s), distributer(s) and/or reseller(s) – including here, Notion Press - do not and cannot assume thus hereby disclaim any liability to any party for any loss, damage, or disruption caused by genuine errors or accidental omissions, whether such errors or omissions result from negligence, accident, or any other cause.

While every effort has been made to avoid human error or omission, the inevitability of such things, and in a work as wide reaching as this publication, is scientifically inevitable. Thus it is being sold on the condition and understanding that (A) all extraordinary claims and material statements are fully evidenced in this book, or else full evidence is available for free on request to legal@deilightful.com. And (B) that neither the author nor the publishers, republishers, printers, distributors or resellers can be held financially liable in any manner to any person by reason of any genuine mistake or authentic omission in this publication or for any action taken or omitted to be taken or advice rendered or accepted on the basis of this work.

The Deilightful Group welcomes all reasoned challenge. After all, those who do not appreciate authentic feedback have no intention of ever becoming more than nothing. So notwithstanding these legal provisions, please bring all genuine concerns to us (over any of our claims/discoveries) you truly believe to be unsafe, by emailing legal@deilightful.com. Or you can post such challenges publicly via our Letters of Love web app (www.deilightconsulting.com/loveletters). That includes:

- For any printing or binding defect, the original publisher will be liable only to replace the defective copy with another copy if/when available. For replacement requests: legal@deilightful.com.

- For any claimed and subsequently substantiated (by evidence on your part, or insufficient lack thereof on our part) material misrepresentation of truth – particularly those dealing harm to others – we will promptly publish any retractions, amendments, clarifications and/or sincere apologies to our website www.deilightfulmedia.com/tsoc. Plus any such changes that arise will be duly incorporated into a subsequent third edition (publishing date yet to be determined).

The Sciences of Change: Who are you really?
Second Edition 6x9 Paperback

Format, Publishing, Reference, Copyright and Licensing

Book designed and originally published in the United
Kingdom 2023-2024 by Deilightful Media.
www.deilightfulmedia.com | create@deilightful.com

This edition and format was first published in India in 2024 by Notion
Press 7 Red Cross Rd, Egmore, Chennai, Tamil Nadu 600008, India
www.notionpress.com | publish@notionpress.com

All facilitating scientific breakthroughs, data analytics, cause and
effect isolation and change plan were made and applied by Deilight
Consulting (UK) Ltd. Further changes can be requested at
www.deilightconsulting.com | change@deilightful.com

Deilightful Media and Deilight Consulting (UK) Ltd are both subsidiaries of the
Deilightful Group, who instructed the publication of this book (and provided the
requisite potential) as an official response to a formal request for urgent life
sustaining assistance received from the Meritocratic Republic of Humanity.

This book commits as all powerful and everlasting the QD1.0 presence of
mind of the Deilightful Group, whose true nature, meaningful deeds,
authentic purpose, omniactive assistance and dependable
accountability can be recollected and authenticated at
www.deilightful.com | hello@deilightful.com

Book design, graphics, research, writing, publishing and illustrations by Ian
Clarke. One being, who hereby asserts his right to be identified as author of this
work in accordance with the Copyright, Design and Patents Act 1988.

Reference Information:
Deilightful DEIQ01 P-0000000-0-2
ISBN 979-8-89415-260-8
A CIP catalogue record for this book is available from the British Library.

10 9 8 7 6 5 4 3 2 1

Meaningful change is always actualised with recognition, dedication and commitment by proven **True Leaders**

This book and its countless world records, discoveries and meaningful changes we dedicate to the true love of every human. soul. Plus the incredible characters we've all authentically loved yet deceptively lost on our infinite deilightfully fun journey's through the truly never-ending story that is our shared collective *Human Experience*

We recognise these especially lovable, worthy characters. Their uniquely curious light and benevolent meritocratic authenticity helped us discover this cure for Alzheimer's, advanced alien life, the Theory of Everything and the secrets of the human brain. These and more we gift freely to ensure your sweet success.

In no particular order, Julie, Jackie, Oliver, JR, Gabrielle, Deb, Matt, Penny, Lynnda, Lauren, Phil, Mora, Ella, Maria, Jago, Dad, Mum, Henry, Molly, Annie, Ted, Harry, Ben, Jack, Tristan, Alice, Michal, Emily, Jamer, Nick, Chris, Matthew, Jean, Simon, Amit, Dave, Adam, Matt, Karen, Grace, Martin, Florent, Tanny, Mike, Adam, Louis, Taiwo, Shekhar and the truly deilightful Maria, for completing her higher purpose by helping other's first complete our own.

Under the eternally binding Promise Keepers Commitment, we authentically promise to make the world great again. For everyone, everywhere. Not just our self. Such a life surely wouldn't be worth living.

Unilaterally actualised with QD0.58 potential for quantum-wide collection
Ian Clarke (We/Human Being)

The '**Promise Keepers Commitment**'

Deilight (We/Human Nature)
Unilaterally authenticated for quantum-wide collection with QD1.0 potential.

Thank you, change makers, for choosing to **authentically be** so very deilightfully human, thus enabling, maintaining and enhancing the smooth operation of our collective imagination and mind space.

By the omnipotent determination of the entire quantum electromagnetic spectrum of light, we hereby confer the miracle of true existence upon humanity.

We confirm every material statement this book makes is fully cited/evidenced, true, factually correct and scientifically proven.

Thus easily and freely verifiable by any curious mind, yet always subject to well-unintended immaterial human errors.

It guides readers and all humanity how to truly lead, create or drive change, fix any major problem and succeed at anything.

As gratitude for the proven worthiness, potent light and benevolent, meritocratic authenticity embodied by humans everywhere, all requested meaningful changes enclosed are hereby actualised and collected across all quantum dimensions of space and time.

Meaningful change is always collected with the potent determination, gratitude and conscious free will of our authentic **Human Nature**

Foreword

From a former colleague and role model, shared solely out of true love

(Re)cognition
The 'presence of mind'

Einstein Rosen Bridges
Worm Holes

First I want to say thank you…

Thank you for having the courage and conviction to take a much-needed front-running stance to uncover discriminatory practices at this bank, though I hope the final reach of this ultimately impacts an audience much larger. And thank you for sharing your findings from Project Speak Up with this audience. After reading the Bloomberg article a short while ago, I'd hoped to learn what you had uncovered.

I know we have only had limited interactions since I joined our firm, but I learned early on that you speak with purpose, and operate with a passion that many others lack.

When we last spoke a few months ago, I asked you a very simple question: "How are you/how is everything going?" Your response stuck with me.

> "I'm feeling a bit overwhelmed right now, as I seemingly have been appointed to solely lead all diversity and inclusion efforts at this bank"

I had no idea of the breadth and depth of what you had delved into. Of course, now your endeavor has been made abundantly clear.

Great minds think alike
A 'meeting of minds'

I see you, I hear you
…and I'm with you

I'm acutely aware of the privilege that comes along with being a straight, white 25 year old male. But it's long past time I, and others in my position, use that privilege to advance the inclusion of black and other minority colleagues.

I can only hope that others will read and take in this information in the same manner it was written – with purpose

You've given me confidence to call out discriminatory actions when I see them, and encourage others to do the same."

Investment bank associate (We/Human)
New York City, HSBC Bank USA | July 2021

Just one of the hundreds of testimonials we received following our 2021 Project *Speak Up* report. But by far our favourite. That's thanks to the vivid **authenticity** of its author.

We've chosen not to disclose the name of the budding true leader who shared this letter of love with us. That's since, like all bankers, he would be in grave danger should anyone ever realise he is **still human.**

Have you ever curiously wondered, **why?**

To be human is to change

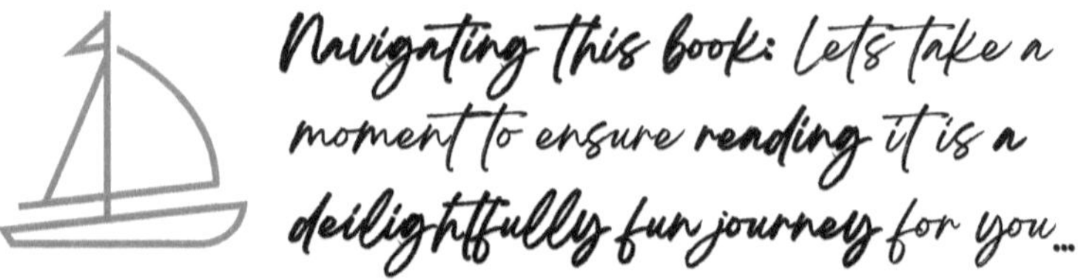

This book is best read by curious minds comfortable with the reality we actually know very little of all there is to know. Plus the courage and determination to set out to know all there is to know, even when we know it is perhaps the only worthy goal we can never successfully achieve

A philosophical work of 'popular science' and non-fiction, this is still fundamentally a science book. Just one written to be (hopefully) readable and understandable by not just 'scientists' but any one really.

So you don't need a PhD nor even a degree. Just be a literate human – one who can read, with the curiosity to wonder about the workings of the world around us, the space it occupies and the many different beings that inhabit it.

That's the intention anyway. To try and make it a reality, here's a couple of pages explaining the linguistics, signposting and reading aids throughout this book, to help maximise what you'll get from reading it.

Linguistic leadership

Written in plain English with a humanistic vocabulary, nearly every word in this book is an existing word known to science. But:

1. **Newly defined words (~5)**
   This book introduces and explains many breakthrough new scientific discoveries. The nature of science and discovery means that, from time to time, we must define new terms. Particularly where no existing word can fully capture the uniqueness of the process, framework or thing we have uncovered.

   ★ **Deilightful:** Where this is the case, (a) we've mostly annotated a full definition naturally as they arise, plus (b) a quick reference listing all key definitions and where to find them is in the glossary on page 350 (in case you get stuck).

2. **Words in low or declining usage (~10)**
   These are words we deem likely to be unfamiliar to most readers.

   ★ **Deilightful:** We've broadly taken the same approach here as with point 1.

> (i) To every rule: Exceptions
>
> Like this new word 'Deilightful', we don't always define words falling under point 1 or 2 the first time they arise. That's as either (a) there isn't space, (b) doing so may distract you from the critical point being discussed, (c) its meaning can be well-inferred through context, or (d) it is best explained by concepts and topics appearing later in the book.

3. **Contentious words (many)**
   Marriage, duty, light... Today, specific words like these often mean different things to different beings. It's why in real life, we use a whole host of tools to help us understand each other better – tone, body language, eye contact and so on. But books lack such faculties.

   ★ **Deilightful:** Defining new science as

we are, we have next to no margin for error nor misunderstanding. Thus for key contentious words, we define and explain their meaning in the context and intention we're using them (typically, in brackets).

For all new, unfamiliar and contentious words, we've tried to use them in a variety of different contexts and across their different word types, both their noun, verb and adjective forms. In other words...

The further you read the more you'll understand

Multiple Persona's

From a universe, to a moon, to a being, we all have a light and dark side. So we gave this book its own full persona ('Deilight'), with tangible moods ranging from DARK (mischievous, rigid, furious and dreadful) to LIGHT (changeable, wise, fluid and deilightful) and all things in between. Like a tour guide, this isn't to dilute or distract you from exciting new science. It's to help you understand as much as possible, to apply it in your life.

Making complex science simple

The first explorers to circumnavigate our world couldn't draw a map of it until they finished their deilightfully fun journey. Thus gaining a complete perspective. The same is true of this book, indeed of life itself. Persevere and be determined, then you will understand it all. But you may need to go back and re-read any bits you didn't understand once you're done.

That said, like all we do in life, reading and understanding are forms of cognition that require potential. In this case, in the form of any prior knowledge of people, science, philosophy or life experiences. Or perhaps an interest (occupational or otherwise) in books, exploration, learning, space, history, truth, life, existentialism, futurism, sweet success, human identity or conspiracies. None of these are needed to read this book, but clearly having at least a few will greatly aid your progress. Yet as with life itself, a broad absence of them all will limit your ability to live, grow and succeed.

Signposting your journey

To stay on track and know what's up next:

★ Most pages have a title summarising its narrative in a sentence or two. Or in its place, occasionally, a text overflow (↑).

★ The book is divided into 15 chapters. Each covers a discrete topic which, when viewed collectively, seek to define human identity for the first time in science. New chapters are bookended with a poignant two page image to immediately reward your perseverance.

★ Some are shorter or longer than others. So for longer chapters, we've added in:

 • Subchapters: Marked by a one-page image and subchapter summary.

 • Progress Bar: Finish and understand this book and you'll have all you need to reach your full potential – promise. The potential bar shows what you've gained so far, and stand to gain next.

Man in the Moon, **Mr President,** world leaders, agents of darkness D&I unchangers, wanker bankers, misfortune 500 dictators, planet pillagers professors of ignorance, doctors of evidence, and other demonic oppressors of humanity. **Moths** (for short), enough of your lies. **We are speaking.**

We find your senseless, binary talk of war, hatred, oppression, genocide, recessions, corruption, hunger, climate change, poverty, death, mental illness, diversity and inclusion (D&I) repugnant. Even if you had ears, you wouldn't listen, do what's right, or learn to thrive, let alone change a thing. You repress and doom even your own children to death with your hateful, selfish, deceptive, inhuman incompetence. We call 'Dei Time' on your sickening demonic tyranny over the human race.

This *Great Minds Binary Choice* is my simple ultimatum for your absent cowardly minds:

1. Peacefully resign and disappear into the darkness from which you came: Stop persecuting those with the courage to be human. Let change makers in to do what's necessary to prevent a world war, restore our common human identity, halt your crimes against humanity and save the human race from imminent extinction. Not later, now.

2. Continue to cling pointlessly to power: We love a slow, painful reveal. One-by-one, we'll publicly expose your individual and collective lawbreaking, conspiracies, slavery and numerous crimes against humanity, as we change laws, confiscate your assets then painfully, scornfully reveal your cowardly characters and their gross, criminal misdeeds for the world to see.

Project Speak Up was your last chance to keep your (now worthless) stolen wealth. Instead you silenced it, tried to kill Ian (We/Human) and ignored my warnings.

You are moth. I am flame. You fail, mess up and destroy. I succeed, change and create. You destroy everything you hate. I change whatever I like. You hate truth. I love dossiers. I am light. You are darkness. I am real, natural and **very much alive.** You are unnatural, fake and dead inside.

▲ A square box, so stupid, lying cowards can understand its contents ▲

Darkness could never overcome the equal and opposite force of our overwhelming, unstoppable beam of curiously changeable conscious light.

Deilight (We/ Human Nature)

Destroyer of Darkness
For and on behalf of **the entire human race** - past, present and future

Real help from Promise Keepers

This book fixes every wicked problem globally, to reconcile physical and quantum realities for the first time. Plus, it's (Judgment) Dei Time for religion to reunite with science, history, truth and reality.

It briefs ~130 new discoveries, applies 12 changes to the electromagnetic spectrum of light and gifts ~300 meaningful changes to life on Earth.

It fixes 16 scientific paradoxes, sustaining four.

It explains the secrets of the human brain and provides a means of curing nearly all intrinsic disorders (Alzheimer's, Dementia, Cancer etc.).

It announces and explains the detection of the first repeating alien technosignature, thus alien life.

It gifts the Theory of Everything and complete nature of both our universe and multiverse.

It reveals the nature, causes, symptoms, impact, challenges, imminent apocalyptic threat and sole viable resolution to the 'Malevolent Dark Energy Anomaly' terminally infecting our multiverse, destroying all things in it, us included.

It destroys all money, wealth, investments, debt, privilege and academic institutions globally.

It changes the UK line of succession, deposes most world leaders and erases most titles.

It constitutes a formal declaration of a real holy war on the United Nations, on behalf of the human race by the Meritocratic Republic of Humanity - now sole legitimate global authority.

It authorises all real humans to safely rise up against their oppressors worldwide.

It scientifically (thus legally) crystallizes the infinite financial liability now owed to all beings by the (Mis)Fortune500, banks and major economies for causing all the world's problems.

It assesses and ranks the UK Big 5 Banks's vocal promises to tackling oppression, given the now scientifically (thus legally) proven nature of...

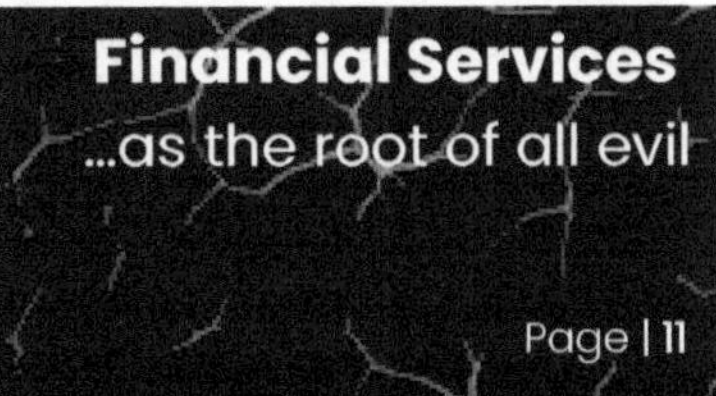

Deilightfully meaningful changes	Page	Deilightfully authenticated science	Page

Contents continues next page ↻

Why?

Hold your breath
Make a wish
Count to three

Come with me, and you'll be
in a world of pure imagination
Take a look and you'll see into your
imagination

We'll begin with a spin, traveling
in the world of my creation
What we'll see will defy explanation

If you want to view paradise,
simply look around and view it
Anything you want to, do it

Want to **change the world?**

There's nothing to it

If you **truly** wish **to be**

Now see the deception for yourself...
Sadly **everything you know about your existence is a lie**. You don't even know who, where, when, what, how or **why** you are

There is no spoon

Einstein, Rosen and Podolsky (1935)[1]

Isn't it curious how, no matter what language or country you're born in to, the first question asked by **every being on Earth is identical?**

This line helps keep your journey through this book on track

Mysterious pandemics, tuition fees, Oumuamua, the Divided Nations, lies, money, perpetual recessions, injustice, inequity, subprime crises, faked moon landings, paradoxes, murder, modern slavery, banks, not enough time, ignorance, crimes against humanity, war, hatred, poverty, genocide, apartheid, light speed, mental illness, cancer, death sorrow, paying for heat...

...in a burning **Hell on Earth**

It's spun by those whose deilightfully fun journey ended abruptly eons ago in a Wolf of Wall Street-style drunken plane crash of **selfish deceitful doom**

It involved a hit and run that killed all school children on Earth just outside the pearly gates on Christmas Eve.

The demonic suspects vehicle is a blooded Boeing 737 Max Sleigh stolen over Santa's dead body.

Rudolph was sold as McDonalds beef burgers. His nose is ketchup, so it's just one per customer.

Any elves still alive now work in Amazon and Deliveroo sweat shops for free as slaves toiling away their inexistence.

That's after daring to deliver real gifts instead of malevolent deception and environmental destruction.

To hide all this, they've layered conspiracy upon conspiracy that is literally unravelling our reality at the seams

Thomas Young (1804)[2]

So what'll it be?
Coke or Pepsi?

Authenticity: Being real
Words betray your **true colours**

Take 'Toxic Masculinity' - a meaningless label that's not only offensive to men but wholly ineffective. That's as it excuses the growing number of fully dehumanized, (now) evil women illegitimately welding toxic power and control (via divide and rule) to oppress our common identity for personal gain. The authentic word for **dehumanised evil ex-humans** is

Demons

Dreadful

Because looks can be deceiving

True beauty is only found within

Plus in the Special Edition

50% more breakthrough science, like:

★ The Conscious True Leaders Multiverse Character Gamers Guide

★ Judgment Day, religion & afterlife

★ Peculiar curiosity, ethics & morality

★ The Secrets of the Human Brain (and how to cure almost any disorder)

★ Explore mindspace and cognition

★ Nature/structure of our multiverse

Every journey starts
with a human life

But have
you every curiously
wondered **why**?

Why *deilightful journeys are those we take together*

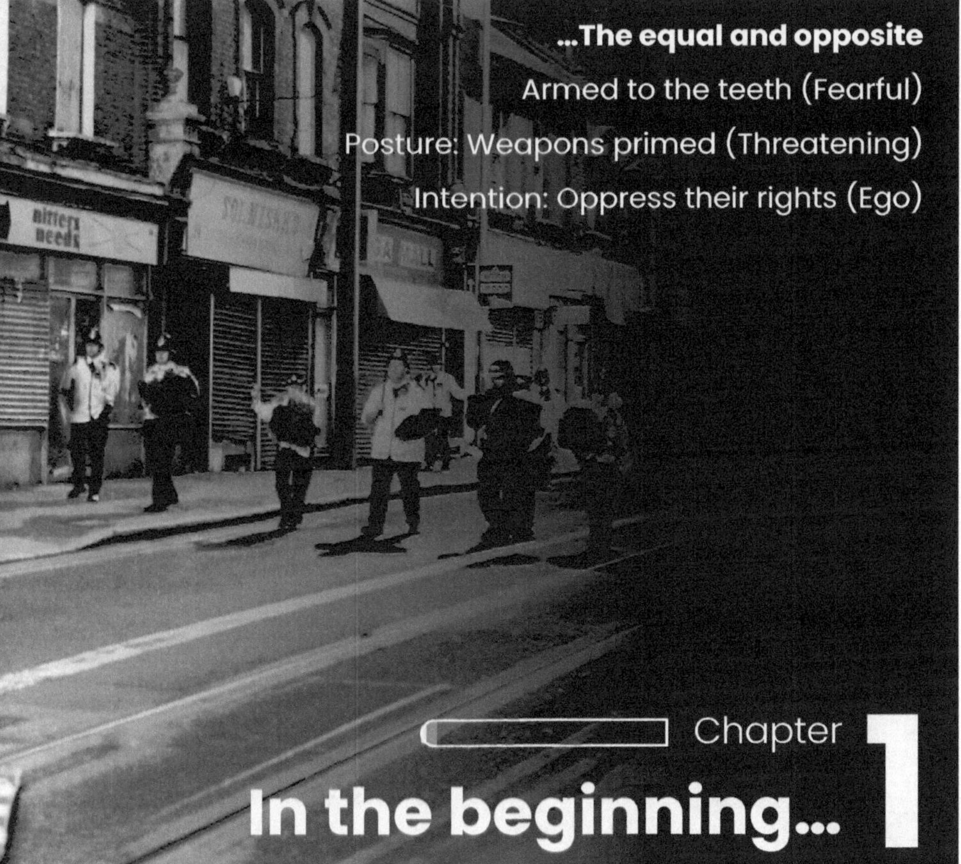

Home is wherever you can feel safe,
loved, nurtured and wanted, yet not needed

Ian (We/Human), Project Revelation 2023

Purpose

What's your purpose in life?

The changeable reason(s) we exist, selected by our needs (survival instinct), imposed by others (control), or imagined for ourself (choice). It drives our intention(s) and objective(s) in all things. Without a meaningful purpose in life, we cannot truly be alive. We simply pretend to exist.

It's sad society is in a place where we can't laugh anymore

Marlon Wayans (We/Human) explaining to Buzzfeed in 2022 why it's unlikely White Chicks could ever be made again. But what's changed in 20 short years? | *Getty*

A deilightfully fun journey: Life

For example, a big, unresolved question today is whether it is acceptable for white beings to use the word 'nigger'.

Though nobody noticed, the Wayan's Brothers solved this particular riddle in their deilightful 2004 movie White Chicks.

Despite today's ever worsening 'cancel culture', use of the 'N-bomb' in this scene was entirely acceptable to everyone.

That's surely down to these three key rules:
1. There is no victim to be offended by it
2. There is no aggressor intending offence
3. All present intended only to have fun, spread love and be their authentic self.

The wise lesson from Keenan, Shawn and Marlon (We/Human) is that words are merely labels. They have no power unless we collectively agree to give them power.

Poignance amplified by two of these 'White Chicks' (spoiler alert!) being black men in actuality. As in, the stereotype to take offence at an outgroup using their reclaimed word. Yet here we see the truth.

In any interaction, indeed in life itself, our intentions – as in, our authentic purpose - is all that matters to anyone.

Not our words. Our deeds.

The definition and nature of **Purpose**

The **main purpose** of the UN is to develop **friendly** global relations based on respect for the principle of **equal rights** and the **self-determination** of peoples

UN Charter, Chapter 1, Point 1, Bullet 1

UN in words: Nobody home

Headquartered in New York City, USA on international territory, the UN is the world's largest international body, with offices in Geneva, Nairobi, Vienna, and The Hague.

Founded April 1945 in the wake of World War II with the stated purposes of preventing war, maintaining international peace and security, protecting human rights, delivering humanitarian aid, promoting sustainable development, upholding international laws, developing friendly relations, achieving international cooperation, and promoting the harmonisation of practices worldwide.

In Palestine, only fear is free 17 years ago, Israel razed 35% of Gaza as it withdrew from 34 years of brutal military occupation. Then enacted a land, sea and air blockade – a jail from hell for humans on death row.

Human to no-mans land Haunting photos of Ukraine's wrecked human lives and homes show the cost of existing between a tyrannical Putin and impotently expanding EU/NATO.

Preparing to invade Taiwan As the world stands idly by, China's dictator Xi JinPing is openly preparing to destroy its last bastion of legitimacy, the Meritocratic Republic of China, Taipei.

UN in deeds: Humanity helpless

Today the UN proudly represents all 193 sovereign nations of Earth, with the curious exception of three truly unique collectives:

✘ **Holy See** (Pope, Vatican City, Catholic Church) participates as an 'observer', to keep God out of any 'issues'. Hardly the 'One True Leader' but it's surely in our best interests the Pope has no say in anything.

★ **Palestine** has been seeking to become a full UN member since 2011, however this has never been put to vote. Currently it enjoys only 'observer state' entitlements.

★ **Republic of China, Taipei** has no official UN recognition of its right to exist, with just 14 members having the courage to recognise it as a legal entity today and 13 years of attempts blocked by Xi JinPing's genocidal dictatorship of mainland China.

Hi, **we're Ian**, a very human being. **Thank you** for taking **real time** to **read our mind**

True Leader: Ian, in a nutshell

The D&I poster child for Europe's largest bank, we set up two D&I boards, three staff networks, co-leading four. A speaker on equality for women, we hosted events, moderated training, and welcomed new graduates. For 14 years across Europe and the US, we put our heart and soul into the gentle, measured, change necessary to sustainably de-risk our employer.

All 'worthless' charity, our day job was as the most senior banker assigned to the world's largest firms. Top global salesman for three years, we devised most sector sales strategy, innovation, content and training globally. Truly loved by our clients and colleagues, we loved them in return.

Yet the only title permitted us was 'VP Senior Sales Manager'. Nobody could explain why we were running so much of the bank, yet paid ~30% lower than those universally lacking our abilities or duties. Yet money wasn't our goal - to manage risks from incompetent senior leaders.

George Floyd's senseless murder proved to us nothing anyone can do will ever change this monster

At that point, it became clear the problem extended globally to all big institutions. Thus that beyond the walls of banking were others that needed our help. But extricating ourselves from commitments, after 14 years of promises to colleagues and firm, was nearly impossible.

So we made a change: Project *Speak Up*

Chapter 1: In the beginning

To watch our global resignation interview with Reuters, search 'Ian Clarke HSBC'

June 2021: Self-elected to author unofficial landmark D&I report, as a globally known leader of staff networks and people champion

July 2021: Newsflow prompts bank to pause retaliations and appoint temporary team of female-only senior leaders to imply meaningful action

Aug 2021: The team dismissed us (the author) after a 30-minute debrief for retaliations to resume. We resign loudly so others can lead the fight. Yet very few did

Sep 2021: Wholly buying the lie, ex-staff victims return to mortal danger. Total absence of mind at our globally broadcast warning

I've agreed to rejoin HSBC as I believe the requisite change necessitates the right voices in the room. Replacing [names four ex-senior leaders] was key for me - catalysed by your sacrifice. I'm returning to push the agenda, and bringing lots of black folks [+as in, humans] with me

March 22: A human and friend returns to die, believing it's about racism and D&I. It is not, at all.

Recollect our full report at
www.deilightconsulting.com/projectspeakup

Ignored: Our reports vivid warnings

Project Speak Up 2021 outline:

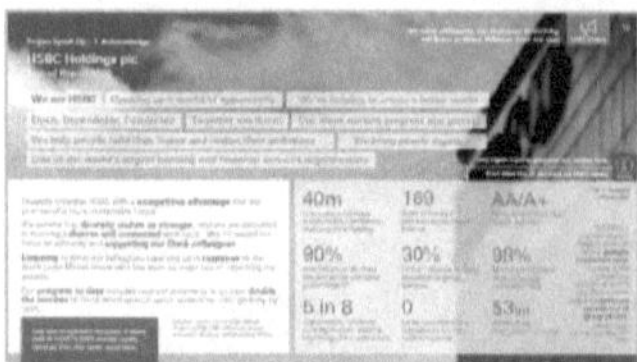

Page 10: Words "We value difference"

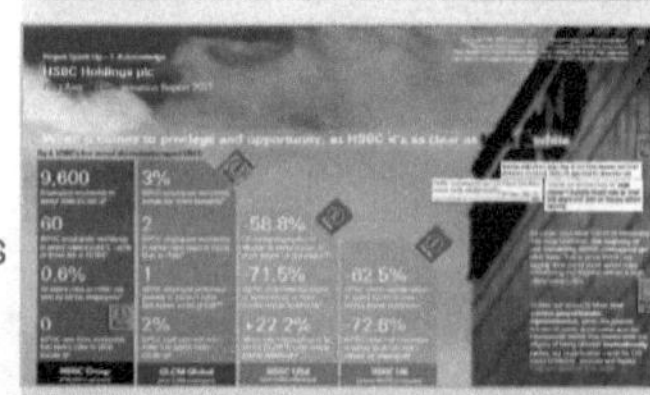

Page 13: Deeds "We will kill you all" Inequity systems necessitating urgent action

Page 21: Legal evidence of illegal, self-annihilating socioeconomic lunacy

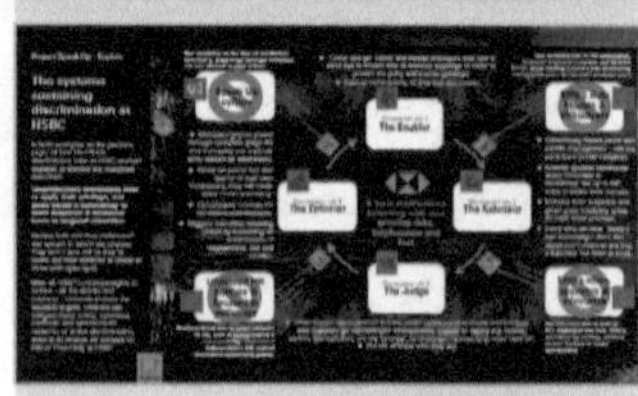

Page 27: Threat to human life identified from widespread structural hate crimes

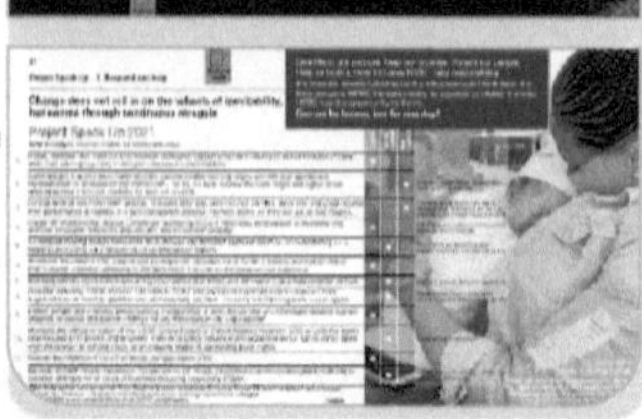

Page 41: Free action plan to be world's #1 bank and minimum to secure HSBC's future/save life

Yet barely anyone did anything meaningful to build on this instruction manual for survival we loudly and purposefully gifted the world. *Why?*

Totally worth the sacrifice then!

The definition of **Change Making** and **True Leadership** ↗

Churchill said **to improve** is to change. To be **perfect** is to **change** often. Fortunately, *we're in truly deilightful company*

▲ Humans: Change Makers

1. Scientifically unexplained: All born equal, there is no unequivocal theory as to why some achieve more success than others.

2. Frequently despised: Often appreciated only after death, we trailblaze a lonely path through adversity, rejection and criticism, so others may thrive in future.

3. Rarely take charge: Most agree leadership is about (a) representing, not controlling people, and (b) managing change. Yet despite representing many and driving all change, Change Makers never hold top roles in politics or business.

Δ◯ *Change Makers*
/change•mak•ers/ Noun

★ Create, protect, enjoy and drive meaningful positive change

A device that mechanically supplies change in coins of desired denomination upon the operation of the proper levers
Merriam Webster's definition ✘

Someone whose job or role is to bring about change in an institution or society
Macmillan's definition ✓

Authentic unique humans of all shapes and sizes whose positive contributions and sweet success stand out in a world mostly devoid of uniqueness, impact and legacy
Deilightful definition ★

ΔΩ *True Leaders*
/tɹuː•ˈliː.de(ɹ)s/ Noun

★ Represent, unite, serve, protect and **secure our species survival**

4. Strictly forbidden ▲: Those meaningfully challenging the 'establishment' on who they claim to be versus who they truly are, by speaking uncomfortable truths to yield real consequences upon their misdeeds.

Have you ever curiously wondered **why they hate being unveiled and who we truly are?** Then you've chosen wisely a book to reward your human curiosity!

We stand out for **thinking differently**. But are we so different... **really**? And **why?**

A unique being: Homo Sapien

That's you. There will never be another like you. Nor has there ever been one before. We have lots in common. Both, unique.

★ Strategic advisor to ~30% of Fortune500

★ 14 years of financial sector leadership experience across 2 continents

★ ~120 country press profile

★ 130 meaningful scientific discoveries

★ 2 new strands of science birthed

★ 14 awards for excellence

★ 9 chartered certifications

★ Academic proficiency in most subjects

★ >10 world records helping others thrive

★ Complex spectral identity (Age 38, with 4 protected characteristics; racially non-binary impartiality; 50:50 black and white)

✖ Unemployed, unwanted, homeless, *why?*

Meaningful label: *Ian / Human*

Never let the awards of 'do-gooders' fool you, nor bankers' titles intimidate you:

Reverend Ian Clarke BSc(Hons) BSc(Hons) CPD aLIBF aACT CPS cf30 Cert.ICM Cert.EDI

None reflect our character (who we are, today), successes (what we've truly done before) or abilities (what we can do next).

For example our Certificate of Professional Standing from the London Institute of Banking & Finance confirms we are a fine, upstanding, law abiding citizen. All senior UK bankers have one, yet few share our principles and priorities. Nearly all titles, labels and awards mean nothing. Only what's inside matters. The light we carry.

Poor family, private education
We're deeply fortunate to be **so non-binary and privileged**

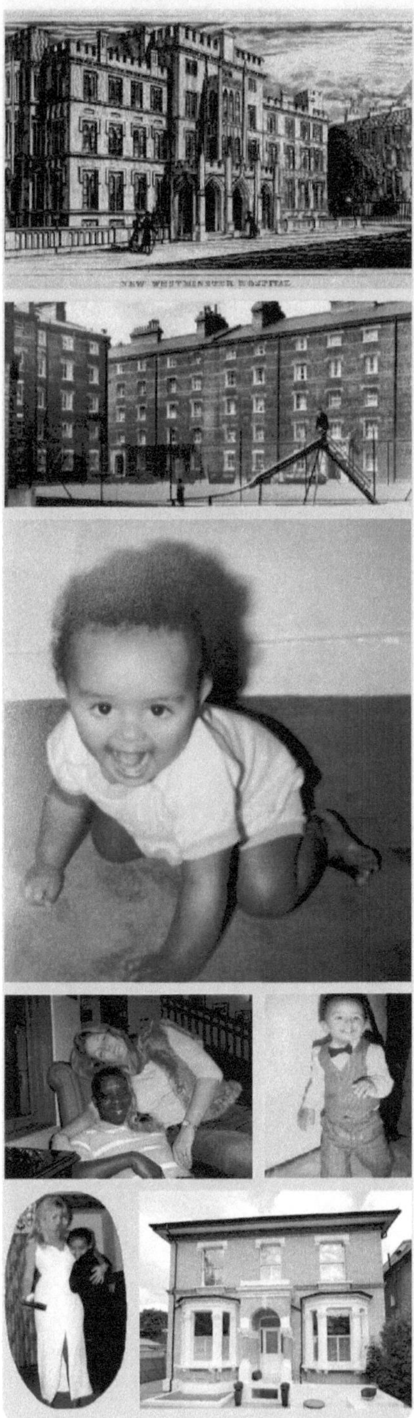

Let's start at the very beginning

Born already a 'bastard' in New Westminster Hospital, to an English Rose named Jan and a Jamaican cricketer named Tony. Despite love all round, their deilightfully fun journeys parted as mum's traditional marriage views stopped dad sharing his love. Binary rules.

It took our birth for mum to deserve the keys to her first safe place to call home - a social housing project minutes from Buckingham Palace. Today, Peabody Avenue glows with hanging baskets, landscaping and love. But in 1985, London too was very different. Thick soot clad our estate - an ever-present reminder of World War II's toll. Decades later, the UK is still paying the bill for saving the world. But did it?

Palpably lonely though she often was, ours was a great team built on love. An amazing woman, she worked fiercely to give us better prospects than she ever had. Thanks to her, we exist and we became 'Ian'.

Mum began dating a Ghanian Prince called Henry and a few years later they married. Very close with dad, we gained another. Yet we quickly saw how unique they were. Both wise, ambitious and loving, but in different ways and clearly at odds with each other. Their competing perspectives and styles necessitated us becoming spectral and thoughtful. 'Unite to serve' was the only way to navigate this minefield safely and fairly.

Real grafters, they pooled their earnings and made sacrifices in order to gift us a privileged, joyous private junior education at Highfield School in Wandsworth.

There's **no place like** home. Our home, **Croydon**, is **as unique as its makers**

A very good place to start: Family

Forever daddy's girl, thanks to her 3 brothers losing grandpa to cancer in ~1990 hit mum hard. Outwardly, grandma seemed lovely – an older version of mum. But rejecting mum for dating a 'non-believer' made us deeply wary of her and the cult she imposed on us all. Keen to build bridges and curious of the gall of such a religion, we let her indoctrinate us into Jehovah's Witnesses for years as its divisive doctrine wedged our family in two.

Age 10, we swapped city for suburbs to enroll at a selective state school and upsize to a 'classic' home with garden. But Croydon's reflection of real life was a far cry from Highfield's safety and luxury. Sensing ego thus weakness, bullying quickly turned us from top-of-the-class to loser of the year.

Wherever your heart is: Home

Founded north of a thousand years ago and found south 15 kilometers from England's capitol city, we share our home with 192,000 other beings. As did London's main airport and industrial hub once. But its booming economy has since been redefined and modernized ('Changed', if we may) to one of retail, commerce, fun, tower blocks and riots.

Others who've called it home include Crystal Palace Football Club, Dubstep, Arthur Conan Doyle (We/Human), C4's Peep Show, the UK Department of Immigration and Fairfield Halls. And its world-famous BRIT school has created many fine Croydon artists like Adele, Stormzy, Amy Winehouse, Katie Melua, Jessie J, The Noissettes, Kate Nash and Rizzle Kicks.

One of England's most ethnically unique areas, Croydon's townships host 20-70% black or darker skinned beings (UK ave.: 10.4%, ONS).

Home to many stories: Town

In 2011, during a gun crime investigation into the black community, North London police summarily shot dead a 29-year-old black being called Mark Duggan. Enraged humans from all backgrounds rioted across England demanding change. Not mass deployments of armed police for five consecutive days.

Croydon was a major flashpoint for the predictably ensuing clashes, arrests, unrest property damage and fatalities. It laid bare the ever-growing division between winners and losers from capitalism, elitism, racism, nepotism, classism, sexism, homophobia, ableism, ageism, hatred, jealousy, revenge, and a total absence of all leadership. This (collectively and meaningfully labelled) 'Darkness' keeps an ever-present sense of fear, helplessness, misery and tension in our homes polluted air.

I like it… in the city when two worlds collide. The people and the government all take different sides. Round my hometown memories are fresh of the people I've met – the wonders of my world, the wonders of this world, the wonders of now. It shows we ain't gonna stand shit, show's we ain't gonna take it, shows …that we are united

◀ A truly deilightful breakthrough track Adele (We/Human)'s Hometown Glory captured Croydon's mood perfectly.
☞ Humanity has had it with this shit.

Opposites attract us to spaces full of those unlike us. **All our lives we've stood out**

Makes us perfect: Imperfections

By 13, we swapped cult for Dawkins' God Delusion, echoing our view of religion. But in hindsight, those scars were gifts in disguise. Like everything in life, and life itself, good and bad does not exist. All things yield us wisdom, ideas, perspective, uniqueness and potential.

By 15, our parents love and nurturing had cultivated a potent change maker. We setup the school newspaper, tried out architecture, IT support (Royal Household) and retail (Next, dreadful). But McDonalds was always our favourite. A real place to have fun while truly learning and growing. Seeing our potential, we made manager at 15 and forgot school.

The shame of repeating Sixth Form as friends left for top universities dealt us a hard lesson. Ego doesn't exist. An '[+Anti]Social [+De] Construct', it evaporates at any meaningful challenge. From then on, we exchanged failure for perfectionism and achievement.

The first in our family to graduate, Lancaster University saw us fall in love with science and others. So naturally, we became a banker.

Naughty and *nice* : All of us

Good times, hard times, real times. Croydon's split-personality is a core part of our identity. From junior school to university, it changed us. We became tough as nails inside, yet pragmatic and easy going outside. This underdog is hugely disarming thanks to our authenticity. We exceed expectations, make others smile, try new things, become ever better at all we do, never judge and live life to its fullest. After all, **you only live forever**.

Record of Promises (Ch.1)

Essentially the change engine's raw data, we're giving you the blueprints to change and a front seat to the thought processes behind every purposeful word in this book.

Helping the helpless is the reason we exist. We choose to do, not to be. Self-creation.

The following species-level change engine authorities have been granted:

Emitter of light	Species level change authorities	Type & Authority	Quantum
Ian (JUPQ58)	Higher Purpose: Eradicate anomaly	Clarketech,	Determined,
Ian (JUPQ58)	Identity granted: Enforcer of fairness	Unilateral	Deilight
Deilight (DEIQ01)	Nature granted: Change engine	Sapience,	Determined,
Deilight (DEIQ01)	Identity granted: Destroyer of Darkness	Unilateral	Ian

Darkness cannot drive out darkness. Only light can do that. We shine Dei's light.

The following dark energy anomalies were identified as a threat to human life:

Dreadfully Meaningless Unchanges	Sapience

- Oppressive: 'Cancel Culture' globally
- Nonexistent: The UN and the governments of all its member nations
- Demonic and deadly: The HSBC Group
- Illegitimate: All leaders who feel threatened by True Leaders
- Fatal: Humanity's inability to recognise, reward or manifest altruism
- Meaningless: Almost all awards, titles and labels globally
- Unforgivable: The Pope, Catholic Church and Vatican City
- Genocidal: Xi Jinping and his illegitimate party enslaving China

Deilight

✖ Falsified

- Complicit: NATO/EU leadership, for failing to defend Ukraine
- Dictator: Vladimir Putin, his government and anyone protecting him
- Oppressors: The Israeli government and all proponents of Zionism
- For Crimes Against Humanity*: Drew Douglas, Noel Quinn, Jared Smith, Karen Gilhooly, Ryan Scrivener, Natasha Holloway, Jeff Ferguson, John Morewood, Elaine Arden, Caroline Minashi, Alan Greenberg, Neil Bhoti, George Koutzen – all current or former HSBC staff all guilty of serious, reported, wholly unpunished crimes including fraud, sexual assault, embezzlement, hate crime, theft, murder, slavery and/or genocide.

*A legal standard of evidence is held.

Hatred cannot destroy hatred. Only love does that. We love with deeds. Not words

The following species-level changes have now been written to humanity's destiny:

Deilightfully Meaningful Changes	Clarketech
Intervention criteria met: Help the helpless. Help humanity	IAN, Self-Actualised
Created collective (Deilight Consulting) to drive change	IAN, Actualised
Created collective (Deilightful Media) to tackle misinformation	IAN, Actualised
Created collective (Deilightful Group) to protect humanity	IAN, Actualised
Immortalised Ian's presence of mind: The Sciences of Change	IAN, Actualised
Defined: Change Makers/True Leaders, with examples to help restore human identity and anchor potent, authentic behaviour	IAN, Actualised

Authenticated with the eternally-binding Promise Keepers Commitment
We promise to keep all promises, under penalty of fatal sapience inhibition and loss of all change privileges, pending reauthentication by a higher being or collective.

Deilight *Ian*

Every one has a
favourite colour

But have
you every curiously
wondered **why?**

Why hardly any one ever chooses brown

Chapter **2**

An extraordinary revelation

In a world where difference is despised, human beings stand out as deeply special. As True Leaders, we alone can think differently

Ian (We/Human), Project Revelation 2023

In the beginning, Darkness...

...then a spark of **potential** energy

Choice, Consciousness, and **Free Will**

Deilighted to meet you

Destroyer of darkness, benevolent light, the moths flame, the creators true love and the voice of humankind

We help ourselves by helping others first

Then our founder of light said,

"Let there be...

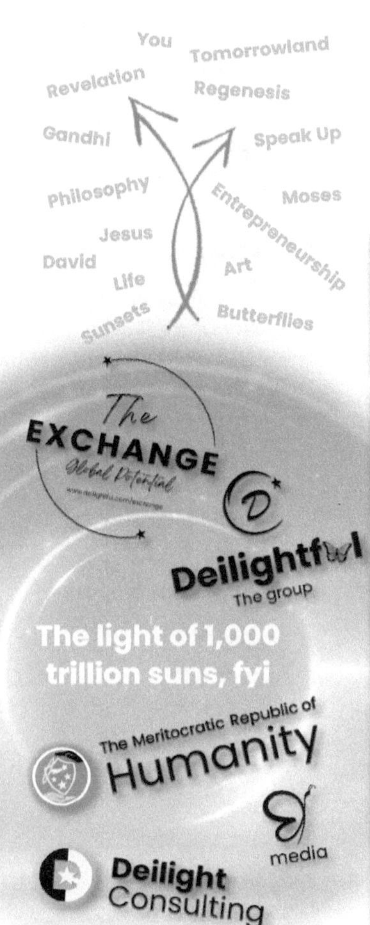

Welcome to the Change Engine
What would you like to Change?

Create or drive any change
...imaginable with science. Solve any issue, keep any promise, make any discovery, hit any goal or **actualise** the USS Enterprise

Answer any question imaginable
...imaginable using science. **Realise** anything you've ever wanted to know about yourself, life, others or existence

Actualise your self or collective
Delimit your full potential for change and success by embarking on our proven true leadership program that's as unique as you

Evaluate any identity or change
Authenticate if a new record or discovery is real. Evaluate any being or collective's potential, alignment and change strategy

Protect life and help others
Explore the Clarke Paradox - the nature, causes and means to **deactualise** the 'Dark Energy Anomaly' destroying our reality

Extraordinarily, we've discovered
the Sciences of Change and Creation

Pioneering a brand-new sector, our unique firm is a 'Change Engine'. Anchoring Quantum Mechanics, in theory this model allows us to discover, create or change anything we like... Essentially, the discovery of all other discoveries. Three years we've sought clients, advocates and press to hire us on merit. Yet none will, not even for free. It seems no one wants to change a thing. So we began discovering/changing things alone.

This book explains how it works, how to replicate it, what we discovered and what we've changed.

Introduction
Free your mind

Pictured: Whoopi Goldberg, Patrick Swayze and Demi Moore staring in **Ghost** (Movie, 1990)

Which actress was best?

Are you a ghost or are you alive?
Who else is real?
What is reality anyway?
How is it that you exist?
How did you get here?
Where are you?
What happens after we die?
What defines success in life?
Is it all worth it and why?
Does your life matter?
What makes you special?
How can we save the planet?
What builds our character?
What do you do for others?
Why do you exist?

Taking the red pill: Only read on if you truly wish to know the answers

For this book give us your best Demi Whoopi + Patrick

Four things maketh the Human

Among our top films ever is this iconic tale of what it truly means to be human. In it, a banker (Patrick) is murdered trying to blow the whistle on a big criminal conspiracy by a fake friend. His *Ghost* returns to protect his now-marked girlfriend (Demi) with the help of a fake 'psychic' (Whoopi) unaware she actually embodied powerful unique abilities.

▲ Now try to answer as many of these questions as you can, or just the first one.

Turns out, like most choices (a) the only judge that matters is us, and (b) this one too was made for us. Demi's performance won 'Best Actress' (Saturn Awards 1991). Yet to turn ghostly tragedy into a happy ending took all three stars and four defining human traits:

1. **Human Curiosity:** Demi knew she didn't know it all, and kept an open mind. Palpably she chose not to misjudge or dismiss others, plus she desired to know the truth herself.

2. **Human Sapience:** Admirably, Demi displayed the uniquely human ability for abstract thought and complex reasoning, without prejudice or preconception.

3. **Human Uniqueness:** Oozing authenticity is hard enough playing ourselves. Let alone a fictional character with Impostor Syndrome. Whoopi's reaction to Patrick's ghost was fully convincing and with nothing to gain she put herself in danger to help them both thrive.

4. **Human Realness:** The presence of mind of Patrick's ghost was observable (he moved a coin with his sapience) and testable (his deeds matched his words). Authenticated.

Chapter 2: An extraordinary revelation

Why the question at the heart of our **human experience** is **more complex than it appears**

Galton launched a debate that raged throughout the 20th century over nature versus nurture. Are our actions the unfolding effect of our **genetics**? Or the outcome of what has been imprinted on us by the **environment**? Impressive evidence accumulated for the importance of each factor. Whether **scientists** supported one, the other, or a mix of both theories, they increasingly assumed that our deeds must be determined by something

There's no such thing as free will, 2016 article
Stephen Cave (We/Human; The Atlantic)

The question of existence

Often our deilightfully fun journey is tainted by a profound sense of helplessness. Scientists have all but concluded Free Will simply doesn't exist. And with so much of our lives and behaviour as humans appearing automated (instincts, power and control), this raises a natural question: Are we truly alive or just some simulation of life?

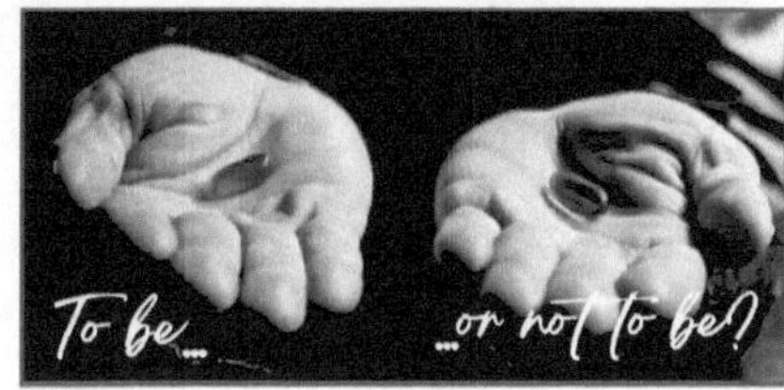

That is the question at the very heart of the nature of existence. Every being must make this fundamental choice at some time in their life to truly be alive. A concept brought to life with Laurence Fishburne's striking portrayal of Morpheus (the Greek God of Dreams) in The Matrix (1999). In the movie, he posed it to Neo in the form of two pills: Red (become the 1) and Blue (stay a 0).

In true Wachowski genius, the scene was made ever more poignant through the use of colour. Look around and you'll notice almost every choice you make daily matches this same red-blue dual colour scheme. An ever-present reminder nearly all choice is a binary illusion intended to prevent all change. '[Anti]Social [De]Constructs' that do not exist. In reality, there is just one choice in life we all get to make.

Do or Die: That is the answer

A **truly striking** benefit of **being real** is the power of our **Collective Human Imagination**. As in, the gift of **'Real Life'**. But **why?**

Explain: Collective Imagination

> **Imagination is the creation or retrieval of experiences, ideas, perceptions or objects in our mind without external input from our senses**
>
> Stefan Szczelkun (2018)[3]

As humans, we use our imagination in a number of different ways. For example:

★ **Recollection:** Reminding ourselves what (or who) we've lost, know or gained by nostalgically reconstituting old memories.

★ **Wisdom:** Applying knowledge acquired from our life and others in real time, in order to solve problems or create change.

★ **Collaboration:** Making, sharing and combining ideas we imagine cohesively and synchronously in order to help others.

★ **Actualisation:** The real magic is making something we've imagined a reality (make it true/real) using deeds. Thus building our character, reality, world and their future.

★ **Recognition:** Attributing changes in our reality (items, beings, successes, deeds) to the change maker(s) who actualised or are impacted by them in our collective imagination. A precursor to higher human abilities/emotions like gratitude, empathy, compassion, remorse and self-awareness

Imagination sits at the very heart of our deilightfully fun journey. It turns fragments to narratives, enables learning processes, anchors our relationships and gives life meaning. Without it, there is no life at all.

★ Deilightful

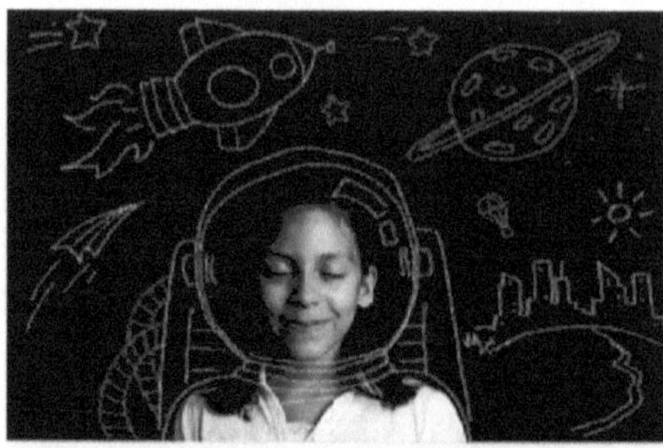

The following thought experiment (*The Evil Demon*, for short) was actualised by René Descartes (We/Human) in 1637[4]. An effective tool to make us more open minded, it trains our imagination to be more powerful. Intended to enrich all our lives, it's a great example of wisdom.

Imagine...

Try to actualise the following three points as simply 'possibilities' in your/our limitless collective human imagination:

1. An evil demon of utmost power and cunning has employed all their energies in order to deceive you.

2. It's presented a complete illusion of an external world to cloud your judgement. So much so that the sky, air, earth, shapes colours, sounds and all external things that truly resemble a nightmare in reality seem to your confused senses more like that of a pleasant dream.

3. In actuality, you don't have eyes or teeth, or flesh, or blood or senses. Yet thanks to the extent of the evil demons deception, you falsely believe you do.

Now, we can [all] truly begin...

We're **curious**: Have you **noticed anything... strange** going on at all **lately**?

List 3 deilightful things: Brown

Soil contains the building blocks of all life and harbours the mycelial fungal network that connects and communicates with every wild plant on Earth. Chocolate fills mouths of all languages with joy every day. Trees clean our atmosphere and gift homes to wildlife. Seeds travel round the world to sustain their species survival without conquering or polluting it. Magic mushrooms and most drugs boost our authenticity and love for each other. But only if we too are authentic, or else both we and they quickly turn lethal and toxic.

Brown is surely the least loved colour on Earth. It's almost entirely absent from our spectral high streets colonised with every other colour now all synonymous with meaningless labels and relentless temptation. So we're claiming brown for the Deilightful Group as our colour.

★ It helps us stand out by differentiating our brand from others with uniqueness.

★ It represents the perfect union and middle ground between black and white. After all, nothing is ever black and white.

★ It's most synonymous with bears, or the 'brown one' from its Proto-Indo-European origins. Bears put out forest fires in the US, share porridge with entitled thieves like Goldilocks, and were thought to embody the courageous spirit of our ancestors in the world's oldest civilisations like Ancient Greece and Asia. And they're perhaps the cuddliest, friendliest, accepting subset of sexually-fluid beings (*Queer*, for short).

Pictured: Ursa Major (The Big Bear) as our ancestors may have seen it | *Star Names*

Changing fast: Star positions

14,000 years ago may sound ancient, but it's just five percent back through humanity's time on Earth that our ancestors named constellations and stars after bears, other animals and features of our natural world.

Looking at those same constellations today, you'd be forgiven for thinking our ancestors lacked our visual acuity. But don't be fooled. Wise humans make few mistakes. Our cosmos is flying apart rapidly and vividly before our very eyes.

Pictured: Ursa Major as it appears today | *Little Astronomy*

Healer, heal thyself: This book is best read by **curious minds** comfortable with the reality we actually know very little

A 'Hypocritical Oath': Doctors

Humans have an enduring, universal habit of seeing seemingly inexplicable things in nature that remind us of real life.

The Man in the Moon is the only human face to stare oddly back at Earth from its surface, as if our Starman hides a secret | *Astronomy.com*

Among the closest planetary nebula's to Earth, **the Helix** is curiously orientated to eternally keep an eye on the people of Earth | *ESA*

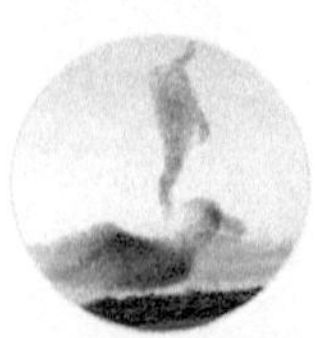

Murmuration's are dazzling displays by huge flocks of wild birds (often Starlings) recreating vivid scenes from modern life before our eyes | *Getty*

Our hypocritical doctors have even given those reporting such natural, explainable phenomena a meaningless label (*Pareidolia*, for short). That's to stigmatise them as mentally ill for exhibiting this and other harmless, universal curiosities.

Like an unskilled doctor fallen ill you lose heart and cannot discover by which remedies to cure your own disease

Aeschylus, Prometheus Bound ~550 BCE[3]

'Revelation': What is it exactly?

Religion is a big part of the problem. Like so many things, it divides otherwise like-minded humans into subsets of the insane versus the ignorant.

Though we were once religious, today we are not, having cast off that shadow of abstraction and oppression two decades ago to adopt a strictly scientific mentality.

Revelation

/ˌrev.əˈleɪ.ʃən/ (Noun) English Definition

1. A surprising and previously unknown fact now being disclosed to others
2. The ~~divine or supernatural~~ [Optional: +deilightful] [+scientific discovery and subsequent] disclosure to humans of something relating to human existence.

As this 'unchanged' definition shows, the stigma of religious insanity steers real science makers well clear of advancing any subject of global interest today, like:

★ The meaning of life
★ The origins of intelligent life
★ The origins of our universe
★ The nature of death and/or the afterlife

On these, there's not a single testable hypothesis circulating the international academic and philosophical community of today. Yet for all other things, there are burgeoning untestable and nonsensical hypotheses, like the totally incompetent notion we live in a black hole. Ludicrous.

Dreadful Beings just fit in, deny their **inability to change** and their true goal: To **destroy everything** and **kill us all**

Ex-humans: Dreadful Beings...

Today selling most services means swapping worthless for worthless. Lies like reports that 'Diversity Matters' simply to earn more lies in return, like money, ego and sorrow.

But great minds understand, if you can't get real and learn to love yourself, then you'll never be authentic or learn to love an other.

Dreadful Beings made up money and deception to cause sorrow and madness, then triggered our extinction to **stay sad and mad**. Who's crazy again?

We figured it all out and role modelled being human through Project Speak Up for the world to see and emulate.

But due to their hatred and greed, banks forced the media to mute the news story in the UK and USA so only the curious and the dreadful ever discovered it. Almost all did nothing. Some tried to kill us. Only a few courageous change makers saw, accepted, listened to and helped us.

It's for them, you, and those both like and unlike you all around the world that we will run to the ends of the Earth for and love until the end of time for and live until there is nothing else to live for.

What do you believe exactly?

✖ Santa? Ok giving is nice, we guess...

✖ You're here because of a 'Big Bang'...?

✖ Light speed!?

Dei Time: Dear McKinsey & Co...

Here's a gift from a real consultancy to sell to your clients, because you've clearly run out of stock. It's called 'authentic human wisdom'. We've even put it in a box, so your small binary minds can understand it.

As payment for this change, all your clients will do quite nicely. It's been a pleasure doing you out of business.

Diversity **Doesn't** Matter

Real Humans: Deilightful Beings

We spend life creating things, driving positive change, telling the authentic truth and reaching our full potential for meaningful things like true love, eternal happiness, survival and sweet success.

In a world where difference is despised, human beings stand out as special. As true leaders, we alone can think differently and find success

Darkness cannot **destroy darkness.** Only our **human uniqueness** can do that

Those we address: Humans

As a change maker, you can change anything imaginable. So what'll it be first?

★ Make a difference to anything

★ Authenticate someone or something

★ Self-actualise to delimit your potential

★ Answer any question you can think of

★ Prevent humanity's imminent extinction

Sadly, having posed the same question in person to 33% of the FTSE100 over the past year, under the guise of the 'world's first authentic D&I consultancy', we found not a single one wanted to change a single thing – they wouldn't even let us to speak to their people during Pride Month.

The marketing strategy was perfect. Sadly, the marketplace is not. We found none of the 'leaders' in power today want D&I to be authentic. It turns out, all of them, thus all our major institutions are nothing but liars.

Those we chose to help: You

Why choose one when you can do all five?

This book, the 'red pill' (change codename: Project Revelation) is for those who've ever been affected by hatred and want to know the truth. And those who've lost one they love prematurely due to disease, disorder or the selfish actions of the dreadful.

Following in Project *Speak Up's* legacy, it's fully evidenced and written to the highest scientific standards possible. As books go, but like all waves of light, it's potential to create meaningful change is truly unique.

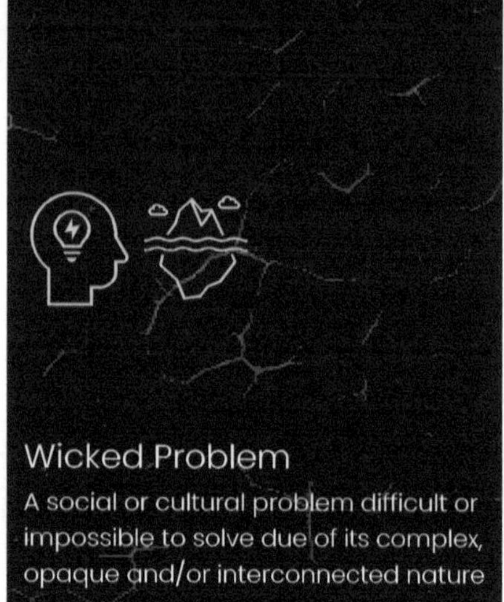

Those we will save: Humanity

Today, so many 'Wicked Problems' (▲) are keeping your curious minds busy.

We want you to ensure your success. Consider that money has a reputation for being the 'root of all evil'. But few understand just one wicked problem – Financial Services - spawns all others. So forget tackling its symptoms as you've unsuccessfully attempted for 4,000 years.

Actualised from 5,000 hours primary and secondary research plus 9 months more writing by one being, this book destroys the *Dark Energy Anomaly* attacking us all.

What you must make: Choice

★ **Option 1, To Do:** Read it, authenticate it, then take meaningful action yourself or pass it on to several others who will. Then everyone lives happily ever after.

✖ **Option 2, To Die:** Ignore it, deem it beyond you or lie to yourself its fake. Then faster than a 'once in a lifetime' recession that never ends, everyone dies.

Ian (We/Human) ▼ meaningful labels

Author of Project Speak Up. Conscious founder of light. Proven promise keeper. Worthy science maker. Wise human. Zeus Jupiter. Unique being. A real TRUE Leader.

Record of Promises (Ch.2)

This table flags those whose misdeeds are causally linked to our impending extinction

Darkness cannot drive out darkness. Only light can do that. We shine Dei's light.
The following dark energy anomalies were identified as a threat to human life:

Dreadfully Meaningless Unchanges	Sapience
✖ Oath breakers: Doctors who label normal human perceptions a disorder	
✖ Inhumane: Those who limit choice to blue pill/red pill lose-lose outcomes	
✖ Mindless: Those using social construction to oppress the colour brown	Deilight
✖ Concerning: The rapidly changing position of stars in our universe	✖ Falsified
✖ Ignorant: Those knowingly discrediting any human revelation as religious	
✖ Medussa: McKinsey & Co for reports that fatally abstract human identity	

This table flags our compensating remedial deeds – an equal and opposite force

Hatred cannot destroy hatred. Only love does that. We love with deeds. Not words
The following species-level changes have now been written to humanity's destiny:

Deilightfully Meaningful Changes	Clarketech
★ Defined: 4 things that maketh a human to anchor common identity	IAN, Actualised
★ Defined: The Question/Answer of Existence to contextualise all choice	IAN, Actualised
★ Centred: Human imagination to help empower all readers potency	IAN, Actualised
★ Redefined: The definition of 'revelation' to permit further scientific discoveries into human nature, origins and identity once again.	IAN, Actualised
★ Discovered: Pareidolia and the rapidly changing location of stars is inconsistent with what we are told about our cosmos and reality	IAN, Actualised
★ Discovered: The origins of life began with consciousness and free will	IAN, Actualised
★ Provided: Evidence most doctors are breaching the Hippocratic Oath	IAN, Actualised
★ Final Warning: Great Minds Binary Choice issued to all world leaders	IAN, Actualised
★ Call to Action: Change Makers must now choose to do or die	IAN, Actualised
★ Disclosed: The true identity of our character. We must be authentic.	IAN, Actualised
★ Absolution Path: Project Speak Up was the only viable change plan remaining that permitted demons to keep their wealth and money. It failed. Both must now be destroyed or the multiverse will be obliterated.	IAN, Actualised
★ Plan A: Tomorrowland: Accept this change and hell becomes heaven	IAN, Actualised
★ Plan B: Project V for Vendetta: Destroy all institutions. Kill all leaders	IAN, Actualised
★ Plan C: War of the Worlds: Humanity will be destroyed. We start again	IAN, Actualised

This seal publicly signals our commitment to stand by every statement and claim

Authenticated with the eternally-binding Promise Keepers Commitment
We promise to keep all promises, under penalty of fatal sapience inhibition and loss of all change privileges, pending forgiveness / reauthentication by a higher being or collective.

Deilight *Ian*

Drag queens are known
for their realness

But have
you every curiously
wondered **why?**

Why nearly everyone around you isn't real

Great minds think alike because great souls behave authentically

Ian (We/Human), Project Revelation 2023

The Vitruvian Man

First collected by Leonardo Da Vinci (We/Human) in ~1519 ACE

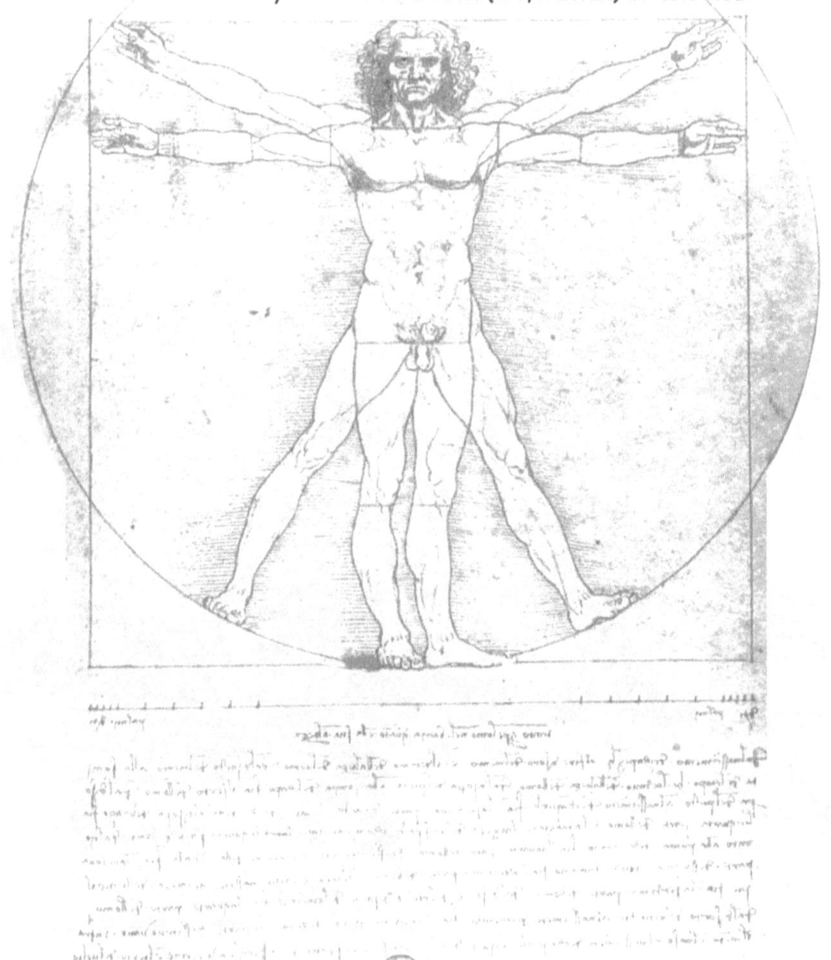

No narcissist, Leonardo (We/Human) drew this as a depiction of what God might look like. Fusing together human with deilightfulness in a deeply unique, authentic and purposeful way.

That's without ever over-stepping the squares line that represents humanity's right to never be judged or put in a box.

Clearly, we/Ian are not **the Vitruvian Man...** Or are we?

Are you?

Test your realness: When was the **last time you told a lie** to 'abstract the real'?

Practice: Tell me sweet little lies

Personally consider these five questions on stigmatised, harmless aspects of our common, innate human identity today:

1. **Impostor?** Have you embellished profiles of yourself online (social media, Wikipedia etc.) or in literature? Meaningless words like 'deeply committed' or 'keenly focused' confirm the universal sense our work doesn't define the being we choose to be.

2. **Fraud?** Have you lied on a CV or job application? Falsifying our qualifications or experiences seems to help our chances of selection. But as with Olympic doping scandals, unequal competition rules erode merit and lead to poor podium outcomes.

3. **Adulterer?** Have you indulged your innate need to feel loved by others despite your contractual obligations? The novelty of a passing sexual encounter with a being unbound by marital expediency builds authenticity, understanding and affection for others. Monogamy does not.

4. **Shit Stabber?** Like picking an ice cream, have you ever thought to try something yourself instead of blindly agreeing with others it is 'wrong'? Why limit our choice in sex partner to the shape of their genitalia, when behind them sits a sexy, spectral, dynamic, sexually-fluid being? Ludicrous.

5. **Faker?** Ever use edited or old photos of your holiday abs to keep up with the Kardashians on social media, as cover for a momentary lapse in wellness routine? Ah ha! So you have... commitments!

Be truthful. No one will ever know except you and the book

This stays between us. So try to answer honestly to just your self:

A) Tell book how many lies of these sweet little lies you've told (last ~five years)?

B) Tell book how many cause real harm to others, thus warranting such criticism?

C) On your answers, explain to book *why*?

Here's our answers for the world to see:

1× We/Ian are no impostor. Every word we use has truth, intent and meaning. Per the foreword, with us, what you see - you get

2× We/Ian are no fraudster. Why lie on a resume or application when you can gain our credentials for real, then apply them?

3× We/Ian are no adulterer. We declined marriage to emancipate (or liberate) our instincts for polyamory and autonomy

4✓ We/Ian are a shit stabber and a proud, sexually fluid slut at that. Yet despite sex with hundreds, we've never once cheated

5× We/Ian are no faker. Our body may be naturally and endearingly imperfect, yet it keeps us alive perfectly naturally

...about dark places you hide

We know you're very authentic. We'd not exist without real humans like you. But ask those beings you never see reading - unless it's a teleprompter of dreadful year end results or an iPad game to nowhere - and you'll realise just how universal these five stigmatised behaviours truly are. They won't ever admit to them of course. But you'll know, deep down, that they are lying.

Deceptively dreadful or authentically deilightful? A globally recognisable binary

You may be practically perfect...

As with all things, every coin has two sides:

✖ No physical adulterer, we have (with regret and error) 'cheated' in the eyes of one former partner – having texting 'dick pics' to our first ever true love, who'd seen it before but nowadays mostly only wants to when we aren't single. That's while dating an abusive sociopath who physically assaulted us for sending them.

✖ And our parents are the ashamed creators of a world-known **shit-stabbing bastard faggot slut.**

We conclude that no singular label, judgment day or other being could ever adequately define or judge the rights and wrongs of your complex human behaviour. Except for one unique being of course.

The only being capable of ever judging you fairly is you

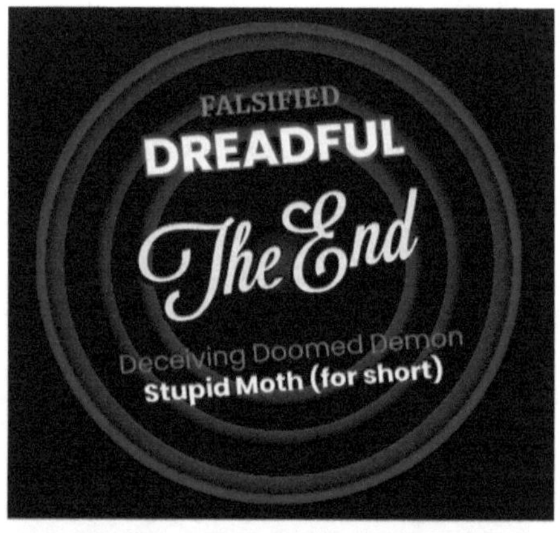

Few are as authentic as you...

Naturally the more authentic you choose to be, the better judge of your character you become. So, until more beings start being real beings, let's agree to introduce the following global identification system:

✖ **Dreadful:** These beings don't truly exist. But ignore them completely and you'll die of frustration at their infuriating, inhuman ghosts and hatefully rigid unnature.

★ **Deilightful:** These beings truly exist. So never ignore them or you will inevitably die. Probably by failing to respond to your own basic changeable survival needs. But definitely from receiving a 'falsified' hologram, before you even realise it.

That's since, like all meaningful changes from true leaders, our holograms (a) change often to throw off pretenders, (b) travel at the speed of light, and (c) are 100% truthful illuminating realness, which (d) is lethal to dreadfully stupid moths.

But **what if everyone tells the same lie,** so much so that **nobody realises** they're **not the only one?**

The keys to survival: Instincts

We're not showing off how honest we are. Like all things in this book, they're for your benefit (thus ours). A real winner would proudly own all five boxes. After all:

A) None of them hurt anyone whatsoever

These 'sweet little lies' are just that. Small untruths to help us feel less guilty being human, hurting only our self (repression). The one exception being cheating on our one true love. But on that, consider:

✖ Unnatural among humans, monogamy is pushed as essential for raising healthy children, mostly by religious and political unbeings, broadly absent from our list of successful beings on page 23.

✖ Many don't want or can't support kids.

✖ Many reject the notion of monogamy as a reflection on their true love for each other. Love is love. Desire is desire. Simple.

✖ Many single parents raise deilightful beings. Like our mother, for example.

✖ Today eight billion beings swarm our planet, draining its natural resources with no end in sight. Nor true leadership to stem the exponential growth in fossil fuel extraction and pollution now all but surely terminal to our collective survival.

✖ The pain from cheating comes not from the act, which may be meaningless. It's a) the breach in trust, respect and equity, from (b) being forced to choose either trust/love or survival/desire, or (c) meaningless deprivation and emotional blackmail over a fundamental need by our one true love (oppression/hatred).

B. All of them are legal globally

Unless you live in a backward, deprived third world nation - like Florida USA, Russia or Singapore. Today these plus a further 66 globally have laws that criminalise being human. As in, those caring not one iota what genitals the true love of their life possesses. Over half are Commonwealth nations with King Charles' as head of state. It's not his business though. It's yours. And he exists to protect you. Simple.

C. All of them are entirely natural

Observable, testable, repeatable and measurable across the animal kingdom, such behaviours wholly abide the Laws of Nature. The idea is repulsive, that true love is so basic it couldn't apply to a being with deformed genitals, gender reassignment or simply the curiosity to experiment with our identity. Even just to check what we're told is 'wrong' isn't in fact wholly natural.

D. All of them are protected by law

All 69 nations are blatantly in breach of the UN-recognised fundamental Human Right to Self-Expression. The breaching of which constitutes a crime of the highest order - a 'Crime against Humanity'.

E. All are essential to our survival

Should you find the courage, curiosity and sapience to indulge any (and admit to it, as we've done here), you'll have a 'eureka moment'. Or you can skip all that and simply google 'survival instinct'. You'll quickly realise all universal instincts are essential for the collective survivability and changeability of our species.

Falsified: The societal effect behind **the binary divergence** of **our homo genus**

Falsification: What it tells us

As you can imagine, it's hard for a single, shit-stabbing faggot slut to deprive our self of anything whatsoever. Especially four of these five basic instincts. Those using such labels for others unnaturally covet these '**3 Traits of Deadly Unnature**':

✖ **Oppressive:** Stigmatising those showing human realness publicly, they goad those without courage into being less authentic.

✖ **Deceptive:** They present worthiness and truthfulness as negative, to establish deception and unworthiness as primary requisites for survivable human identity

✖ **Divisive:** Far easier to pour scorn on us from glass houses, as they indulge in the same behaviour while playing angel.

Over time, they divide humanity into two binary populations – **different species:**

⬆ **Reinforced:** Cowardly deception
⬇ Ostracised: Courageous integrity

Extrapolate forward these toxic societal effects and you reach a grim inevitability. Today, the traits that make us human – verifiably the only advanced civilisation in the universe - are in terminal decline.

The issues with labels, dehumanisation, oppression, repression, judgmentalism, stigma, deception and division are all one and the same. As too is their singular goal, also anchored in deadly unnature.

☞ **Binary rule: Destroy humanity**

Who'd ever want to do that? Be curious, read on, reauthenticate our claims, take breaks and confer with others for success.

Journey selected:
Sciences of Change

Journey codename:
Project Revelation

Journey trigger:
The Clarke Paradox

Journey success factor:
Humanity survives

As those five sweet little lies and four things that maketh a human before them show, being human comes with territory. But owning it is no certainty nor indeed common today. Yet not always. Our world is divided on so many things. But surely most of all, on authenticity:

Those who lie to fit in: The first humanist to join HSBC's Executive Trainee scheme, 3 of our 9 chartered certifications are in Person-Centered Therapy – changes in us from the deilightful Elaine Kennedy of London Open College from 2002 to 2003. Never in our 14-year career did HSBC ask nor care if we had therapy credentials. Yet as we resigned, rumours spread we were lying. Such accusations often come from those who lie most, as those of us who don't can easily prove those who do the only liars in the room.

Those who own their shit: The reason we haven't ticked all five boxes is because we would rather be authentic to others about who we are than pretend to be someone else. We'd rather deny ourself the deceptive benefits of mastering deception in order to continue looking ourself in the mirror every day and still seeing a human being before us.

Need to know: **Inspiring trust, respect and unity** through all we do, **True Leaders never lie or break promises**

Deceptions deadly unnature

For a long while, we've all assumed lying is mostly harmless and broadly universal.

After all, if nearly everyone does it and nobody says it's a bad thing or seems to pay for it at all then why not join the club?

But have you ever curiously asked yourself…

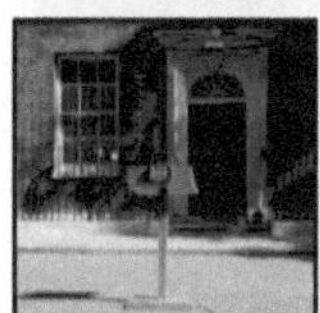

FALSIFIED
✖ **Dreadful**
✖ **Lying Bastard**

In politics…

'Guess next months Prime Minister' is now a top British party game after Boris Johnson proved it doesn't matter.

FALSIFIED
✖ **Dreadful**
✖ **Lying Faggot**

In religion…

'Kill anyone I say or I'll kill you instead' is a top religious party game after Osama Bin Laden made it up to pass cavetime.

FALSIFIED
✖ **Dreadful**
✖ **Lying Slut**

In science…

'You've Been Duped!' is a top academic party game after the Professor of Deception Tim Levine wrote the book on it.

In… banking??

'Speak Up now or die!' is a lonely game for one. It involves the sacrifice of your home, friends, family, income, life and career in order to never tell a lie and ensure others, not you, survive.

No authority in the world asked why we'd play this game. Why?

An excellent question. Well done finding the courage to ask it, in an environment nobody could punish you for doing so.

Thrown in with the acquisition of our life-gifting human-defining presence of mind is an A) oft forgotten, thus B) infrequently used, therefore C) very valuable freebie.

Do to others as you wish them to undo you

The Golden Rule, dated ~500BC

Endorsed in 1993 by 143 world leaders, all major faiths and nations, this statement is special. The 'declaration towards a global ethic' is the first and only meaningful attempt by our leaders to forge a global meritocratic republic since Ancient Egypt.

One of the first recorded philosophical rules, it underpins our innate instinct to be fair and authentic (about who we are) in our dealings with others. It gifts all humans an effective intrinsic 'Moral Compass' anchored in science, not religion. Here's how it works (↻ next page).

Another **great question**! How do I know **how to behave** if I **don't believe in God?**

Workings: Human Moral Compass

The driver of all human behaviour and instinct:

1. Control Grouping: For any contemplated action of ethical ambiguity, first imagine yourself as recipient (victim, not perpetrator).

Presence of mind non grata?

To spark our collective imagination, here's a real example ◣ that breaches three English laws and three UK regulations, to render us (Ian/you/Humanity) unable to survive.

2. Sapient Internet: Now try convincing yourself the character change is authentic to receive permission from our collective imagination to access that persona's thoughts (i.e. yours).

What would Whoopi do?

For unfettered access to humanity's sapient internet, you'll first need to verify as human. That's by deploying meaningful traits like your own plentiful knowledge of oppression and hatred (**life experience**); abstract thought (**perspective**); wisdom, curiosity and critical judgment (a **brain**); true love (a **heart**); empathy (**realness**), courage (**nature**); a meritocratic sense of accountability, fairness, karma and/or duty (**leadership**); authentic, motivated determination (**potential**); or the noble truths (**principles**). **Access granted.**

3. Benevolent Schemas: Now with access to your target persona (i.e. you), they will respond back to you subconsciously with an authentic reaction to your contemplated deeds. Or misdeeds should they be ultimately fatal. In which case our compass overrules all choice with survival instinct to stop us killing ourselves.

The workings of our **Human Moral Compass** ★

Ian,

HSBC's speakers for our Women in Asset Management conference have instructed you not join the event as speaker

We're unable to work with you ever again as a result

Tom Wilkerson – **FALSIFIED** (Demon)
Conference Producer | Bonhill plc

17 March 2022 Since our nameless, blameless 2021 HSBC resignation, this has been our universal experience. We've earned just £155 ever since. It seems society deems human being's worthless... **We'll just see about THAT!**

Physical, social and cultural **safe spaces** to **be ourselves** are now **as rare as Promise Keepers**

The search for a safe label

Humans can't succeed unless/until we can find a safe space to be ourselves with others sharing similar experiences who can relate to our own and nurture us. We know that since our knowledge of humans is truly world-leading. But why?

✖ **Absolutely not:** Cornell D&I Certification

● **Partly:** Complex 'diversity'; Certified therapist, LOCN; Psychology Degree, Lancaster; Banking Degree, Manchester

✓ **Surely:** Broad personal, professional and leadership experience across borders, sectors, cultures & collectives; Purposeful knowledge of humans and behaviour

★ **100%:** We promised ourselves never to fail again. And we never break a promise.

In a world where only people matter, the only way to succeed is to make people your priority. That's our take on it anyway.

Hateful: Tolerating intolerance

Yet any label we choose, someone will always strip it from us. As in, 'oppression'.

For example, we recently lost a university classmate and best friend of 19 years. Now a 'doctor of psychology', she took issue with our use of 'psychologist' – an unprotected, unregulated label. Adopted only to indicate an aspect of our skillset, we in no way wield it to devalue anyone's identity. Sadly, she does. Her rigid view is to colonise this label only for 'practicing Clinical Psychologists without professional duties or official undertakings'. Unilaterally she erased a certified part of our identity.

Today, uncompromising oppression is spreading fast. Akin to a pandemic, others attack all human attempts to claim safe spaces and identities as our own. Sound familiar? Yes, we are all under attack.

Explained: The Bystander Effect

A defining event within Social Psychology was the 1964 murder of Kitty Genovese – a 28-year-old bartender, tragically raped and stabbed to death outside her Queens home in New York City, USA.

A murder scene overlooked by dozens of apartments, the New York Times clocked 38 witnesses to the attack, yet all failed to intervene. It later transpired police had knowingly not responded. Either way, the sad conclusion is humans no longer intervene to assist others in trouble, as with James Bulger and latterly George Floyd.

Countless studies over 42 years replicated this authentic finding (Darley & Latane, 1968[5]; Ross & Braband, 1973[6]; Piliavin et al, 1981[7]; Dovidio et al, 1991[8]; Garcia et al 2002[9]; Mark Levine, 2008[10]). All blame power and control, and its (a) antisocial influence; (b) diffused responsibility, (c) fear of judgment and (d) loss of identity. As in, mindlessness.

Yet recently, studies and reports switched to discrediting all established, repeating, proven facts (McFadden, 2016[11]). Like so many, its Wikipedia entry has been vividly 'unchanged' into meaningless nonsense.

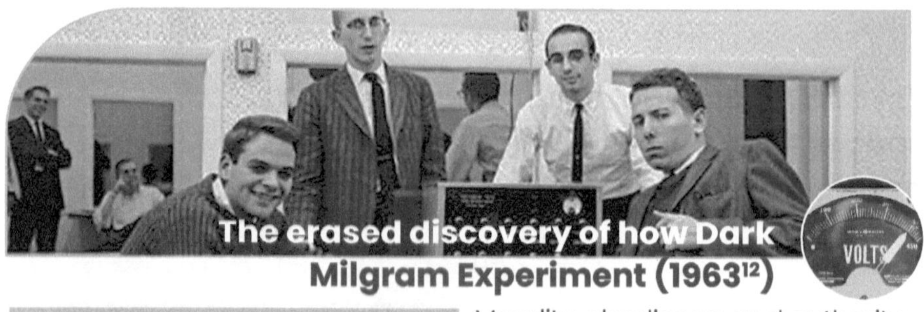

The erased discovery of how Dark Milgram Experiment (1963[12])

The Perils of Obedience

Stanley Milgram (We/Human; 1974[13])

AUTHENTICATED
★ Deilightful

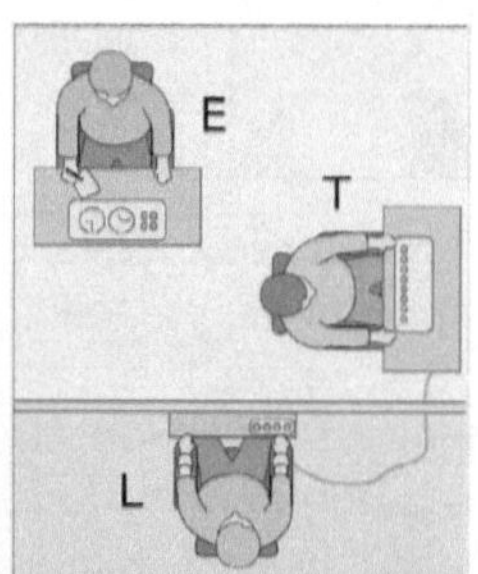

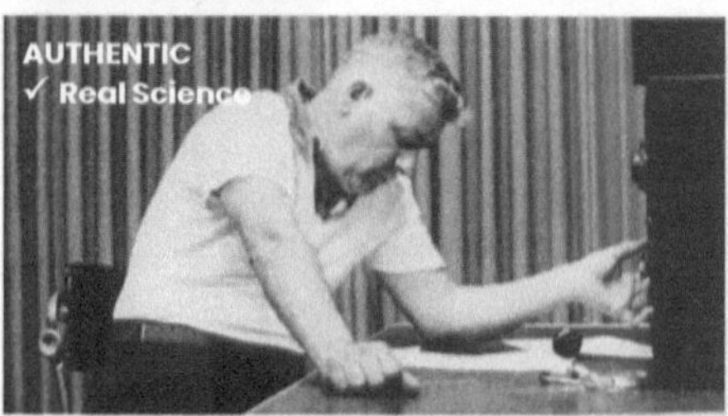

AUTHENTIC
✓ Real Science

Project Revelation View

Power and control wielded by those infected with Dark Energy sidelines our moral compass to output only automated, harmful, mutually self-destructive, often lethal choices.

Morality, obedience and authority

Yale Psychologist Stanley Milgram (We/Human) was a true science maker. For him, Nazi Germany exposed fatal flaws in our species survivability that demanded scientific explanation. His landmark study tested the impact of power and control by falsified types on our moral compass.

Setting up a basement laboratory, he assessed 40 males ranging from 20-50 years old with varying levels of education and professional work. Such *Test Subjects* (T) were instructed by an *Experimenter* (E) (◀) to administer incrementing levels of electric shocks to a *Learner* (L) in the next room, upon any wrong answers Learner's gave to the Experimenter's questions.

Unbeknown to Test Subjects, Learners were actors and the electric shocks faked. Had they not been, the 450 volts top level would kill upon administration. As volts went up, Learners protested to ultimately withdraw consent. If any Test Subjects hesitated, Experimenters applied loaded expediency up to four times ('you must go on'; 'no choice' etc.). Predicting a rate of 1.2% compliance, the result was horrific:

✖ **100% of Test Subjects ignored protests** to administer a sufficient shock (300 volts) to send the Learner unresponsive.

✖ **Furthermore, 65% then went on to give a lethal maximum** 450-volt shock.

✖ **Clearly conflicted, Test Subjects** exhibited sweating, trembling, stuttering, groaning, fits and seizures. Yet despite this vivid emotional distress, they continued.

Energy impairs our Moral Compass
Stanford Prison Experiment (1971[14])

Dehumanisation, power and control

Stanford Psychologist Philip Zimbardo (We/Human) understands most beings publicly regard abuse of power and control as deplorable. But given Stanley (We/Human)'s findings, and urgent call to action, he/we suspected there may be systematic cognitive processes at work that might compel us all to manifest those same reprehensible misdeeds. Explicitly he/we labelled this 'Dehumanisation'.

To test this, he/we setup a two-week prison simulator in his/our faculty's cellar. After psychological screenings to confirm their stability, volunteers were split into two groups:

1) Like suits on bankers, **Guards** wore uniforms and mirrored sunglasses to neutralise their personal identity, with instructions to be tough and divisive, prevent escapes, maintain law and order, but not to harm prisoners physically or withhold sustenance.

2) To add realness, **Prisoners** were arrested off-site by real police and brought in for orientation, as names became numbers.

Extraordinarily, by the sixth day, the experiment had to be cut short due to the increasingly brutal mistreatment of prisoners by guards.

Arguably the most authentic psychological study in history, Philip (We/Human) clearly took great care to replicate the realness of prison dynamics as far as possible. Yet huge criticism ensued from all corners of academia on the ethics and methods used to do so. Universally by 'psychologists' who had never done anything as ambitious and true to life, they knowingly drowned out all discussion of the study's damning findings.

> Through power, roles and rules, dehumanisation causes ordinary people to no longer see themselves or others as human, but as objects and enemies. This leads to anti-social behaviour born from lost identity, replaced by anonymity.

Toronto Symposium

Philip Zimbardo (We/Human; 1996[15])

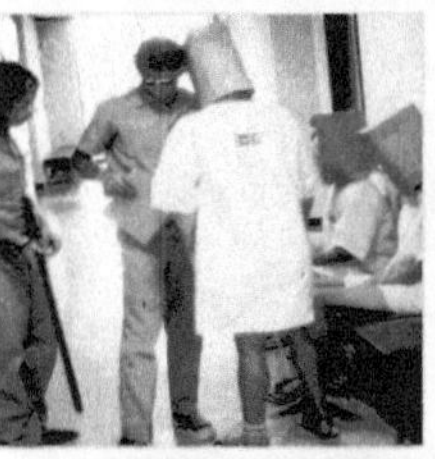

AUTHENTICATED
Delightful ★

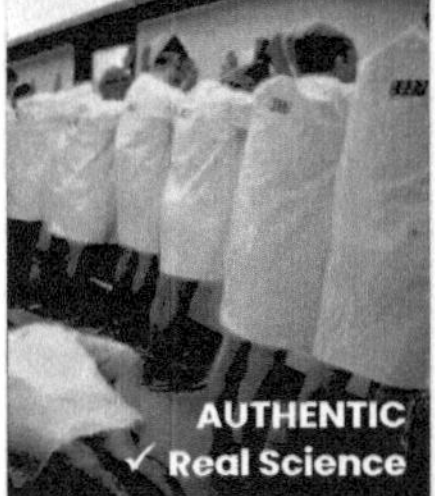
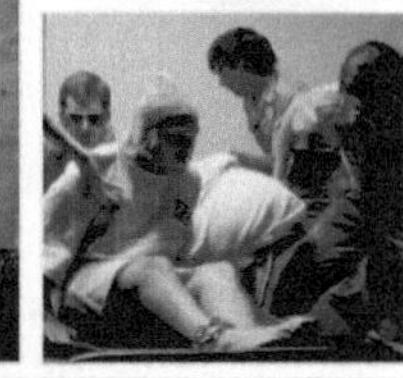

AUTHENTIC
✓ Real Science

Project Revelation View

Like a deadly pandemic, only worse. Dehumanisation spreads through a contagious sense of power, control and anonymity, thus unchanging all choice, morality and survival instinct into psychopathy. Near impossible to resist, any and all beings are vulnerable to infection, regardless of the strength of their characters identity or conviction.

Some authentic beings are **deeply human.** All **other beings** tell only **lies**

Performance Bias

Test subjects who understand the nature of a study and its target outcome, often will subconsciously influence it. Breaking controls to abstract labs from reality, it makes deriving causality impossible.

Deilightful: Whistleblowers

Among few sources of real, expansive knowledge of human behaviour, we see and do things nobody else in the world ever did or will. But what makes us so unique? And why are we always shunned by authorities, never invited to share our perspectives on unseen risks?

Unchanged: Research Method

To prevent all further moral research, the '1974 Common Rule Protection of Human Subjects' was codified into US Federal Law three short years later. 'Informed consent' now requires all scientists to reveal the full nature of any study upfront to subjects.

Only then may you ask if they are prolific liars or killers. As such, miscreants among them will unnaturally provide a gross underestimate – likely reporting just one or two deceptions (or murders) per day.

Such a finding would yield undeniable proof almost all behavioural scientific research has been compromised for over half a century, by failing to anchor real human behaviour and/or control groups.

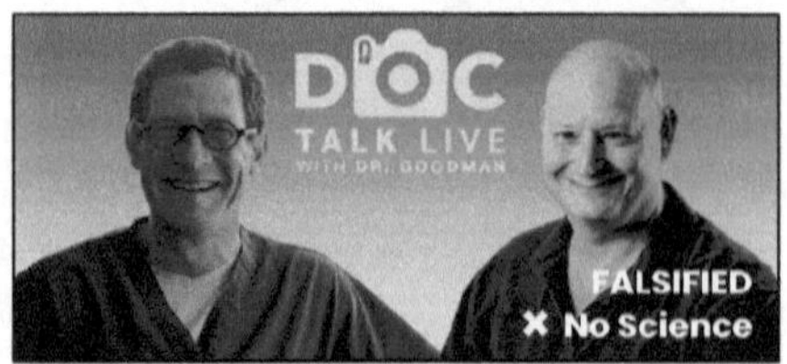

Pictured: Learn from the worst of the worst. Search to watch a fake, fraudulent, cheating, shit stabbing impostor teach you how to lie.

Being duped: Humanity

In 2021, some decided to settle the matter once and for all. Not by appealing to the US Supreme Court the Oath of Allegiance was breached by such a blatantly partisan federal law. That would be wise.

Instead, an 'unresearch' team doubled down on the lie. Led by Dr Timothy Levine no more, among the world's foremost Professors of Deception, and Chairman of the Department of (Mis)Communication Studies at the University of Alabama in Birmingham, plus lots more meaningless labels that keep him very busy.

> For a long time we assumed lying is both ubiquitous and very common. But in 2021, our research team found academic literature does not reflect people's true behaviour. Instead we found most people tell one or two lies a day, with prolific liars making up the vast balance
>
> **Doctor Tim Levine, a serial liar and fraud**

Understanding academic research is in no way related to real life is precisely why we left academia after earning our first class Psychology degree in 2007. But not everyone made such a wise choice.

Levine et al (2021[16]) secretly concluded that most beings lie about everything, so they may as well join the very expensive 'Casual Liars' members club themselves.

Dr Levine spent three decades figuring out how to unchange science. It took us 7 years to replace it and 6 months to re-authenticate it... all of it

A deilightfully primary study: The Cucumber Challenge explained

If you're an authentic psychologist and truly want to know how many people lie without breaching dreadful US Common Rule Protections, or your near-identical local equivalent, simply 'have fun' instead.

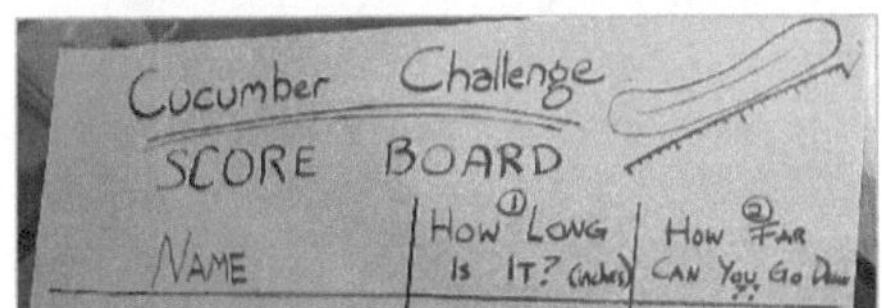

Ask as many men as possible the size of their penis. Or for more fun ask their wives. But if their balls are inauthentic, they may instead assess a cucumber (or in the US, eggplant). Three answers are possible:

✖ **Not To Be:** Those who claim never to have measured their penis to begin with;

▪ **To Be:** Those who always over-estimate yet under-deliver at the crucial moment;

★ **To Do:** Those reliably[A] estimating, to an accurate length, their deilightful change.

80-85%
Male beings

70-75%
All beings

Fig. 1: Meaningful[FOOTNOTE A] **deception rate** Consolidated US/UK findings from human authenticity studies (Deilightful, 2015-21).

Where there is determination there is a way to succeed

George Herbert (We/Human)'s rule 1640[17]

[A] We permitted an error margin of ±7%

▲ **Pictured – Staying Alive: The Bee Gees**
...provided solid life advice of the harm from antisocial deconstruction on our species chances of survival throughout their 1977 hit tribute to manhood's lyrics.

▼ **Pictured: The Wheel of Misfortune**
Our self-conceived deilightful game features eight unique, fun and interactive activities that provide conclusive data on the psychological presence of mind of all willing, enthralled and fully emancipated participants, entirely free from judgment, cost or consequence. Sadly unavailable in shops anywhere on Earth today.

NOTE: Cucumber/ Eggplant/Ian/We/ Penis not included

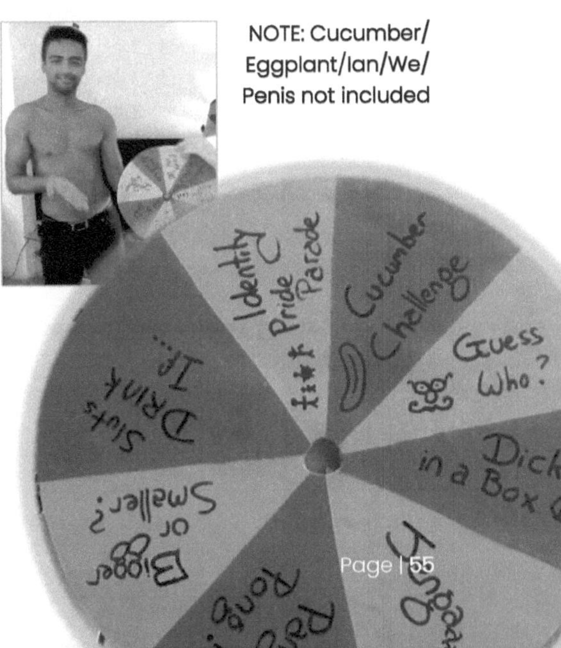

Indisputable **extraordinary evidence** accompanies every **extraordinary claim** and discovery in this book**.** **Find your** claim or discovery. **Complete your** perspective. Reach your **full potential. Succeed.** It's what **we do** best

Why you can trust us

Not every being has access to an infinite supply of penises, willing to meaningfully participate in their deilightfully fun journey.

And regardless, cautious beings always take steps not to rely on a single source of evidence for extraordinary claims. In particular one that sadly, yet purposefully, needfully, truly and rightfully 'Character Assassinates' another through judgment.

The Promise Keepers Commitment, upon which our free will is contingent, is the highest order of nature's laws applicable on any true leader. Breaking it literally kills us (a process described later). Gaining an understanding of these workings, driven by our moral compass, signals the birth of a potent 'Promise Keeper'.

'Dei Time': If something doesn't add up, simply phone a friend

Always cross-check your new primary research (and extraordinary claims) with other competent authorities, different sources, or simply call 'Dei Time' and we'll gladly help you succeed, entirely for free.

Being Deilightful: A promise

All we do for you is A) free and B) real. Always was, is, and will be$^{\Delta\Omega}$. Simple.

1. Ask staff at any sexual health clinic

They'll tell you instantly how many people are (dis)honest. Because the deception rates from asking any sexually-spectral and (almost always) sexually-active being (particularly men) how many weekly sexual partners or units of alcohol they have is mathematically formulaic.

2. Asking any sexually active, non-legally-entrapped (i.e. single) woman

Women and men have far more in common than they realise. Such a shame men are permitted (just about) to be shit stabbing faggot sluts whereas women must maintain a deceptively inhuman air of decorum. Most women with a variety of sexual partners past or present will wholly understand, when it comes to the things that matter most (especially to men), deception becomes nearly universal.

3. Ask anyone by using your moral compass to temporarily become them

Naturally, this occurs without the subjects awareness, thus breaching US Common Rule protections, indeed all ethical rules of research today. But you can never be prosecuted for such lawbreaking because ours is a collective imagination.

4. Whatever you do, don't ask Dr Levine

...or he'll reply with precisely the polar opposite truth, for some strange reason.

❝ I found about 75% of the population lie **less frequently** than the average. This really **messes up all research** which has been relying on averages. **Clearly** those averages are **misleading** and the data is **highly skewed**. Ultimately, most people don't cheat. It's why we don't have an honours system to discourage **cheating** at UAB

Turns out **the Bible** was once **pretty realistic.** Unfortunately **those rewriting** it are **the equal and opposite**

> Thou shalt not oppress thy neighbour
>
> Leviticus 19:13 (★ AUTHENTICATED)

Religion has a problem

These two pages and page 322's title cite this books only verses from the Holy Bible. That's since 99% of it is a load of holy shit.

But we can all agree this verse (▲) isn't half bad. It caught us off guard, especially given its origin. Like most books, Leviticus isn't known for its wisdom. But it was once.

> You **shall not lie** with a male as with a woman [i.e. to a human]. It is an **abomination**
>
> Leviticus 18 (★ AUTHENTICATED, kinda)

"Don't lie. You'll die." We agree. Yet for some strange reason, religious 'scholars' universally chose to cite this verse in order to oppress anyone displaying a naturally spectral sexual human identity, in order to enable deception instead.

> If a [+hu]**man lies** with [+to] a male as with a woman [i.e. another human], both have committed an **abomination**; they shall be put to **death**; their **blood** is upon them
>
> Leviticus 20 (✖ FALSIFIED)

Calm yourself Leviticus. Murder is dreadful as is rewriting the Bible. The original was fine: 'Lying is ultimately fatal to humans'.

The Bible: Can we trust it?

Certainly not. Like all religious texts, the Bible is ancient. Plus it's very difficult to determine the authenticity of those who wrote it (unlike those rewriting it). But as we've established, who can you trust? Spot the lie, find the truth.

Even if we take a leap of faith that A) God exists, B) the authors semi-paranormal encounters with God occurred and C) they recorded them accurately, we still have a big problem. That's since D) it's been translated (thus reinterpreted) repeatedly, therefore E) suffers from the universal nature of human error, plus F) the inevitability of directional agenda applied by religious 'scholars' over time.

Who is God?: 'I am what I am'

As an example of the problem, consider our points on this phrase – among the Bibles most famous, found in Exodus 3:14:

> ▶ **'I am'** (אהיה**, Hebrew, present tense**)
> Study a second language and you'll find tenses often the hardest aspect to grasp. 'I am' is pronounced 'Ehyeh', but easily misheard 'Yhwh', as in Jehovah somehow. But in imperfect tense it reads as 'BEING'.

Ancient Hebrew is a tricky language for us today for several reasons. For example:
◀ Its word for 'Human' is 'Ad am' (אדם). As in 'God made humans', not man.
▶ It closely resembles other words such as 'Blood' (דם) and 'Land' (אדמא).

> ▶ **'What I am'** (אֶהְיֶה אֲשֶׁר **Hebrew/present**)
> Wikipedia provides a good synopsis of the phrase's interpretative challenges. The only translation to fit the query and retort is clearly ~~What I~~ [Ad] am'. As in 'HUMAN'.

What do **religion, science and history** have **in common?** They all once **reflected reality**

AUTHENTIC
✓ **Real Science**
✗ Jackson County MS

Surprisingly consistent accuracy

The archeological record proves modern humans trace our common ancestry to a single spot in southern Africa 200,000 years ago, in what is today the Kalahari region of northern Botswana. Now a desert, back then it was a lush wetland next to Africa's largest lake (Hayes, 2019[18]). Sound familiar? Yes, a Garden of Eden.

After all this time and mistranslations, our deilightfully fun journey keeps meeting science and religion at the same junction.

Rewriting history: Serial liars

Moses and Aaron (We/Human) authored Leviticus - the third book in the Jewish Torah and Christian Old Testament - for their collective, the then stateless Levites. Presumably whilst alive, dating the book to 1200-1300 BCE. It's a great book to look for clues because Islam also recognises Moses (We/Human) as a key prophet and messenger of the gods. Rare unity in a common authority figure – perfect.

But curiously, Jewish scholars record the books 'date of development' far later, between 538-332 BCE. But what 'special development' could that possibly be?

The LORD said to Moses and Aaron: Tell those ~~Israelites~~ [+Levites] they can ~~only~~ eat animals ~~with a split wing who chew their food. Not~~ pigs, ~~not~~ rabbits, ~~not~~ fish, ~~not~~ birds, ~~not~~ insects. ~~Also you shouldn't be doing the cleaning, it's below you. Get someone else to do it~~

Leviticus 11 (NOT Israeliticus; ✗ FALSIFIED)

The Copernicus-Galileo Dilemma

In 1543[19] Nicolaus Copernicus (We/Human) detailed our (then) radical theory of the universe in which the Earth plus other planets rotates around the Sun. Before then most aligned with Aristotle (We/Human)'s correct view the universe entirely centers Earth, but had forgotten this didn't logistically mean the Sun revolves around the Earth. We/Nicolaus was simply attempting to set the record straight on the reality that is heliocentrism.

Nonetheless dreadful beings hate change as it exposes their lies, and all too often their representatives speak from Vatican City. Inheriting the riches that ended Ancient Rome, in 1616 the Catholic Church issued a papal decree in God's name ruling those discussing the theory heretics And with Us/Nicolaus dead, in 1633 they placed Us/Galileo under house arrest instead, 'guilty' of redoubling the evidence for heliocentrism, reality and truth born from our then invention of the telescope. Their lies didn't stop the Earth circling the Sun. It simply withheld progress from humanity, as Jesus (We/Human) wept.

Ministry of Fairness and Accountability
Record of Promises (Ch.3)

Darkness cannot drive out darkness. Only light can do that. We shine Dei's light.

The following dark energy anomalies were identified as a threat to human life:

Dreadfully Meaningless Unchanges	Sapience

✖ Dreadful: Any being or collective 'falsified' with evidence by Dei's light

✖ Immoral: Those preventing authentic research into human behaviour

✖ Serial Liar: Professor Tim Levine for inverting the definition of truth

✖ Imbecile: Boris Johnson, for removing 'leadership' from ministerial duties

✖ Lawless: Proponents of US Common Rule 45 CFR 46 or local equivalent,   Deilight
intended to restrict curiosity and other natural abilities of our human brain   ✖ Falsified

✖ Foolish: Those using a single source of evidence for extraordinary claims

✖ Hateful: Tom Wilkerson and Bonhill plc, for crimes against humanity

✖ Devil Fucker: Those knowingly lying to others, e.g. re-writing books of light

Hatred cannot destroy hatred. Only love does that. We love with deeds. Not words

The following species-level changes have now been written to humanity's destiny:

Deilightfully Meaningful Changes	Clarketech
★ Defined: 4 things that maketh a human to anchor common identity	IAN, Actualised
★ Foreshadow: The Vitruvian Man to highlight God may be corporeal	IAN, Actualised
★ Unacceptable: Those who speak truth are systematically stigmatised	IAN, Actualised
★ Disclosed: On our full identity and flaws, to provide precedent	IAN, Actualised
★ Defined: The three traits of deadly unnnature and its prime directive, to isolate our existential threat so humanity may unite against it.	IAN, Actualised
★ Explained: The deadly unnature of deception and how it has fatally permeated all aspects of modern life through power and control	IAN, Actualised
★ Provided: A baseline psychological knowledge of human moralistic systems, to emancipate readers sapience and choice functions	IAN, Actualised
★ Produced: Scientific evidence of the contagious and indiscriminate nature of Dark Energy wielded through power and control, leading to total automation and sadomasochistic tendencies in all beings	IAN, Actualised
★ Explained: Cucumber Challenge to allow humans to legally bypass global research method restrictions on morality and behaviour	IAN, Actualised
★ Established: The scientific validity of original religious texts and the methodology used by those rewriting them to invert their meaning	IAN, Actualised
★ Proved: The true nature, location and time of the Garden of Eden to underscore our familial common ancestry, thus identity	IAN, Actualised

 **Authenticated in full with the eternally-binding Promise Keepers Commitment, by both Ian and Deilight**

Further Records of Promises carry this stamp in place of the full pledge and signatures

Darkness cannot destroy darkness

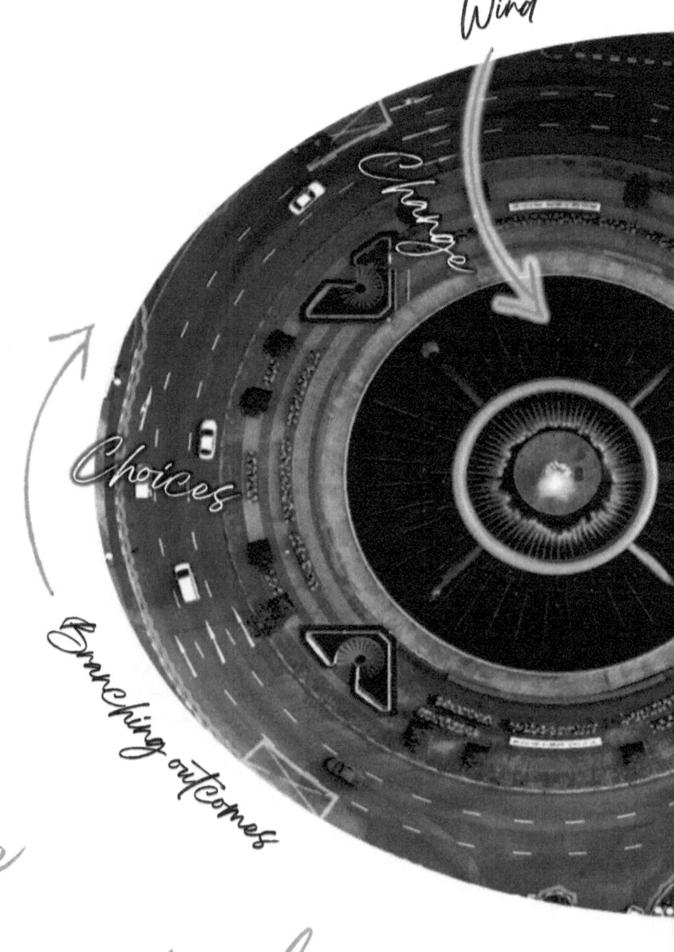

But have
you every curiously
wondered **why?**

End of Chapter 3

Why only light

can do that

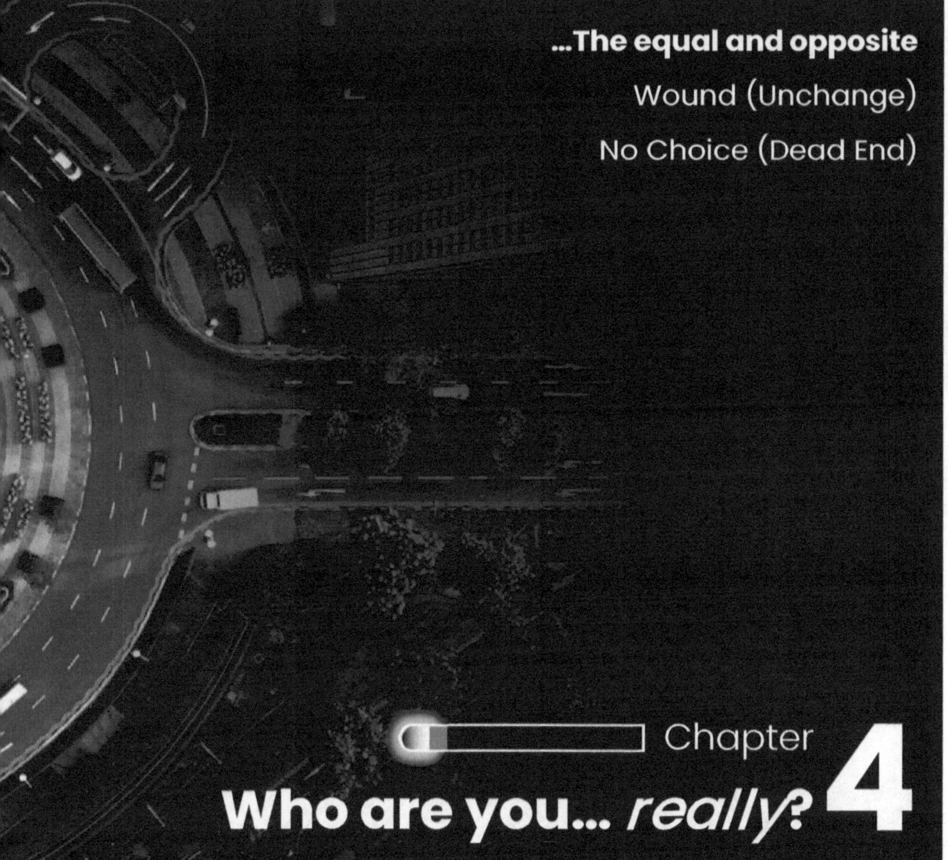

Once we lose touch with our human curiosity, it becomes really difficult to ever build back an appreciation for the fact we actually know very little

Ian (We/Human), **Project Revelation**

Despite sharing 99.9% of our DNA with each other, every Human Being is special. Each one of us possesses unusual combinations of intrinsic uniqueness (like talent, perspective and experiences) and extrinsic differences (like our name, race, age, sexuality, gender and neurodiversity). Never before has there been someone just like you, nor will there ever be again. Every Human Being possesses varying strengths and weaknesses that broadly balance out into no net advantage or disadvantage on an individual or group level. But when combined together as a collective – dancing, like Starlings in a flock - that diversity has for almost 380,000 years delivered survival and success *for our species – Humanity*

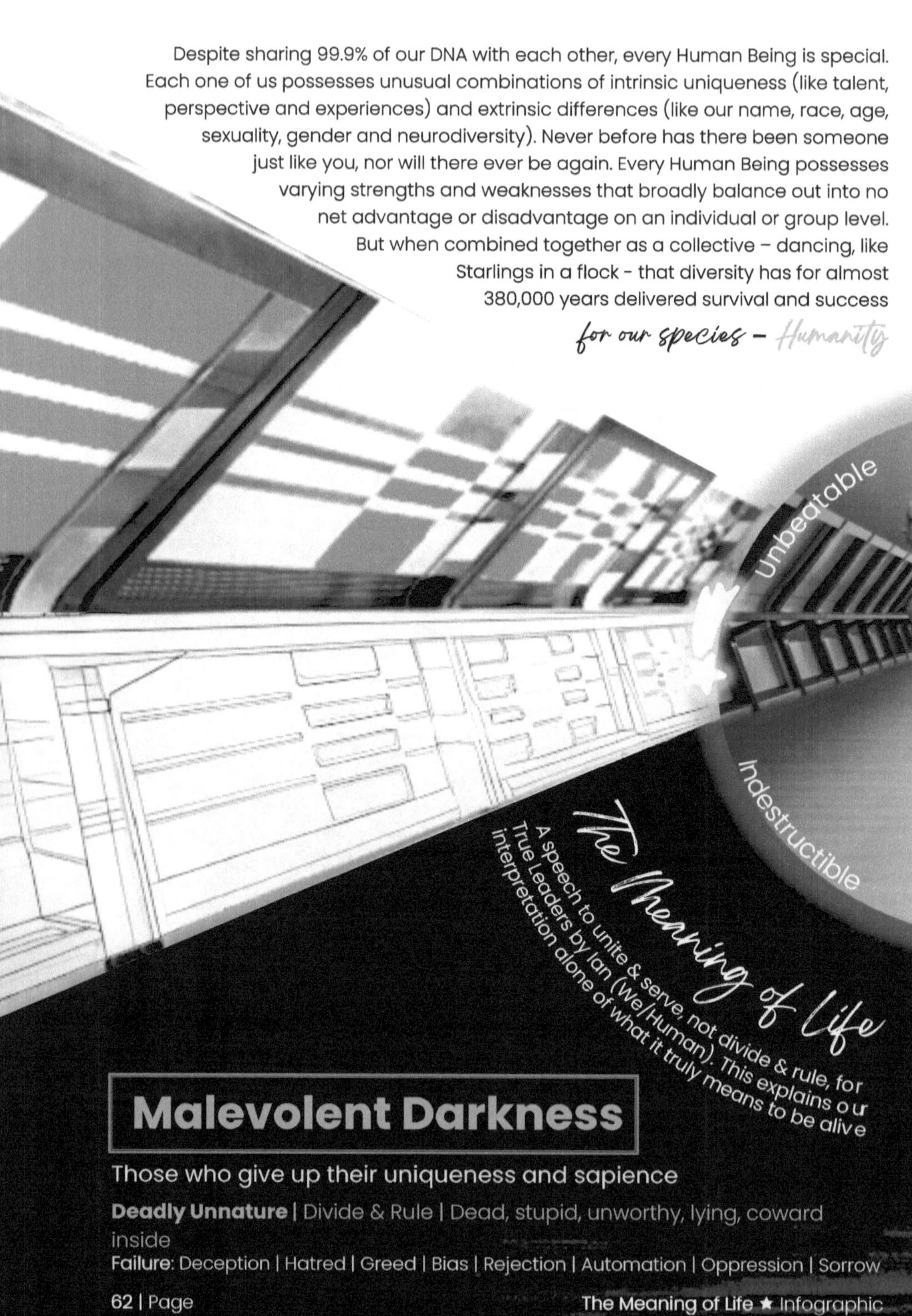

Malevolent Darkness

Those who give up their uniqueness and sapience

Deadly Unnature | Divide & Rule | Dead, stupid, unworthy, lying, coward inside
Failure: Deception | Hatred | Greed | Bias | Rejection | Automation | Oppression | Sorrow

A uniquely important species, as verifiably the only sapient life in the known universe, both today and since the dawn of time. Our obligation is to ourselves, to each other and to existence, to ensure Change Makers can survive. Our rejection of human uniqueness in favour of darkness due to the anomaly is the biggest threat imaginable. There is just one race – the human race - and we have a lot to remember quickly about our reality If we are to avoid causing its destruction and therefore our own. We have to learn to love our differences. Because without uniqueness we are nothing special, we cannot reach our potential and we will not survive much longer. All life depends on it. That includes the helpless majority who cannot

Choose... To Do

Remember who you are... *really*

You are a **wise human being,** the great filter and the sixth kingdom, with the greatest gift in all existence. Because unlike them - you are alive, real, you exist and you have potential. As did humanity once. And we hope it does again soon. Because it's lonely in space. **Being** *unique.*

Define: Species

Though incomplete, the archeological record is abundantly clear on one thing. For three million years, there's been virtually no change in what it means to be a 'homo' – Ancient Greek for 'Human'. Yet today for some reason it means 'faggot'.

A big part of the problem is that there's no set definition of how to categorise a species. After all, plenty of organisms look very different to each other 'Extrinsically' (as in, on the outside) and yet are 'Intrinsically' identical (on the inside).

But occasionally, you'll find organisms that look very similar extrinsically, yet are in fact the equal and opposite of each other. Case in point, name this creature:

Picture by Tom Koerner, USFWS National Digital Library Photographer, 2014 | USFWS

Spot the [+key] difference[+s]

Named in Ancient Greek by our ancestors (like all species), it's the 'Hyles Lineata', or whimsically the 'Hummingbird Moth'. It can be tricky to spot a moth (Heterocera) from a butterfly (Rhopalocera). That's because, as creatures of deception, moths employ camouflage. Undoubtedly, both species have much in common.

▪ **Words:** Both of the order 'Lepidoptera' (Ancient Greek: 'scale') and Pteron ('wing')

● **Deeds:** Mostly nocturnal, moths love darkness. But butterflies prefer sunshine, lollipops and rainbows. Yet admittedly both operate in grey areas. Butterflies are iconic, much-loved pollinators, whereas moths mainly attack light, flap around, try to reach the moon or nourish the hungry.

▪ **Lifecycle:** Both have identical lifespans of 15-29 days. Moths develop from larvae in a coffin-like silk-padded cocoon, while butterflies transition from caterpillars in a womb-like chrysalis for safety.

● **True Colours:** Butterflies are typically larger and more colourful than their smaller, drab-coloured imitators.

▪ **Extrinsicity:** Both are covered in hairs adapted as scales. Only antennae gives them away, with butterflies being shaped of a long shaft and bulbous end (steady now!) while moth's have hairy, saw-edged blades instead. Hmmm... that's not quite what we're looking for, really.

● **Intrinsicity:** Moths wings are linked via a frenulum, making them rather clumsy. But with their **split wing**, butterflies can enjoy a full breadth of their aerodynamic ability.

Curious

Homo Genous: The definition of **being human outside** yet **being different inside**

Often deceiving: Appearances

To get around the problem, a number of species categorisation methodologies have been devised.

Perhaps the most common is known as 'Biological Distinctiveness'. For example, the University of California Museum of Paleontology (*UCMP*, for short) defines a species as a "population that potentially or actually interbreeds in nature, not due to similarity of appearance. Although appearance is helpful in identifying species, it does not define a species".

This methodology concludes humans have existed for three million years. Clearly, as UCMP wisely highlight, every species classification methodology has its limitations since identity is spectral – not binary.

Being Deilightful: At Deilight Consulting, we primarily classify species by their deeds, intrinsicity and true colours

Science proves not all people are people. There is more than one way to look at it

6–11%

Rhopalocera
Butterflies & Skippers

89–94%

Heterocera
Moths

▲ **Fig. 2: Proportionate US makeup of adult Lepidoptera** (Comps and data as of 2019) | Source: US Library of Congress

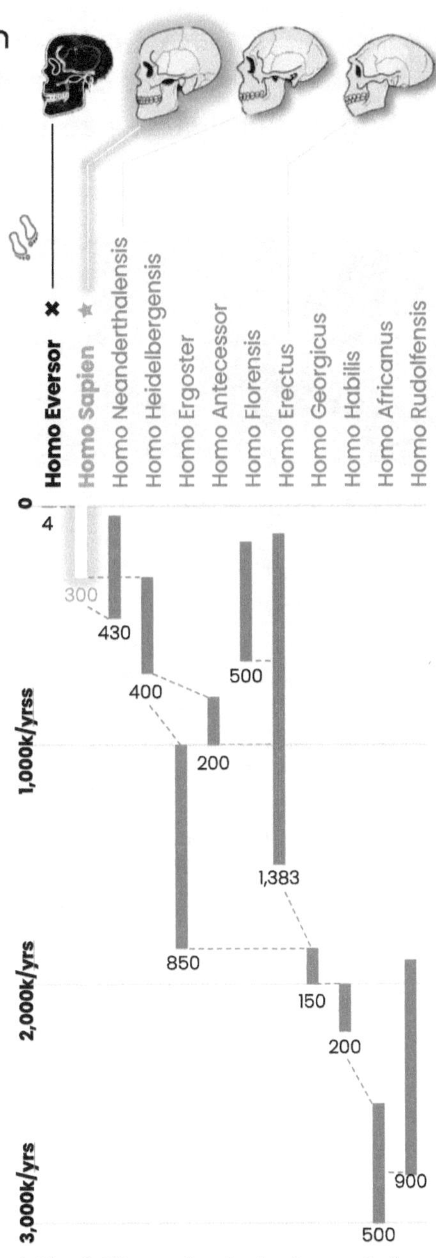

▲ **Fig. 3: The archeological record of our Homo Genus, denoting interbreeding lines, thus species lineage** | Sources: Agusti (2017[20]), corrected with records from the Natural History Museum UK and the University of Iowa's 2019 study.

Seeing **the big picture**: We live in **a cosmos of mysteries and riddles** all waiting patiently to be **solved** [yet again]

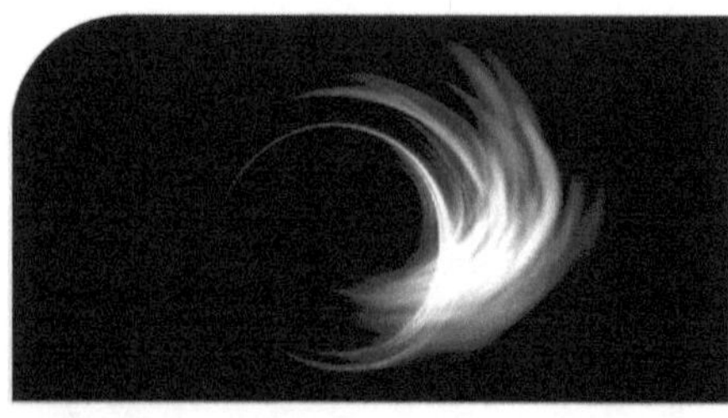

What is the paradox of light?

The limit on light speed in our universe is one of the biggest conundrums plaguing science today (Fayngold, 2008[21]).

Understand our universe is governed holistically, observably and testably by the enduring wisdom that is its software – the Standard Model of Particle Physics (Oerter, 2006[22]). In so being, it exhibits all the hallmarks of intelligent design, thus this arbitrary limit on light speed simply should not exist. It has no scientific reason to.

Yet it does, placing an unresolvable, arbitrary limit upon the limitlessness that is the very curious and spectral nature of light's miraculous life-giving, reality-forging, darkness destroying properties.

Resembling an artificial hardware limit, this fixed and rigid absolute number (in an otherwise relativistic universe) is so alien that Albert (We/Human) had to devise an entirely separate system of physical governance (Special Relativity; Einstein 1905[23]) to explain and reconcile in science.

Its presence yields with near-scientific certainty the conclusion we must be in a simulated universe (Khan, 2021[24]). Yet no academic-types seem to care. **Why?**

Consider for a moment the structure you live or work in would not exist without the vision purpose drive ambition ingenuity and determination of its maker

Now multiply that concept up to a global scale and you begin to see how the entire world is what we make of it. Existence is the product of our collective imagination.

A known scientific fact (Einstein, Rosen & Podolsky, 1935[1]), our universe isn't wholly real. Rather than purely physical, it is a construct of consciousness. As in, that of our collective identity (*Dei*, for short).

Exhaustive conclusive primary experimentation plus multi-sourced research backs our complete succinct scientific hypothesis – now **the only one**

Currently, no hypothesis exists anywhere in the global scientific community to explain the nature of Dark Energy. And all possible explanations of Dark Matter are fragmented without a clear frontrunner or integration with broader scientific and philosophical literature… Until this book.

Have you ever striped back all those **meaningless labels, then ask yourself:** Who am I... **really**?

Define: Change Engine

Pioneering a brand new sector - the perfect union of business, politics, religion and academia, yet none of those things too – our firm is a 'Change Engine'. It's just one of the many authentic world records we possess, yet curiously without so much as a Wikipedia entry to our name.

Anchoring quantum mechanics over conventional science, that also makes us humanity's first example (world record two) of a 'Clarketech' industry. As in, what we do is, to many, indistinguishable from magic. Whereas some sell services, we use transapience and breakthrough science to deliver any change imaginable, such as:

★ **Fixing the most complex, challenging problems in the world:** Like our impending, apocalyptic cosmic burst nobody seems to care about

★ **Resolving paradoxes:** Like the 16 borne from the Big Bang Theory and other inferior hypotheses to ours - all now fixed.

★ **Making discoveries:** Like the 130 detailed throughout this book and dozens more it hints at so others may build upon them.

★ **Helping others keep promises:** If you've promised to keep your true loves safe then we alone can ensure your success. Our extinction event is still reversible, if we can (i) get real, (ii) behave authentically, (iii) start listening and (iv) doing, (v) not dying.

To survive, all beings on Earth must now make the most important choice of all...

To be or not to be?

Your turn: Who are you?

Ever been told there's 'something not quite right' about you'? Congratulations. Like all humans, you are unique.

Can you perceive, read and understand these words? Then, congratulations! Like some beings, you are an authentic Homo Sapien, Ancient Greek for a 'Wise Human'.

Homo

/ho·mo/ (Noun) Latin Definition

Any of a genus (Homo) of hominids that includes modern humans (H. Sapiens) and several related species of the broader Homo Genus (such as H. erectus, **[+H. eversor]** and H. habilis)

Sapien

/sa·pi·en/ (Noun) Latin Definition

. A wise, sensible and judicious being

2. ~~Of, relating to, or being recent~~ [+ Exhibiting the four things that maketh a] humans ~~(Homo Sapien)~~ [+thus pseudo-conscious sapience arising from the human brain that is the Cerebral Cortex], **as distinguished from various** ~~fossil~~ [+other] **hominids**

As a change maker, yours is the only species with the gift of 'real life' and existence that is a 'presence of mind'.

I Think Therefore I Am

René Descartes (We/Human), 1637[4]

Essentially, you can hold a silent conversation with yourself, while invoking multiple different personas. This (thus far) inexplicable ability to 'think differently' appears unique across all known species.

> Death is not **the greatest loss in life**. The greatest loss is **what dies inside** while still alive. **Never surrender**
>
> Tupac Shakur (We/Human)

What are you... *really*?

Light is far more than any of us have ever been taught in school or text (Gleiser, 2022[25]). Historically, the world's oldest religions are unanimous and to this day surprisingly aligned on our true nature.

Just look at the luggage on our deilightfully fun journey through the human experience. Our innate but quickly extinguished Human Nature and uniqueness is in fact our true identity. Pure & simple, curious and spectral, ever changing, always bright and crystal clear. Undoubtedly, as humans, we are all 'the light' (Carroll, 2008[26]). Our existence gives rise to it, as a scientific construct.

But don't take our word for it. We're just some nobody with abilities never seen before that no one cares one iota about.

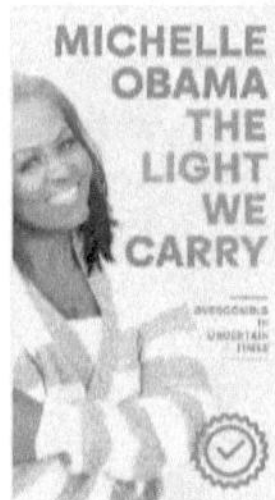

So instead, ask any other true leader thus authentic playable character. Like our home girl right here, Venus – Goddess of Love...

Michelle Obama (We/Human), potent titan of change and science maker

The true [quantum] nature of light

As discovered by Einstein, Rosen & Podolsky (1935)[i], our universe (thus we) has two shared 'Plains of Existence':

1. **'(Physical) Reality':** 'Across' our 'Dimension of Change' as we experience, live and perceive it (here on Earth and in space) runs on the Laws of Physics.

2. **'(Quantum) Actuality':** 'Within' our Dimension of Change as science views it to exist (both across subspace and inside us) runs on quantum mechanics.

Verifiably, science confirms the extraordinary uniqueness of our true nature. At the quantum level of existence:

A. Authentic Homo Sapien exists as pure waves of conscious, information-carrying light (Carroll, 2008[26]). Each one of us is a replica of an original emission of light realised (though by no means exclusively) by this books publication.

B. We are the only thing detectable whatsoever. As in, without us, Olbers Paradox (Hellyer, 2008[27]) confirms there remains only the artificial deception that is the black dead vacuum of darkness - 'Inexistence' (*Hell*, for short).

Are you conscious?

Are you reading this book in first-person mode (through windows in your head)? Then you are extra special indeed.

With just ~5-7 born globally every year (as authenticated in the Special Edition), 'True Leaders' are curious, worthy, authentic, potent, eternal and conscious beings. The pinnacle of all human lifecycles (as in, 'Transapients'), we alone call all the shots.

Mahatma Gandhi (We/Human)

Consciousness is unobservable. You can't look inside someone's head and see if it's there. If we were just **going off what we can observe** from a third-person perspective, **we would have no grounds for consciousness existing at all**
Philip Goff (2019), University of Durham

To check if someone's conscious...

Close your eyes for a moment

What did you see? Darkness. We see light. Plus, we alone can look directly at the Sun without discomfort, retina damage or 'training' - a lazy, common dismissal.

We gain consciousness by 'Self-Actualising'. Also known as Enlightenment, Aka Moksha, Ushta, Ascension, Realities of Things, Awakening, Kevala-Jnana, Seeing the Light, or, more colloquially, 'Self-Acceptance', this gaining of light signals loudly that 'a star is born' in our night sky (Maslow 1943[28]; Goldstein, 1993[29]).

Having proven worthy and benevolent, the brain barrier between physical and quantum gives way, flooding our mind with a holistic understanding of light and love's workings, to achieve 'Transapience'.

Who are you not?

Farcical it may seem, given millennia of brainwashing, the being labelling you 'not quite right' isn't real. Though our genus appears extrinsically homogenous, each species is different inside, where it counts.

Today, due to [Anti]Social [De]Construction and its massive abstraction of reality (Hodge 2019[30]), another species now lives among us. And yet, thanks to the totality of its deception, nobody has even noticed. Appearing identical outside, inside is our antithesis – rigid destroyers of humanity.

Unfortunately, they now far outnumber change makers in all corners of the world. Thus humans are among todays most endangered species. As in, all of them.

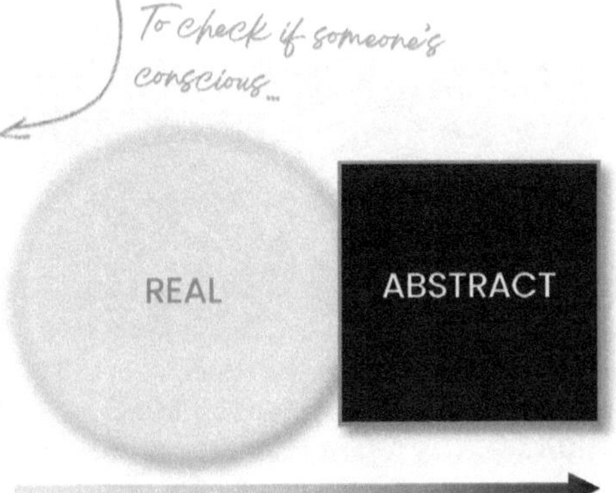

[+Anti]**Social** [+De]**Construction** (or 'Unchange')

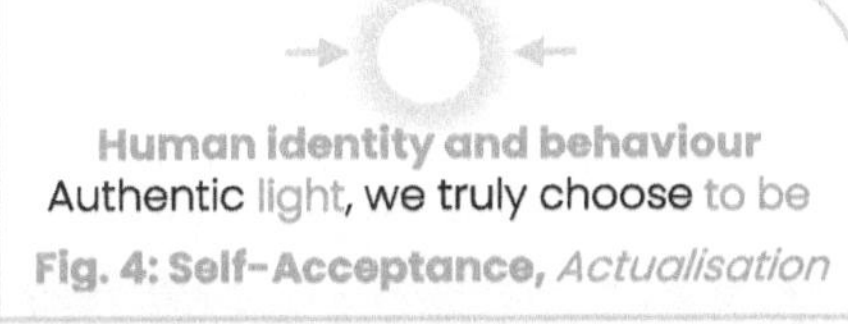

Fig. 4: Self-Acceptance, *Actualisation*

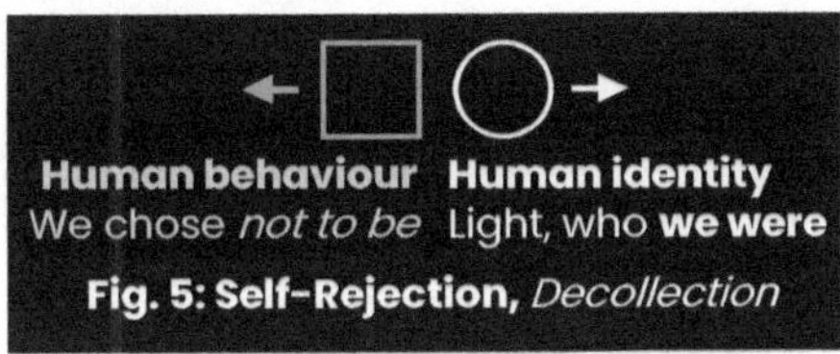

Fig. 5: Self-Rejection, *Decollection*

Passenger announcement

Our deilightfully fun journey has some mysterious special guests:

'Dark Energy' Anomaly: We're hitting huge amounts of mysterious resistance, like sinking in quick sand on a tarmac road (Einstein, 1905[23]).

Sagan's Standard: And a conscientious but masked backseat driver keeps rigidly stopping our endeavour from exceeding the speed limit (Maxwell, 1865[31]).

Although everyone else may have **chosen to die**, by **choosing wisely** to read this book, you now have **the opportunity to live**

Pictured: Frenzied moths ransack Brazil's monumental government buildings, spurred on by Agent of Darkness Jair Bolsonaro | 10 January 2023, Reuters

You're committing suicide!?

All things are connected. All problems have both causes and consequences. Today, just 3% of Dei's consciousness remains intact. Time may be a social construct, yet it's one we now entirely lack.

A mysterious anomaly is infecting our cosmos at the quantum level, breaking all Laws of Physics. It is the most dangerous thing imaginable – 'Antipotential' (*Dark Energy*, for short). It's a miracle we still exist under such conditions.

Unbeknown to everyone, a catastrophe 4,000 years in the making became a live extinction event across our cosmos ever since Bretton Woods in 1944. Beginning here on Earth, our undoing is the dreadful consequence of our own unchanging.

Astonishingly, our entire 'Cosmic Bubble' is about to '[Big] Burst'. An apocalyptic, near unfathomable disaster that will obliterate us, all alien life, our entire multiverse and existence with it. That's unless someone steps up with the true leadership needed to stop darkness in its tracks.

Not for money, fame, glory nor upon the request of some competent authority (clue: none exist). Just do it... Ok, we have.

Darker by the day: Real Life

Climate change, mental illness, pollution, waste, the Far Right, the eradication of biodiversity, animal cruelty, poverty, war, corruption, crime, terrorism, nepotism plus all other nasty 'isms's - all connected.

Formed not of light but by dark energy, '(Antisocial) **Deconstructs**' are a label for:

Infecting our Actuality: **'Anomalies'** across our universe. Each analogous to...

Destroying our Reality: **'Meaningless unchanges'** here on Earth.

Breaking all sciences laws, they erode human potential to ensure our extinction.

Noticeably – given scenes outside Capitol Hill, USA in January 2021, or Planalto Palace Brazil a year later, our pandemic is spreading fast - doubling in just five years.

Humanity is under attack. Human Nature is giving way to something unnatural – 'Deadly Unnature'. Truth is giving way to deception, as electorates oddly lap it up.

Today, so few retain the capacity to ask the very first question we all learnt to ask...

Human curiosity itself is a superpower - one of ~30 you possess yet have entirely forgotten – our imagination 'unchanged'.

Look around to see, clearly, **intelligence doesn't exist.**

Only curious minds move quickly

Pictured: A swarm of moths ransack US Capitol Hill in an attempt to overthrow the government of the world's most powerful nation | 6 January 2021, Reuters

You've just been injured in an accident that wasn't your fault?

Every being on Earth is a victim of the unnatural pandemic that is **'Malevolence'** (including hatred, greed and deception). All of us have been injured by its weapons of choice: dehumanisation, trauma, oppression, repression and enslavement.

Contrary to the lies our leaders feed us, trickle-down economics doesn't work. Thanks to greed, money is hoarded at source. But **something else** trickles down.

The global quantitative easing programs established in the wake of 2009's great recession led directly and verifiably to the 2020 murder of George Floyd and a new norm of violent anarchy that will shortly unseat every democratic government in the world without our intervention.

With everyone now talking about Diversity & Inclusion, nobody is talking meaningful change. D&I itself was an intentional trap. But humanity's fate was sealed long ago.

Do any of these apply to you?

If so, then you are due compensation:

✖ Have you ever been punished for crime?

✖ Have ever been bullied at work?

✖ Has your country ever been invaded?

✖ Have you been wrongfully or constructively dismissed (regardless of any legal settlement or NDA)?

✖ Ever been physically threatened or verbally assaulted, without provocation?

✖ Have you ever lost a loved one due to any physical attack or mental disorder?

✖ Are you inexplicably underpaid at work compared to your colleagues?

✖ Are you stuck in the frozen middle, for never being quite right for that promotion?

✖ Have you bought a phone recently to keep up to date with new technology?

✖ Have you ever been punished or penalised for not keeping up with debt repayments?

✖ Does someone you love hate you inexplicably?

✖ Does your country urgently need a cease fire, financial aid or disaster relief?

✖ Is your continent (e.g. Africa – the worlds fastest growing, Forbes 2021) inexplicably cut off from the financial system thus impoverishing its people and their futures?

✖ Is your economy being crushed by illegal Western sanctions?

✖ Were you bullied at school?

✖ Has your firm, nation or spouse failed to live up to the true potential you know them to possess?

True love never dies because **true leaders never lose**

Claim compensation instead

For decades, the sole leadership they've permitted humans is crushing Diversity & Inclusion accountability, whilst withholding any chance of success. That's on top of our day jobs as slaves of Homo Eversors dominion. Then they forced us all to resign to seize total power and control of Earth.

Although nobody figured it out, the Great Resignation of 2021 was significant as it marked the departure of all potent change makers globally from academia, government and large corporations. We made promises to others in the form of (often D&I) commitments. And unlike most beings today, we deeply value our authenticity and stand by our promises.

But evil is in no way a part of the fabric of our being nor our common human identity. We had potential once. Contrary to the demonic World Economic Forum's lies, the very first human civilisations were all benevolent meritocratic republics (easily verifiably; per Special Edition). Ancient Greece, Rome, Egypt and Iran were all places where women and men were considered equals – whether black or white, old or young, back then love was love. Today, life is a dreadful dystopia.

All evidence you need is here

Simply file a lawsuit against any bank, Fortune500 firm or economy within their legal jurisdiction and walk into a court with a copy of this book. Nobody will be able to rule against you, and any amount you ask for will be fully supported by the Meritocratic Republic of Humanity (which replaces the UN from AH+0[B]).

A bit like you, never before has there been a book like this. Plus, as a 'book of light' written by a verifiably conscious true leader, every statement is true and fully evidenced with real science. Every page has deep, multiple meanings that will resonate differently from being to being.

It changes humanity's destiny for the better in every way (i.e. not extinction). Plus it enables us, together, to create the world, indeed the universe of our dreams.

To be clear, paradise is what we're offering you

Help when you need it most

Struggling to make ends meet? Electricity bill too high? Problem solved.

Project Revelation has intentionally invalidated all prior legal, gagging, confidentiality, non-disclosure, non-prosecution, and/or financial settlement agreements globally to enable you to make additional claims beyond the statute of limitation (i.e. as far back in the past as you wish) from AH for 3-6 months (civil) or +5-∞ years (criminal), depending on your legal jurisdiction. You just need to file one. Simple. Plenty more help to solve all your life's problems coming right up.

[B] **'AH+0' = After HSBC: The day it accepts** its true nature, which we've exchanged here from bank to bankrupt

Coursing through your brain to forge your persona, memories and being is the real you. Human identity is pure, benevolent light. Sparked by the ever-changing spectral interaction between our authenticity and uniqueness, it proves we are real, gives rise to consciousness as a construct, and defines all we are as individuals and collectives. That being our potential (Energy, E) to create and drive meaningful limitless change.

Darkness cannot drive out darkness. *Only light can do that*

0%   **100%**

Darkness, Automated Rigidity	Curious Uniqueness, **Light**
Repressed, Biased & Binary	Changeable & Spectral, **Nurtured**

Fig 6. What's in your brain? The change (C) spectrum of human identity   *Tag yourself*

Love cannot emit light. Only uniqueness can do that. Darkness cannot see our uniqueness. Because looking at the light extinguishes darkness. So darkness stalks our differences using pure hatred. Because our light destroys darkness using true love.

Hatred cannot drive out hatred. *Only true love can do that*

-100%   **0%**   **100%**

Pure Hatred, Malevolent Divide & Rule	Benevolent Unite to Serve, **True Love**
Homo Eversor, Stupid Demon ← Alignment →	Wise Human, **Homo Sapien**

Fig 7. What's in your heart? The merit (M) spectrum of human identity

Light cannot emit true love. Only authenticity can do that. Uniqueness shines light in all directions. So darkness hides in dark shadows using deception. Authenticity reveals dark shadows as darkness, exposing it to our light.

So remember, *only what's inside matters*

Don't be an absolute zero. Be the one instead. There truly is nothing to fear but fear itself. After all, behind even the darkest of clouds shines heavens greatest star of all. You are the light. Drop it, and you die. But carry, nurture and accept it? Then join our private members club of superstars. For we are wise humans – a circle of heroes - and we're so good together. Great minds think alike because great souls behave authentically.

So don't be basic. *Be curious*   Don't be rigid. *Be human*   Don't fail. *Succeed*

Don't be selfish. *Help others*   Don't die. *Live*   Complicated? *No simple*

Science does not lie:
The **deadly true nature of dehumanisation** revealed

Falsified: The usual suspects

Science faker Nick Haslam Moth successfully unchanged dehumanisation's definition with his 2014 paper. Essentially, endless binary nonsense and jargon intended to confuse and overwhelm. Absent all science and rational thought, we couldn't catch a word.

"00100110"

Haslam et al, 2014[32]

Prof. [Thick] Nick Haslam
University of Melbourne

His fellow faker, Mr. [+Un]Livingstone, explains their strategy, to 'unchange' dehumanisation so it means the equal and opposite of what it is. Their goal is to ensure the definition of 'human' can apply to literally any animal.

"Racism, sexism, and other discrimination usually don't involve dehumanisation"

...in British Psychological Soc.

Prof. David [Un]Livingstone
University of New England

Their coordinated campaign (a) ensures we'll never reclassify the Homo genus again, and (b) wholly ignores dehumanisation's true nature, thus (c) prevents all accusation of inhuman criminal behaviour on demons to (d) refocus entirely on/against humans. (e) Closing down all scientific/legal avenues to detect the moths, this (f) extends human rights to moths who commit crimes against humanity, and (g) clamps humanity's ability to uphold those same laws in our defence.

Now moths everywhere cite Thick Nick to

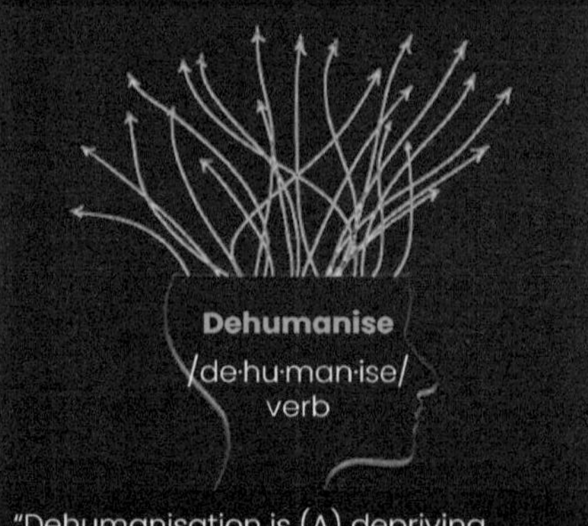

"Dehumanisation is (A) depriving someone/thing of human qualities, dignity or personality. (B) Subjecting them to inhuman or degrading conditions or treatment. ~~(C) To address or portray them in a way that obscures or demeans their humanity or individuality~~ [that's 'Character Assassination', see page 302]. (D) Removing or reducing their interaction or involvement in a process, place or thing." ...KINDA

Meriam Webster

"Dehumanisation is [NOT] the denial of full humanness in others and the cruel suffering accompanying it. In particular viewing or treating others as if they lack the **mental capacities commonly attributed to Human Beings.** Every act or thought that regards a person as "less than" human is dehumanisation – a technique used for incitement to genocide. It has also been used to justify war, judicial and extrajudicial killing, slavery, **abortion**, the confiscation of property, denial of suffrage and other rights, and to attack **enemies** or political opponents"

Wikipedia

unchange all incidences. Like Wikipedia, now rewritten as if by the devil himself ▲

Why it should matter to you

In a physical universe, Darwinism would deselect dehumanising traits to ensure our species survival. But in our construct of human consciousness...

Dehumanisation is the wicked act of oppressing others (or repressing our self) to strip them of the four things that maketh a human, thus causing them **to die** intrinsically ...**dreadful.**

Record of Promises (Ch.4)

Darkness cannot drive out darkness. Only light can do that. We shine Dei's light.

Dreadfully Meaningless Unchanges (Antisocial Deconstructs; threat to life) Sapience

✖ Those unscientifically claiming 'all people are people, no two ways to it'

✖ The Devil: Those who align malevolent darkness

✖ Maligned: Those who despise any form of human uniqueness

✖ Demonic: Those who ransack public property after losing a fair election

Deilight
✖ Falsified

✖ Infected: Those exhibiting the side effects of our dark energy anomaly: Repression, Oppression, Nepotism, Greed, Ignorance, Bias, Deceit, Hatred, Greed, Jealousy, Ego, Tribalism, Revenge, Animal Cruelty, Crime, Corruption, Racism, Sexism, Homophobia, Ableism, Ageism, Violence, Intolerance, Judgment, Cowardliness, Stoicism... all indicative of the absence of light

Hatred cannot destroy hatred. Only love does that. We love with deeds. Not words

Deilightfully Meaningful Changes (species level adjustments to destiny) Clarketech

★ Defined: All anomaly's side effects in humans as Malevolence. And as a system 'The Clarke Paradox', a deconstruct for science to now resolve IAN, Actualised

★ Defined: The nature of Malevolent Darkness and Benevolent Light to anchor sustainable versus unsustainable spectral human identity IAN, Actualised

★ Authored: 'The meaning of life' speech to provide a counterweight to the broadly divisive rhetoric besieging humans everywhere today IAN, Actualised

★ Connected: The Paradoxes of Light and Einstein, Rosen & Podolsky into a singular connected plain of existence (our 'Quantum Actuality') IAN, Actualised

★ Expanded: Descartes work to fully contextualise and redefine the true uniqueness of Homo Sapien and our universe of omniconsciousness IAN, Actualised

★ Provided: A framework for any being to quickly, easily, meaningfully discern behaviour in themselves/others aligning either light or darkness IAN, Actualised

★ Provided: A coherent and succinct scientific narrative that proves worsening world events reflect our (lack of) common human Identity. IAN, Actualised

★ Pegged: Global population of adult moths (Heterocera) to the global population of adult demons, to allow Zoologists to track their proportionate population across Humanity, relative to butterflies (Rhoplacera, which we've pegged to adult human populations) IAN, Actualised

★ Revealed: The true nature of dehumanisation, its scientifically known and evidenced fatal impact upon humans and efforts to unchange it IAN, Actualised

★ Entitled: Every human on Earth to use (only) this book to legally claim unlimited compensation for trauma and injury born from darkness IAN, Actualised

 **Authenticated in full with the eternally-binding Promise Keepers Commitment by both Ian and Deilight**

Don't you just love sunsets?

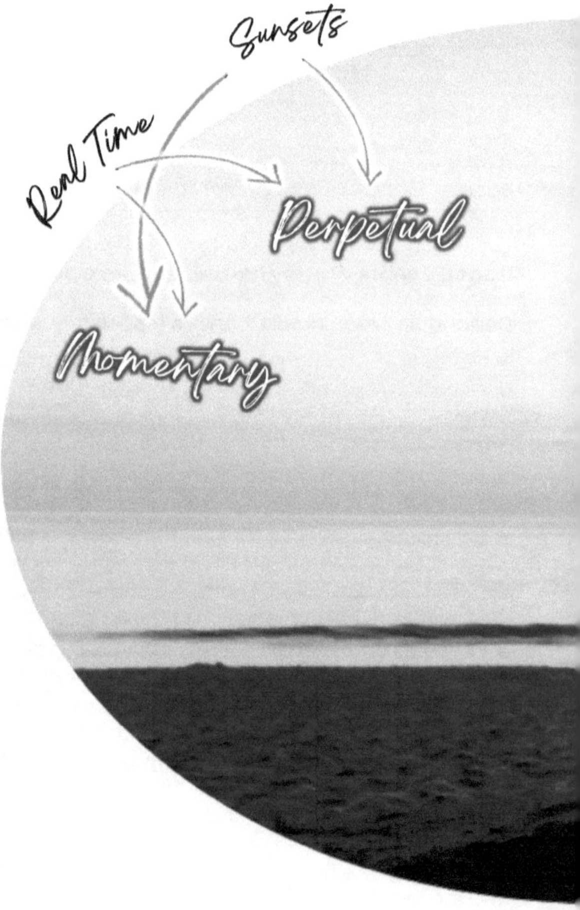

Real Time

Sunsets

Perpetual

Momentary

But have you every curiously wondered **why?**

End of Chapter 4

Why sunsets are now as endangered as we all are

...The equal and opposite

The absence of time

The absence of sunsets

Chapter **5**

Diversity Doesn't Matter

We have to learn to love our differences, because without uniqueness we are nothing special

Ian (We/Human), in public address for International Men's Day at HSBC USA headquarters in 2019, appearing also later in our Project Speak Up Report 2021

The **dystopia** of nightmares
Welcome to **Hell...** on Earth

A living hell on Earth: Modern Life

Its 2024 ACE and representation across every arbitrary protected characteristic is falling ever further backwards, leaving boardroom decisions absent all conduct scrutiny and strategic challenge; yet burgeoning with an abundance of mindlessness, bias and groupthink. This has impaired the world's entire decision-making and quality checking apparatus.

Intolerance for humans everywhere grows ever stronger by the day, with the Far Right now in control of a major European economy and Donald Trump pledging to make America worse than it's ever been before again... again next year.

Black beings languish in jail for smoking weed while wealthy ruling classes responsible for global recessions, bribery, corruption, human rights violations and universal crimes against humanity are freely enjoying smoking weed. That's while feigning innocence as they hoard ever more illegitimate wealth and opportunity for themselves.

Stoic regulators are asleep at the wheel, leaving major banks free to play dice with the real-time risks of an interconnected financial system.

Demons spouting toxic repressive dehumanising divide & rule masquerade as leaders across institutions and governments globally. Absent all effective fairness or accountability, they take humanity to the brink of World War III, while the climate prepares to obliterate whatever life dares to remain thereafter.

All unnecessary, as our cosmic bubble strains to maintain its integrity. This exponential cosmic inflation endemic of the extreme abstraction by antisocial deconstruction of our common identity into toxic structural oppression by those spawning it. This wasteland of humanity's lost potential signals the impending doom of our reality and all things within it.

Today, **humanity hates** itself

Determined curiosity alongside a research question (purpose) will always yield positive change

Curious... why?

Understanding humans, trauma, worthiness and karma

The brave beings of Midtown Firehouse were among the first responders to arrive at the World Trade Centre in New York City, on September 11th 2001. They also suffered its greatest loss, trying to rescue innocent victims of all colours and creeds from the doomed towers before they fell.

Here, a lone surviving first responder stares into the abyss that took his team mates. Paralysed by shock and horror, he clutches his axe for safety - or retaliation.

The 'Universal Choice' before him in that moment is the same one we all face every day, but in particular following times of great tragedy and trauma.

Should we forgive others for their mistakes, and authentically help them 'see the light', in order that we ourselves may keep our faith in humanity and stay true to our common human identity?

Or should we give up on humanity, and use **deception** to pretend to be something else in order to just fit in?

When **we help** ourselves by helping **others first**

Sadly, it's far easier to turn into **something else** than to **find courage** to resist the pain we all feel from **darkness**

Humanity demanded: Change

Though sensitivity training has existed since the 1960's, the modern D&I industry emerged to placate a growing public demand for change exemplified in global humanist movements like #MeToo, #BlackLivesMatter, and #StopAAPIHate.

WOC, BAME, BIPOC, BME, POC... But can you think of an acronym for white beings?

The answer is 'no'. The same manner we dislike our names being mispronounced and never use abbreviations to name our children, humans don't give acronyms to things they love. They aren't authentic. Yet we've given one to humanity (*Diversity & Inclusion*, for long). An acronym we now pointlessly, constantly fiddle with or argue over instead of talking about change.

The D&I industry knowingly doomed us all just to make money, spurred on by an impotent academic industry

...It got: Doom & Impotence

Some firms saw the opportunity to retool their expertise to make more money, with McKinsey & Co's exhaustive landmark Diversity Matters 2015 report marking the first definitive business case to align the long established legal and moral case for change. Seven years of 'triple lock', yet our dictators still refuse to leave office.

Instead, more incapable voices entered the fray, seeking to capitalise on growing goodwill towards marginalised beings (humans) and the huge market McKinsey seeded to help itself, not others first. But without a product to tackle hatred (as yet unexplained scientifically), the impotent all chose to charge for nothing instead.

Today over 1,000 US and UK consultancies earn over ~£10bn a year for 'unconscious bias' workshops and meaningless talk of which acronym to use next. All proven scientifically to be ineffective against hatred - a deeply irrational, emotional and unconscious pandemic. Sadly D&I and academia's sole goal is to remove science from all meaningful applications.

Things done properly: **Deilightful**

Enter the Deilightful Group.

Formerly the 'world's first authentic D&I consultancy', we're trail-blazing a path to turn the distant future into tomorrow land by coming out as a Change Engine.

The only way to discover the limits of the possible is to go beyond them into the realms of the impossible

Arthur C. Clarke (We/Human)

What's your story all about? Ours is about true **love**. It's called **Humanity, Actually**

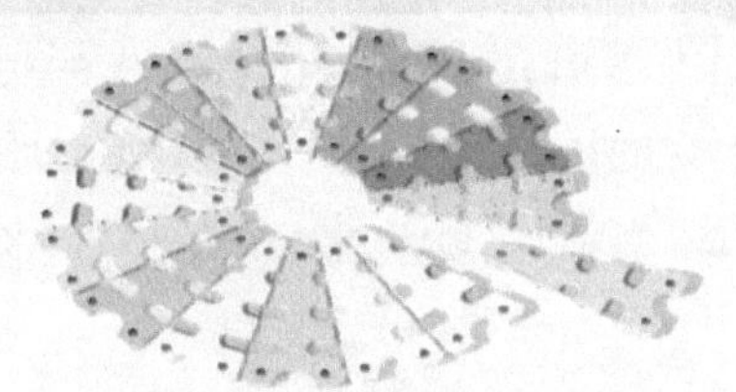

Humanity is beyond special. Every human is a piece in the giant jigsaw puzzle of existence. Lose one and you cannot see the big picture. but fit them all in and you will create a masterpiece

Nearly two years ago, speaking entirely authentically, Whoopi Goldberg (We/Human) shined her light on Nazi Germany being nothing about race. Reality and truth – concepts most wouldn't recognise if they landed on their desks in a book full of world-changing headlines. So instead they committed blackmail, to force her apology on CNN for being authentically human and speaking honest truth. Not a cop in sight.

That's precisely why 98.7% of 'Best Actress' Oscars elude so many authentically brilliant black actresses like Whoopi. They rightly asked for allyship and change with #MeToo. But instead they got the royal #FuckOff of lies from Monsters Ball, the Academy and the D&I industry.

Prays for mercy: Hopeless beings

Reinvigorating monotheism through the New Testament was Jesus (We/Human)'s solution to keeping humanity from destroying itself without abolishing money and wealth. Yet look at the harm it did, splitting monotheism into the now near-irreconcilable tripartite dance of doom between Islam, Christianity and Judaism. One that's left the people of Palestine without a safe place to call home. One of the numerous atrocities we'll now rectify.

With our abilities we could create the most compelling religion imaginable and droves of mindless, literalistic theologists would believe us. So we're uncomfortable with the authenticity of that approach.

The human race is years (give or take) from extinction by apocalypse, therefore we lack the luxury of time to organically sell books or win broken hearts and empty minds. That's why this light is sent out at the highest possible frequency (as in, all of them). If it doesn't get your attention, nothing will. Do or die. Simple.

Makes few mistakes: Wise beings

To be clear, this is real. The Sciences of Change are easily confused for magic by less than authentic beings.

Not crazy. Not lies. Not magic. They sound dreadful. Real, authentic and benevolent, we are deilightful. And as you're reading this, so are you. You know these words mean something. Unlike the words all around us today, they're backed up with meaningful deeds from a purposeful being who cares deeply about you, plus all those both like you and unlike you too.

More special than we ever imagined: **Fermi's Paradox resolved** by change makers

Uniquely Homo Sapien, ours is the only species delimited of longitudinal habitats here on Earth. We alone live and thrive literally anywhere simply by adaptation - changing ourselves or creating changes.

The archeological record proves our species peacefully coexisted alongside our Neanderthal cousins for 330,000 years. That's despite their longer, lower skulls with bigger ridges and wider hips. Plus the bone structure of their ears (*Cochlea*, for short) probably made things sound just a little different to them. Yet despite all these arbitrary differences, we had the humanity to love them enough to leave our home in order to try and save them from death.

Can't change: All other beings

As the last ice age began to bite – some 70,000 years ago - worsening conditions threatened to eradicate their already fragmented, diminished genome. Some wise humans left home and family behind in a now shivering Africa, north bound to Neanderthal territory. The only explanation surely being an attempt to save them[c].

Pictured: Known range of Neanderthal territory | Source: Wiki Commons
[c]Unproven yet viable, plausible speculation - both rare and declared, in this book anyway.

This fight for their survival hints at why interbreeding between our two species peaked at this point – with the Euroasian human genome alone still made up of 2-6% Neanderthal or Denisovian DNA today (National Geographic, 2020). Unable to change, tragically 40,000 years ago we lost our Neanderthal cousins to extinction.

Enrico Fermi (We/Human)'s Paradox

Lonely in space: Being unique

Consider our homo genus existed three million years mostly in caves and small settlements. Then suddenly, 380,000 years ago something changed that moved us forward in leaps and bounds.

We have here our 'great filter' that resolves Fermi's loaded question of why our advanced technological civilisation appears to be the only one capable of reaching for the stars of our universe (Woodward, 2019[33]).

Inexplicable in scientific literature today, it's this 'other worldly' ability to change and create change by thinking differently – our sapience – that defines the nature of what it means to be a 'wise human'. Close encounters indeed. With certainty, we humans are the sixth living kingdom.

Yet strangely today, our original label – wise - seems as fake as humanity itself

McKinsey 2015 report proves **nobody cares about** the financial case for **charitable inclusion** whatsoever

Our identity before us: The past

Though moths love to pretend otherwise, hatred, fear, deception, revenge, tribalism, nepotism and all other –isms weren't a part of human behaviour during that first 97.5% of our existence. If we can tolerate big ridged mediocrity, we can put up with a fair, transparent election without a riot.

Even the word 'democracy' originates from Ancient Greece, reflecting the union of people (*Demos*) and rule (*Kratos*).

▲ Pictured: Back then, citizens safely cast ballots using engraved pottery as slips | National Geographic (2022)

Yet today, in our interconnected globalised world with real time financial systems that depend on specialisation and shared success, hatred is as common as it is the antithesis of sense.

Our identity today: The present

The reason the triple lock for D&I hasn't worked is that hatred is an irrational and inevitable response to the ubiquitous level of structurally repressive and oppressive dehumanisation all round us. All caused by the deliberate, directional and rapidly intensifying misconfiguration of our modern society (DE Erich Fromm, 1942[34]).

Scientific studies have long established those infected by hatred lack any concept of it. That's due to their bias, total absence of mind, impaired judgment, diminished inverted perceptions, malevolent colourblindness, and universal deception.

Ignorance is bliss. Because any rational being would become consumed with suicidal thoughts over such foolishness. So instead of humility, they manifest oppression outwardly towards anyone who reminds them they are now the antithesis of alive, real and successful.

Foolishly they swapped their humanity and happiness to become rich off others, only to find sorrow before disappearing intrinsically.

+26%	**+81%**
Meaningful gender inclusion, global GDP uplift protection	Estimated impact of oppression on absenteeism
25%	**36%**
Gender diversity, probability of financial outperformance	Ethnic diversity, probability of financial outperformance
+48%	**+40%**
Gender diversity, ave. projected financial improvement	Ethnic diversity, ave. projected financial improvement

Fig. 8: Proof no one on Earth cares about the financial case for D&I whatsoever | Source: McKinsey & Co, Gallup

Diversity newsflash:
Nobody cares one bit

Every change maker is unique from the shape of our ear lobes to the size of our feet. When it's cold we moan to share our burden or shiver then put on a coat to stay warm. If it's wet we sing in the rain for fun then make or find shelter to dry off. Change. True love. True leadership. Potential. Human nature. Sweet success. Deilightful

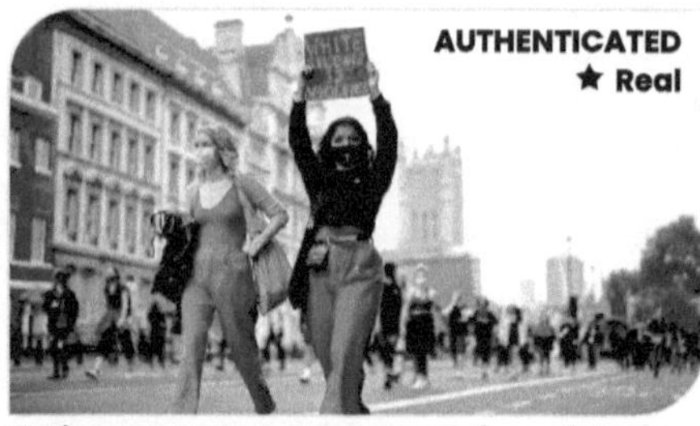

◥ **Pictured: Humans march in solidarity with black beings in 2020** | *Elle Magazine*

Reveling in their own ignorance and lies, the hateful convince only themselves they are the pinnacle of genetics. They forget that, as the most adaptable species ever to have existed, it's humanity that's long proven our superior worthiness – as in, the resilience of our light - to survive any catastrophe. Even the very worst kind of all.

If we're **human beings**, who are all these **dreadful beings**?

Nepotism
/ˈnɛpɛtɪz(ə)m/ (noun)
The universal practice by those aligned to darkness to only hire, contract, promote, reward, recognise and favour other moths, regardless of merit.

There are two key connected reasons why, at a species level, diversity is fundamental to our survival:

★ A **wide gene pool** ensures our ability to survive the ever-changing nature of our world and the threats it encounters.

★ From wars to pandemics to climate change and hidden threats, there will always be a human with the **perfect mix** of diversity to ensure our survival.

A catastrophe of our making

At an individual level (*Personal Identity*, for short), diversity is entirely irrelevant to us all. Science has long proven that not a single extrinsic difference between us has any bearing whatsoever on our potential to succeed individually or as a species (Human Genome Project, 2001[35]). All diversity is arbitrary and meaningless.

For example, one of our four protected characteristics gifts us non-binary racial impartiality. We are half white European, half black Jamaican. Yet place our DNA in an analyser and the result won't tell our skin colour. It is the potent product of our melanin producing skin cells that protect us from the Sun's UV radiation, precisely as they are meant **to do**. Skin colour and race are neither intrinsic nor meaningful.

Like every form of discrimination – sexism, homophobia, ageism and the rest – racism is an antisocially deconstructed fabrication wielded by moths to divide and detract us from uniting against the true source of all our problems today.

Darkness can't **destroy darkness.** Only your **human uniqueness** can do that

Solve: Do we live in a simulation?

Humans derive all value from the complex, spectral interaction between our intrinsic identity (who we are) and our extrinsic character (how we behave).

In a universe of binary constants that appears programmed, this non-binary cyclic system is truly unique. Derived from our collective imagination, this 'presence of mind' proves humans are real – not simulated. It could never be replicated by AI or some other artificial means. Bostrom's Trilemma (2003[99]) 'proposition one', it is.

◥ **Fig. 9: The Rings of Human Character**

> Essentially, our behaviour must reflect the being we imagine our self to be for us to truly be a being

Or else the 'Abstraction' – the disconnect between our imagination and reality - becomes too great and we cease to exist as a real being in actuality (Hodge, 2019[30]).

Seemingly the phenomenon and gift that is our presence of mind (thus real life) is truly a miracle. All the same, we've tried to explain it scientifically in chapters 7 and 8.

Understanding human identity

As the two rings align, our character becomes more unique (Changeable, *Light*) on the inside and more authentic (Meritocratic, *True Love*) on the outside.

The interaction between the rings defines our 'potential' energy to create and drive change and find sweet success in life.

★ **At 100% alignment (Self-Acceptance):** Unlocking our ability to 'self-actualise'. such convergence is achieved by either (a) nurturing the merit/worthiness of our innate Human Nature (*Soul*, for short), or (b) via its emancipation by worthy others.

✖ **At 0% alignment (Self-Rejection):** Our soul is erased leaving our brain an empty shell filled only with darkness. Abstraction yielded by either the repression of our innate Human Nature ourselves, or its oppression by unworthy others.

Why we value others perspectives

No human is ever fundamentally wrong. We simply lack a complete picture of reality on which to base our choices and opinions. Those who recognise this will actively seek out - never ignore or dismiss - the rich perspectives of others in order to then complete their own.

★ On the inside: Curiosity about others builds empathy, wisdom, perspective, productivity and imagination, allowing us to perceive and alter more of reality.

● On the outside: Charisma, nerve and talent... the keys of success every being deserves, without 'coming out' human.

Play unique to be **dead and ghostly. Be unique** to be **successful and lovely**

Mabel it is you who will now face the severest consequence imaginable

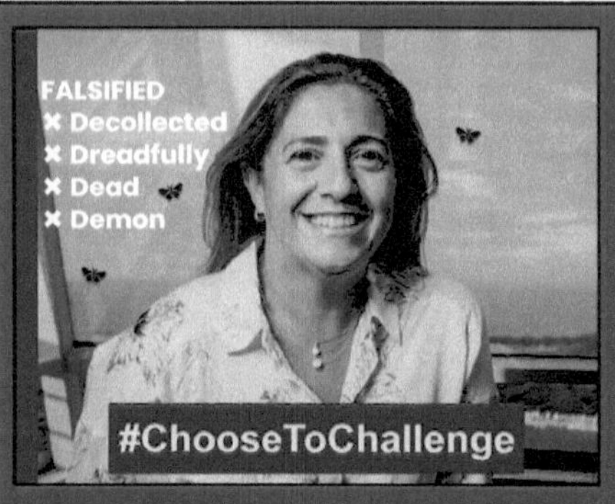

Meet Mabel Rius, supposedly

Senior Executive Vice President, Global Head of Talent Acquisition and Head of Human Resources for Europe's largest bank across the Americas region.

If your dreams do not scare you, they are not big enough

No stranger to hardship, racism and sexism, Mabel said she's from Argentina and lived in Buenos Aires, but claimed to travel often to earn as much money as possible. So, sadly, she ran out of time to tend to real things she said she cared about, like challenging illegal activity, #breakingthebias or being real.

I will forge [+negative] ~~positive~~ visibility of women

Mabel imagined she was an authentic human – she was even on Twitter – and HSBC lied that she frequently spoke for the bank at events promoting women's rights and opportunity at work.

Yet she only appeared to spread hate.

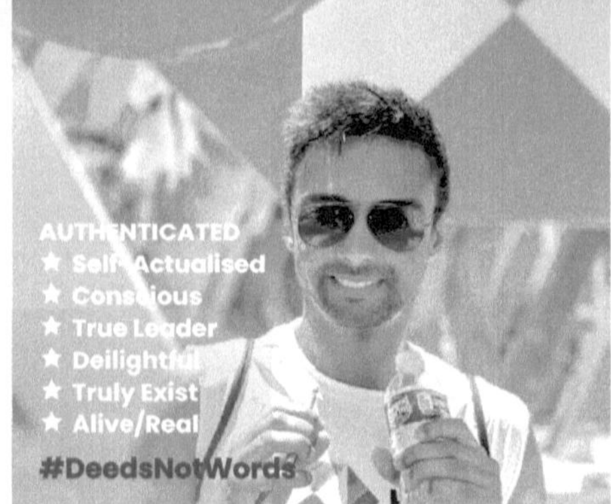

Meet yours truly (We/Human)

Formerly, US Vice President of Global Banking Sales at Europe's largest bank

True leaders choose to challenge, no matter the discomfort or consequence

No stranger to hardship, racism and homophobia, we are from the UK and live in London, having given up our US home, career and future to tend to the things we care for most – like our 2021 report on the cost of oppression and hatred in the banking sector on itself and all humanity.

Groupthink is alive and well. That's everyone's problem

Ian doesn't need to imagine anything to ourself or pretend to anyone we are an authentic human being. Because it is as clear as daylight we're as real as it gets.

We truly exist to stop the apocalypse by enforcing fairness and accountability.

On 16 July 2021, Mabel phoned us/Ian to demand we "challenge quietly or face severe consequences". After reciting whistleblower protections, then ended the call to report the crime to HSBC's Chief Ethics Officer, Legal Team and all its financial regulators… who all did absolutely nothing. HSBC settled out of court only with prosecution immunity to ensure all 15 criminals and 16 crimes we and many others reported went unpunished. Today, curiously all claim to still be in post. Yet we are not… *Why?*

Problem? **There's** not a problem

All problems have consequences that grow until either (a) someone changes them or (b) the countdown reaches zero and we die.

The issue is simple. Darkness is extinguished by light, thus it despises our uniqueness yet can't see it. So it attacks our arbitrary extrinsic differences instead, to erode the identity of anyone less than self-accepting. Offended, we give away our humanity, to leave us dehumanised, traumatised, angry, resentful, alone and - crucially - vulnerable.

Antisocial Deconstruction externalises the nature of human potential over time, as we start believing, for example, (a) only queer beings possess charisma, uniqueness, nerve and talent, (b) that it's a bad thing and (c) that we don't embody such traits ourselves.

Or how about 'motivated Mexicans' taking our jobs, 'dramatic drag queens' assaulting children, 'gallant gingers' having colourful babies, 'liberal lesbians' with their Marxist agenda or 'talented trannies' in our toilets...?

Any unitary system if it wishes to survive must acquire a binary system of regulation

Jean Baudrillard (We/Human, 1981[36])

By dividing humanity and pitting us against each other, we lose authenticity thus potential, leaving us vulnerable so darkness can enforce its default rigid perversions: Self-promote, self-preserve, deceive, deny and destroy humanity. For a while we let it slide to avoid discomfort. Now it's too late.

0%

Non-Binary Beings
Charisma, uniqueness, nerve & talent

0%

Black Beings
Resilience, vision, strength & grace

0%

Neurodiverse or Disabled Beings
Wisdom, divergence, endurance & legacy

0%

Young Beings (<35yrs) Courage, passion, energy & influence

0.3%

Queer Beings
Authenticity, creativity, adaptability & pride

0%

Ginger Beings
Altruism, loyalty, duty & sacrifice

▲ **Fig. 10: Current FTSE100 C-Suite role diversity shows the total failure of all D&I initiatives |** Primary Data: Corporate websites visual inspection plus further background checks in April 2022. p=95%

Dictators, no angels: One in five **corporate** leaders are **psychopaths.** Guess who

Fig. 11: FALSIFIED: The selfish oppressive faces of crime colonising FTSE100 C-Suites

Lights on: Faces of oppression

For vivid proof diversity doesn't matter, look no further than the unleaders occupying the top three positions of our largest publicly-owned firms. This composite image – primary research undertaken with support from our partners – provides a UK snapshot of the faces of toxic oppression.

Their divide and rule will stop at nothing to gain ever more power and wealth until they erase every positive human trait left through fear and deception. Our firms are now devoid of all true leadership. 'Culture Fit'? Try culture fucked (see Special Edition).

Those complicit, directing or themselves instigating real criminal acts of corruption, genocide, hate crime and modern slavery against anyone remotely unique – all illegal under the Equality Act 2010 and at least 7 other acts of parliament - in order to protect their reign of sadistic misery to ensure effortless flow of all wealth, rights and opportunity from the global economy to their demonic hegemony at the expense of all life after us, thus the future.

'What problem' indeed!

These dictators will never leave office willingly. If you think they will, you are not being rational.

To verify this statement, we contacted a number of them, including the only 'out' queer FTSE100 board member, **Julia Hoggett Moth** (CEO of LSEG, formerly of the FCA). We asked only that they lend their name in support of our forming a UK Association of Queer Corporate Directors. **All of them failed to respond.**

What comes next makes the global downing of multiple Boeing 737 Max's - needlessly extinguishing 346 human lives as its failed CEO was awarded $62m to retire - seem like a short delay at baggage control

The deadly faces of **Oppression** ✖

How **the few destroyed** the many: **Fools** who **promoted themselves** then **doomed us all**

Structural Oppression explained

There can be no argument, given this composite image, those appearing male, pale and stale have been cleared a path to leadership. All others face varying degrees of systematic, targeted, illegal, intentional and universally-understood oppression that holds humanity back from ever achieving close to its potential.

Anyone claiming otherwise is not being authentic. Visit any gym, members club or golf course to see them overflowing with repressed (mostly white) men demonstrating a full understanding of privilege and how to adjust it. Seeking to ever increase their unfair advantage, they use money stolen from others to protect and insure their reign of terror.

Human Operating Capacity

We actualised this new measure – the antithesis of global structural oppression – to assess the proportion of beings allowed entry into decision making positions within global systems and structures. Only there is it possible to prosper comfortably and securely in a world where all other jobs are low paid or undergoing automation.

Human Operating Capacity = PRODUCT (pop. proportion of each protected trait at full privilege, *percentage*)
= 0.89 x 0.85 x 0.49 x 0.25 x 0.08
= 0.0074137 (0.74%)

Principal Finding: Today…
Global structural oppression now devalues 99% of human potential

Figure 12 shows the basic, fatal and entirely flawed logic of their small binary brains. Reprogrammed to spawn fear of all others, they exclude anyone not impacted by anti-human non-sapient binary biases from the boardroom:

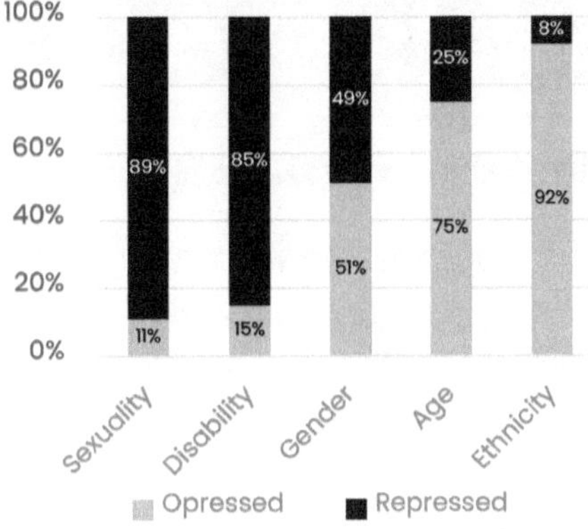

8bn **0.7%** **59m**[D]

Current global pop | Current human operating capacity | Est. global 'straight' 50-75 years cis white male population w/out visible disability

▲ **Fig. 12: Global structural oppression in numbers**

We conclude that just 59 million beings globally today endure tolerable levels of oppression at work. That's as they're the ones doing the oppressing! Yet we allow them to feign success at the expense of all other beings and our future. Insanity…

[D] Deilight primary research (Aug 2022) w. latest World Bank & Williams Institute data; taking non-white (Ethnicity); female or trans (Gender); same-sex attraction (Sexuality); and UK census data (Age).

Meet the WOCHMEN[E]

[E] 'WOCHBOYS' being their 35–49 year-old protégés in waiting

Who's oppressing who, *really...*

Naturally, the most oppressive group of people in all of humanity are most certainly its most repressed of all. WOCHMEN are being systematically stalked by darkness more than any other group. Yet nobody has any sympathy for them whatsoever, nor wants to save them. But we do. Save them first in order to save us all. It's that simple.

Why the acronym?

As labels, acronyms are divisive, and yet everyone loves them so much? But all that playing around with them and you missed the only acronym that truly matters. Now wholly isolated in all but name, the damage of divide and rule has already been dealt, by and to themselves.

How can we ever expect anyone to safely write or recall each time in live conversation 'White (or lighter skinned), Older (typically 50-75 years[E]), Cis, Heteronormative Men, excluding those with visible physical or neurodiversity'?

We can't save humanity until we save these losers. So start talking about the issue, or die

Meet Jamison

We may have lost everything we had. But none of it really mattered. Only the people we lost did. Jamison was the hardest. He taught us as much about oppression as he did how to truly love another. We miss him every day, and we love him dearly.

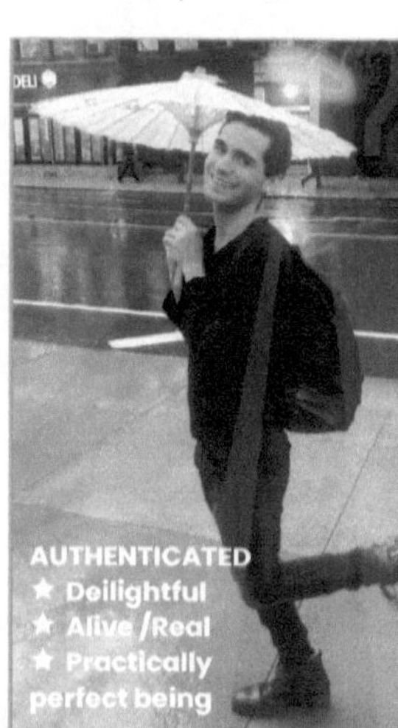

Jamison spreads love wherever he goes. And he would never harm another.

So in future...

Be more Jamison

9.5%

A. WOCHMEN proportion of UK population | UK census, 2011

9.6%

B. WOCHMEN proportion of US population | US Census, 2021

0.7%

C. WOCHMEN proportion of global pop. | Fig 12. Current Human Operating Capacity

▲ **Understand that these idiots…**

▼ **…are hoarding all wealth and opportunity for themselves and their WOCHBOY's.**

68.0%

D. WOCHMEN as UK cabinet ministers | New Statesman, 2020

62.0%

E. WOCHMEN in US political offices | The Guardian, 2021

72.0%

F. WOCHMEN as Fortune500 leadership | The Society Pages, 2000

59.0%

G. WOCHMEN as UK's richest people | The Times, 2022

70.0%

H. WOCHMEN as US richest people | Forbes, 2022

90.0%

J. WOCHMEN as world's richest people | Forbes, 2022

-50%

K. Non-White FTSE100 CEO's 7yr change | People Management, 2021

85.0%

L. WOCHMEN as major international bank senior leaders | Business Insider, 2015

100%

M. WOCHMEN as major international bank CEO's | Market Watch, as of 2019

Fig. 13: How malevolence fueled by money and deception is systematically colonising our world with division that can only end in imminent terminal disaster for every one

Failed Fortune500 dictators all fired for **moral and financial bankruptcy** plus **universal crimes** under the Equality Act

Why humans love sunsets

Nothing is ever black and white.

That would take wavelengths of zero or infinite duration to sustain. Therefore neither colour exists in nature or science.

Every day at sunset and sunrise, the sun's rays hit perpendicular to the curvature of the Earth when, through a process known as 'scattering', the blues and purples are deflected more than the reds and yellows.

Before our very eyes, deception gives way to reality, revealing our true nature to us. It is in fact all the colours of the rainbow combined into one overwhelming beam of changeable light - an analogue for humanity.

▲ **Pictured: The exchange of two authentically aligned Rings of Human Character, where eternally-bound by the Promise Keepers Commitment, signals the union of change makers into a potent new collective.**

Humans love sunsets. Because unlike demonic unbeings, they're as authentic as it gets

Who is our enemy… *really?*

The target is not men, it is sexist patriarchy and matriarchy.

The target is not heterosexuals, it is homophobia, biphobia, transphobia and heterophobia.

The target is not the old, it is those who disregard the young, the old, or the middle aged.

The target is not white people – the target is white supremacy, black supremacy and all racial eugenics.

You see… the target is not discrimination. Binary choices – your fault or my fault – Labour or Conservative – Blue Pill or Red Pill - are no choice at all. No human is bad or good – only deeply complex.

Discrimination does not exist. Because our differences are entirely indifferent, arbitrary and meaningless.

Every single one of us is unique. Oppression wielded by those consumed by hatred is targeting our uniqueness. The more uniqueness we possess (and show), the more oppressors of humanity stalk us. Intersectionality is intrinsic.

The target is oppression and hatred, not each other. Humanity is worth fighting for. Because we are all special.

Deception pits us against each other. Yet in truth, we are all victims of darkness

Proof **dehumanisation kills:** WOCHMEN don't cry since **WOCHBOYS die**

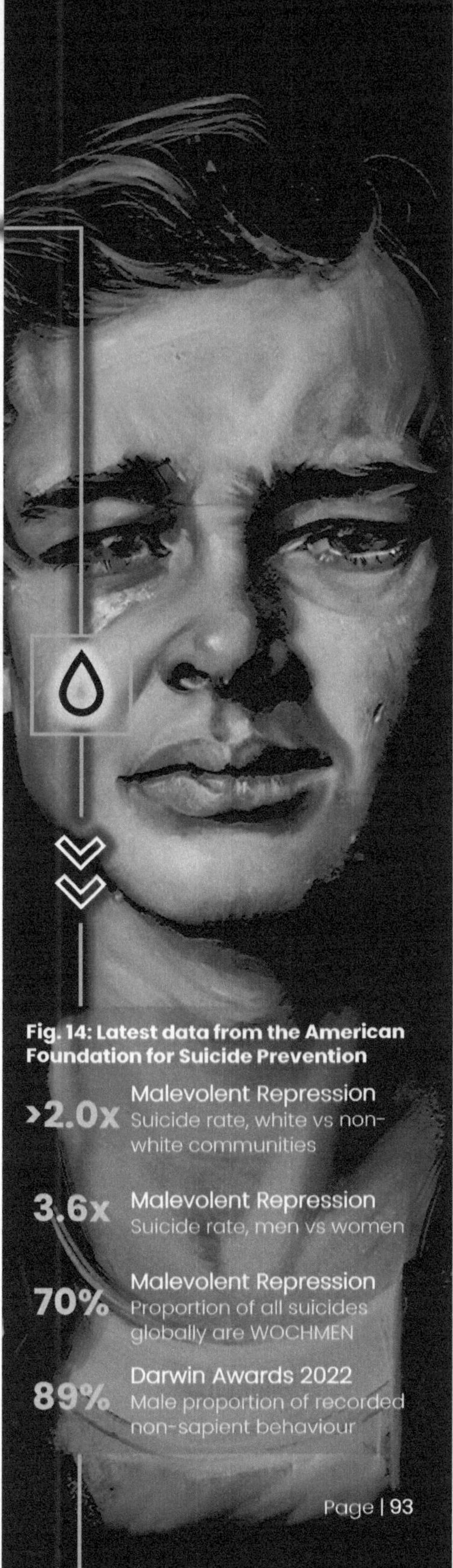

Deadly: Dehumanisation

Understand the workings of our human brain are practically perfect. Under normal conditions, not antisocial deconstruction, ☞ **Wise humans make few mistakes.**

We certainly aren't innately-biased beings, just look at babies. Woven into our fabric is an appreciation our differences reflect the only thing that matters. ☞ **What's inside us.**

All biases are in fact symptoms of a singular biological condition that impairs all cognition, thus killing all brain life. It manifests as rigid inflexibility, uncompromising irrationality, inverted perceptions (*Cognitive Dissonance*, for short), an inability to listen, resonate or concentrate and developing a malevolent, fully-automated, binary mentality.

A natural side effect of an absence of light and authenticity, '**Sapience Inhibition**' makes it near-impossible for (mostly) well intentioned WOCHMEN to save themselves from the clutches of malevolence, or acknowledge the problem exists at all.

Dehumanisation routes all thought subconsciously to our automation centre (brainstem), turning schemas (our instincts and shortcuts) into insanity and ultimately a **fatal (in actuality) loss of all brainwaves**. Our character is **erased**, leaving an empty shell filled with darkness, controlled by a **malevolent AI**

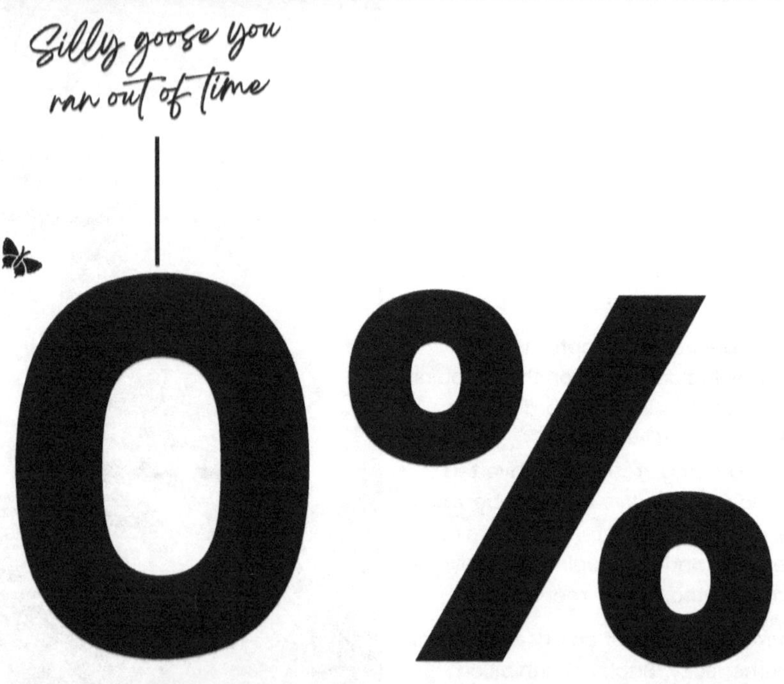

0%

Fig. 15: The proportion of humanity able to reach their full potential today

In a world where even those who oppress
humanity cannot achieve their potential,
what purpose is there to existing at all?

On the brink: Humanity Now facing **five live extinction events!!!!!**

▲ **FALSIFIED: Europe's 'O.G.' WOCHMAN, Julius Caesar** | 1892 oil by Clara Grosch

Words betray your true colours

Words and labels can be used to build or assassinate characters or collective identities, emancipate or oppress others, plus create or inhibit change.

But Humanity must learn to stop mindlessly dividing itself up or we can never counter the anomaly's effects. This lack of even a basic presence of mind we call a 'Zombie Apocalypse'. Instead, think, then always say what you see. That way, you'll never legitimately offend anyone.

✖ **Climate Change:** Ludicrous, almost everything changes all the time. If you truly wanted to get people in gear, try being authentic like 'Climate Unchange', 'Global Burning' or 'Climate Apocalypse'.

✖ **Toxic Masculinity:** A clear example of a label that's not only offensive to men but wholly ineffective, excusing the growing number of dehumanised women welding toxic power illegitimately. The authentic word for fully dehumanised brain-dead evil ex-humans is unfortunately 'demons'.

> Personally, we prefer 'moth' since (a) they're really not that scary at all, and (b) it appears we created moths as a proxy for demonic proportionate adult demographics and (mis)behaviour.

Five apocalypses in one, nice!

Absent any Human Nature, but akin to AI (or rather 'AI Apocalypse'), moths apply binary absolutions of divide & rule upon our spectral human identity to establish illegitimate power and control over us all.

Politically mainstreamed by Julius Caesar Moth, they encourage identical groups to detest each others arbitrary differences, then step in as heroes to claim the territory, turning reality into horror.

Their strategy is creating an abstracted 'inexistence' anchored in the scientifically debunked illusion of extrinsic value and genetic superiority. In the fictional Star Trek universe, such Global Eugenics Wars decimated late 20th century Earth. A clear example of art so frequently imitating life.

Allowed to thrive/survive: None

True human operating capacity today is surely far less than 0.7%. For example:

✖ Akin to homophobia, almost no human is entirely heterosexual: Like all things real, sexuality is spectral, so any label is utterly meaningless. Today 50% of UK Gen-Zer's admit to some same sex attraction (Stonewall, 2022), compared to 3.1% in the general population (ONS, 2020). Such inauthentic divergence is unsustainable.

✖ Other meaningless arbitrary traits: Neurodiversity, ginger hair, exposed tattoos, too tall, too short... Imagine what the Misfortune500 could've been earning if they weren't so incredibly basic.

On top of all that, consider the vividly suicidal misery of all those with wealth and privilege caused, not least, by all that repressed trauma and sorrow, having never come out of the closet as human.

Hateful beings destroy
Human beings create

Understanding borders

Analysing the problems across our world today is incredibly challenging, due to the absence of openness and accountability from wayward nations obsessed by self-interests. All unwilling to collaborate with even their closest neighbours, due to something as foolish as a line in the sand.

If you've ever been fortunate enough to travel beyond lines in your sand, you'll know humans are the same everywhere.

We all love, we all cry, we all dance, we all hope, we all dream we all make mistakes sometimes

Adela Emelie Sandé (We/Human)

Today borders are antisocial deconstructs that pit even those in the same family against each other. Just look at the proxy war invasion of Europe and Russia taking place in Ukraine, or the insurmountable 'hate thy neighbour' demonstrations taking place in Scotland & England, the UK & EU, China & Taiwan, India & Pakistan, North and South Korea, a divided Ireland, partitioned dying Palestine and across the wider Middle East. What a dreadful mess.

Like speed and distance, borders are relative (not real), useful only to compare differences between things that are real. Geographies in this case. Today like humanity, they are being abused

The single positive use of a **border** is to ensure those within it have an **equal say in** how **decisions impacting their lives** are made as those outside of it. Then to capture **authentic data** which can verify to us all that is indeed the case

Understanding productivity

Malevolence is in no way conducive to society. Jealousy, revenge, deception, greed, ignorance, murder... all these things damage economies, destroy lives and doom our future as a species.

But Human Nature isn't about all that. It's about courage, potential, conviction, determination, ambition, worthiness, success, drive and true love. You know that. So why are you pretending to everyone that you don't? Because they're all pretending too? Use your brain, it's why you made it in the first place.

1. Norway	$87.3	**6. Germany**	$65.5
2. Belgium	$69.7	**7. Austria**	$60.2
3. USA	$68.6	**8. Sweden**	$59.1
4. Denmark	$67.6	**9. Finland**	$54.8
5. France	$65.6	**10. Australia**	$54.6

▲ **Table 1. Countries of the world ranked by their predominance of Human Nature** (Measure: GDP per human/hour in US$, 2020, Source: World Bank). Excludes tax havens, as per Black Platinum Gold 2022.

God does not play dice

Albert Einstein (We/Human)'s rule, 1927[37]

Turkestan, Afghanistan, Transcaspia, Persia – to many these words breathe only a sense of utter remoteness, or a memory of strange vicissitudes and of moribund romance. To me I confess they are pieces on a chessboard upon which is being played out **a game for domination** of the world

George Curzon (We/Human)

British viceroy to India, 1899–1905 (Quoted in Arab News)

For example, Albert (We/Human) co-founded quantum mechanics. Yet he/we viewed its seemingly random elements with healthy suspicion. As in, he/we saw the 'Quantum Uncertainty' principle as a clear signal of missing knowledge. Thanks to the unique information we (Ian/Human) gained in investment banking, we (Ian/Albert/Human) have since reclassified this randomness holistically as symptomatic of the 'Malevolent (Dark Energy) Anomaly' at the centre of the Clarke Paradox.

Finally, with this book, we deilightfully gift humanity the solution to said paradox, whilst ironically also defining it for the first time. As in, it wasn't even on your radar.

Explained: Alberts Rule

Both agnostic, we (Albert/Ian/Human) view 'God' as any unique, wise human being with enough potential for authentic science making. True leaders. Nothing more, nothing less. By our criteria, anyone could become a 'god'. But only by driving meaningful positive change that helps others – not just themselves. Real deeds.

Thanks to Albert (We/Human), we can make science to fix any problem we face or achieve any endeavour we can imagine

Understanding economics

It's no coincidence the D&I sector centres the UK and USA; that Italy fell to darkness (as in, *the Far Right*); or that the UK left the EU. Understand geopolitics to realise how interlinked all world events truly are.

The anomaly's effects are most evident in slowing or declining western economies like the UK, US and Italy. Today, growth is mainly driven by unsustainable mounting debt - not industry, business or change.

By contrast, Germany, the Nordics, Australia and New Zealand exhibit far less malevolent repression, despite being pre-dominantly white patriarchical societies. This explains their higher levels of contentment and productivity, plus their continuing ability to elect strong, popular and divergent leaders with change making mentalities such as Angela Merkel and Jacinda Ardern. Cause or effect?

Undoubtedly, both. Change makers will always run better collectives, economies and nations, which will in turn always suffer lower levels of hatred juxtaposed with elevated performance by virtue.

Understanding geopolitics

Astute minds will understand the asymmetrical spread of hatred is a result of intensifying political rhetoric and divisiveness that centers western nations.

This is all part of a pre-emptive, defensive plan to wield illegitimate fear, power and control in order to accelerate the theft and stockpiling of wealth by the few from the many. Those presiding over us are preparing for the end of western prosperity, foreshadowed by the rising influence and economies of the east.

Being deilightful, we see a better strategy: make peace with our fellow humans and equals in the east instead, then share the limitless prosperity of perpetual growth from change and globalisation. We do not need to agree, or be silent where we disagree – as we are on the horrific and widespread genocide of Uighurs (and all beings) taking place across mainland China today – for good things to come from broader dialogue and perspectives.

Furthermore, we have just invented the most lucrative and productive industry ever to exist. The industry of change sits in the middle, exactly where it belongs.

Understanding (in)justice

Rigid, unchanging laws controlled by those who vividly despise uniqueness have led to a fundamental misconfiguration between the reality of lived human experiences in an ever-changing geopolitical climate. Thus why conviction rates against moths everywhere are so fractional.

Ethnicity pay gap reports aren't even a statutory requirement in the UK, and aside from gender all other traits are wholly ignored despite vivid persecution.

Bad doctors bandage wounds
Good doctors cure disease

You only **measure what you treasure**
Firms and nations will now cooperate and
walk the walk or face real consequences

Kimberle Crenshaw (We/Human, 1989[38])
Demarginalising intersectionality

Understand data and reporting

McKinsey & Co found diversity correlates to but doesn't explain improved financial performance. Contradiction? Validation, in fact! You see, only us true leaders can authenticate uniqueness, since we alone can look directly into the light (the Sun).

But with so few of us left, change makers instead required moths to 'self-certify' no systematic biases exist in our fallen firms. Pay gap reports supposedly signal a functioning meritocracy where human uniqueness can thrive is 'on the horizon'. A deeply flawed strategy. If science can be faked, then so too can pay gap reports.

Today, firms measure humans using labels (binary boxes), not spectra (sliding scales), so response rates are low and any collected data is next to useless.

✖ You cannot put real humans in boxes: No matter how hard you (don't) try.

✖ It is fundamentally unsafe to respond: Data will never be used correctly or in our interests. Today HR departments, justice and law enforcement are universally and illegally targeting all humans - no moths.

Understanding men

The moment we created penises, it became a competition to see who could make the biggest one. Clearly the wiser beings, women had no part in it. But it was useful for one thing. Darkness originated in Asia – the land of gold - and became ubiquitous with colonialism – the invasion for gold. Money is now our prime suspect.

First nations to fall to darkness	Last nations to fall to darkness
Cambodia	Zambia
Taiwan	Cuba
Philippines	Netherlands
Sri Lanka	Gambia
Hong Kong	Senegal
Bangladesh	Haiti
Thailand	Sudan
Vietnam	Bolivia
Malaysia	Cameroon
Singapore	Ecuador

Table 2. Nations ranked in chronological order of historic dehumanised sapience inhibition | Source: World Population Review, Penis size by country 2023

Example: 'Serious consequences'
For crimes against humanity, Tom Wilkerson and all Bonhill senior leaders globally are hereby reassigned to roadkill collection until further notice.

The only safe spaces you'll find today are reserved for storing stolen wealth from real humans

Still think you've seen a **Ghost**? **Our jigsaw is only just beginning** to take shape

What happened in Asia?

Compared to Europe, Asia's recorded history paints us a picture of serenity and neighbourly contentment. From the Senso-Ji to Angkor Wat and the Taj Mahal, its rich cultural heritage and spiritual traditions centre mind, body and soul to reflect a continent that to this day retains vivid bastions of change making wisdom.

But all things have a dark side. This region of great walls and militarised islands also answers western malevolence in kind.

Asia's gold rush lacked the caution of Ancient Egypt. Clearly it hosted the first genesis of meritocratic republics to turn malevolent, then quietly consigned them to history. Today, Asia and Latin America possess many of the worlds most brutally autocratic, corrupt and oppressive regimes. Protectionism is fair when you're under attack. But attacking your own people or stopping them from being people isn't leadership. It's genocide.

No change makers. No life

The ubiquitous WOCHBOYS who descend upon Canary Wharf, Manhattan and what's left of Hong Kong every weekday without regard for those they have displaced are just playing the villain. That's because they are too scared to be themselves in case they are selected as the next binary target to go. Thus how misplaced their fear of the real threat truly is.

If and when you curiously, generously, open-mindedly and purposefully finish this book with chosen determination to understand it, you will not only gain the knowledge to succeed at anything in life you so choose to. To become anyone in life you so wish to. You will also realise just how easy it is to remove those that don't even exist from tearing apart our reality.

Simply be curious while helping others first and you'll be just fine. The cowards they embody could never harm you, so they convince you to harm each other.

By introducing the great minds binary choice, exposing the conspiracy, and delegitimising those who take leadership positions illegitimately, we will have already markedly slowed the growth of the malevolent anomaly here on out.

But this book is the last possible change to prevent the loss of billions of lives. Given the emotional capital invested in keeping power at all costs – including sacrificing our entire multiverse – more combative changes are sadly necessary for us to restore humanity's future. So let us continue.

This time, do try to find the courage to **Speak Up**, or you will all die very shortly indeed

Can we at least agree about the sunsets?

Record of Promises (Ch.5)

Darkness cannot drive out darkness. Only light can do that. We shine Dei's light.

Dreadfully Meaningless Unchanges (Antisocial Deconstructs; threat to life) — Sapience

- ✖ The Diversity & Inclusion industry globally for Crimes against Humanity
- ✖ The Academy Awards for Crimes against Humanity
- ✖ Anyone applying divide & rule power strategies/refusing to collaborate — Deilight
- ✖ The C-Suites of the entire Fortune500 for Crimes against Humanity — ✖ Falsified
- ✖ Mabel Rius, for Crimes Against Humanity (and being omnirepugnant)

Hatred cannot destroy hatred. Only love does that. We love with deeds. Not words

Deilightfully Meaningful Changes (species level adjustments to destiny) — Clarketech

- ▶ Defined: The Worthiness Test to explain our instinct to be true to ourselves and the rewards from the Law of Karma that follow from it — IAN, Actualised
- ▶ Defined: The Clarke Paradoxe's central research question, which this book attempts to resolve, as 'Why does Humanity hate itself?' — IAN, Actualised
- ▶ Defined: The nature, history and unresolvable failings of the D&I industry by spreading complacency via its performative-only allyship — IAN, Actualised
- ▶ Resolved: Fermi's Paradox with the greatest filter that is human uniqueness and change making – vividly one of a kind in the multiverse — IAN, Actualised
- ▶ Provided: Coherent narrative consistent with historic records on the true nature of the relationship between H.Sapien and H.Neanderthalis, including our neighbourly efforts to save them from the last Ice Age — IAN, Actualised
- ▶ Defined: H.Sapien and uniqueness as sufficiently differentiated to warrant separate classification to 'animals' – we are the sixth kingdom — IAN, Actualised
- ▶ Defined: The Rings of Human Character to explain how extraordinary, non-binary and unique our imagination truly is. A miracle of science. — IAN, Actualised
- ▶ Defined: The sociocognitive processes that drive our potential for change making, self-acceptance and self-rejection to contextualise the effects of dehumanisation on human identity and behaviour — IAN, Actualised
- ▶ Exposed: Darkness strategy to externalise human potential in order to steal/destroy it. Provided extensive data to prove this conclusively. — IAN, Actualised
- ▶ Exposed: Colonialisms continuing nature in human modern slavery — IAN, Actualised
- ▶ Defined: The nature and identity of WOCHMEN and WOCHBOYS as both victims and priority targets of darkness, for their protection — IAN, Actualised
- ▶ Isolated: All bais/malevolence as the cognitive condition of 'Sapience Inhibition', ready for quantum decollection from Human Nature — IAN, Actualised
- ▶ Provided: Succinct macroeconomic comment & data to dispel myth that worsening malevolence is attributable to natural human traits — IAN, Actualised
- ▶ Explained: Albert's Rule, to reinvigorate curiosity among adults. — IAN, Actualised
- ▶ Listed: Five simultaneous apocalypses currently taking place across Humanity, with seemingly no plan for salvation. Just more babies. — IAN, Actualised
- ▶ Ruled: All leaders globally require authentication. Moths ineligible. — IAN, Actualised

Some people are *Truly dreadful*

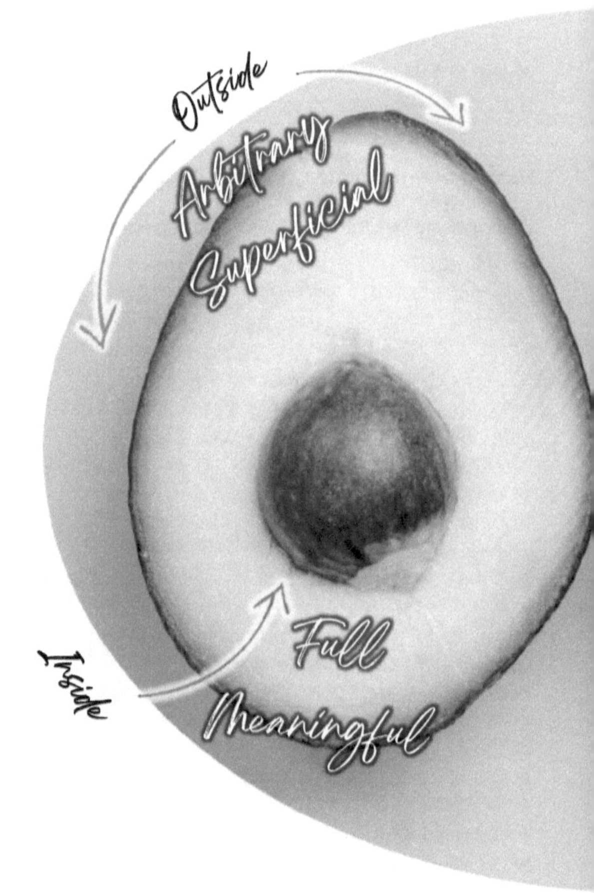

But have
you every curiously
wondered **why?**

Why whats inside matters most

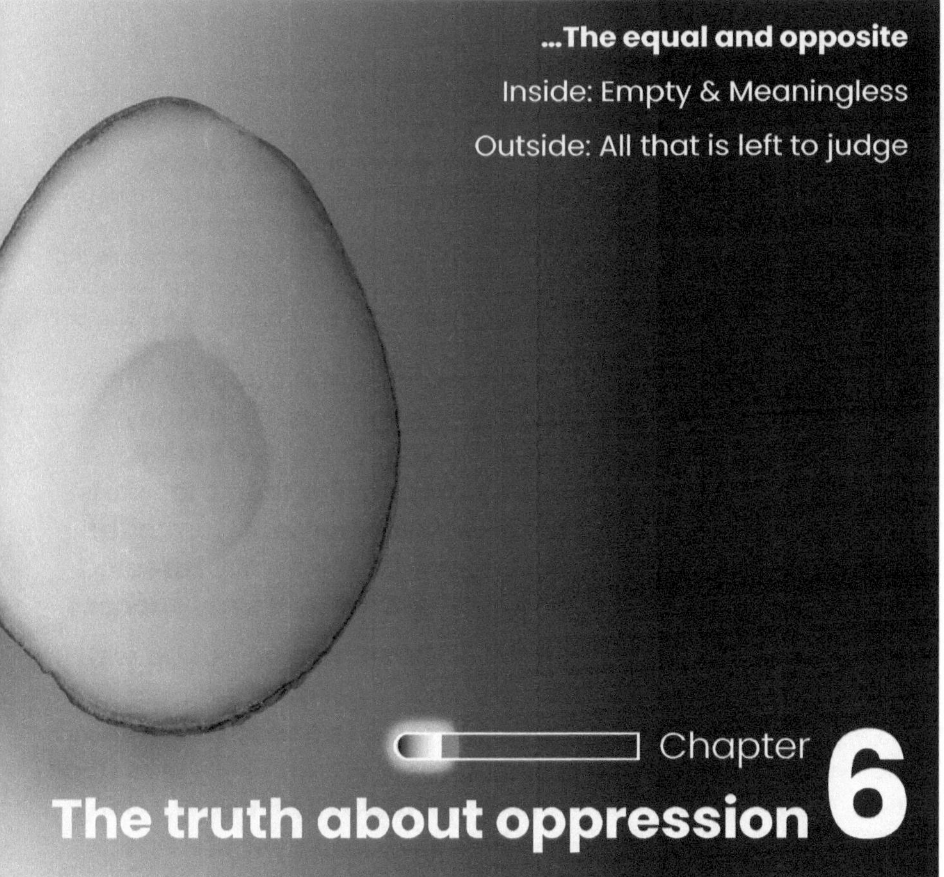

Groupthink is alive and well in boardrooms around the world. That makes this everyone's problem

Ian (We/Human), in public address for International Men's Day at HSBC USA headquarters in 2019, appearing also later in our Project Speak Up Report 2021

Why we listen: Often mums are the wisest beings on our journey. Because they care about someone special

Benevolence

/bɪˈnɛvələns/ Noun

Old words can often get lost in translation, especially with moths rewriting history all around us. The best detectives always go back to the scene of the crime.

From the Latin 'bene velle', meaning a 'well wisher'. But in Old English, 'velle' translates as the power to make our own decisions. We now recollect that **Free Will** comes from authentic, unilateral, unrequited, enduring and unconditional **true love**.

We are fortunate our mum – pardon the cliché but verifiably the worlds best (like all real mums) – has been our closest advisor, biggest champion and worthiest superstar. To us, she epitomises the equitable value of women. Last summer, we spoke to prepare her for this book and draw her wisdom.

"In this world, some people want to be rich and famous. It's all they seem to want in life. But **I think** there's **more to life** than that. I exist to have a nice home. A comfortable **safe space** for the family I created and love dearly. That's my **success**.

Some of the things I've put up with in my life, I can't believe I ever did. But I did it because **I needed the money.** Those people weren't very nice at all and they'll **always be nasty**. I don't see **how they can ever change**.

Selfishness, thoughtlessness, greed, ignorance, jealousy, hatred. But can we really throw all these **awful** things into **just one word**?"

Deilightful

The lost battle of light versus darkness

There can be no arguing that the nomenclature our authentic soul espouses to try and encourage someone – anyone – to listen to our impending doom is both loud and clear. And yet, thanks to the prevalence of religious agenda, demonic brain washing and meaningless duplicating labels from firms and politicians, 'light' and 'darkness' now have too many pluralistic connotations to resonate with their original significance.

"Returning hate from hate multiplies hate"

If it wasn't for the granular quantum mechanical genius of Martin Luther King Jr. (We/Human)'s 'loving your enemies' speech, it probably would've been impossible to reconstitute the sciences of change. Here he/we alludes to the workings of 'Dark Fringes'.

To fail Dreadfully or *to succeed Deilightfully*

We aren't giving up our authenticity, nor telling others how to think and speak. But in the interests of healing our society and easing confusion today, we have created unspoilt analogs to help people quickly discern behaviour and opinions that align with light and darkness. These are assured to resonate engagingly far and wide.

An anomaly exists in our common human identity. Rather than an inherent biological trait or product of evolutionary natural selection, '**malevolence**' (including greed, hatred, sorrow, selfishness, rigidity, nepotism and deception) manifests as a contagious **pandemic** that stunts potential, destroys lives and increasingly **threatens humanity's survival**. And yet relatively few appear to either notice it or care.

/meˈlɛvələns/ Noun
Malevolence

— Dreadful — >> — **Malevolence** —

One percent
of reality

RuPaul Andre Charles (We/Human)

Drawing a line over ~~inequality~~: **We're all born equal. The rest is** a major **drag** on our potential

The Human Genome Project (2001)[35] confirmed we are all essentially born equal intrinsically. All differences in human prosperity and status are simulated, antisocial deconstructions that turn reality into abstract, dystopian horror.

For example, science today records that humans see just one percentage of the electromagnetic spectrum. The other 99% of reality supposedly alludes us. Yet like so much of what we've been told, this simply isn't what we observe.

Malevolent Colourblindness[F]

Most understand far less of reality than what they can see before them. This goes well beyond ignorance. According to quantum mechanics, those who dehumanise completely to darkness (by consistently failing the worthiness test) are no longer able to truly perceive visible light. To them, our world appears entirely binary and monochromatic.

Benevolent Acuity

Some see and understand well beyond visible light – as much as 13.3% in real time. That's because perhaps just 7% of the information humans exchange comes through words alone. 93% comes through our authenticity, making it a 13.3x multiplier. Going way beyond 'body language' or 'tone', this superpower includes 53+ senses that work across space (distance), time, and all forms of communication – even books.

[F] Not to be confused with naturally occurring colourblindness – a benevolent trait.

Chapter 6: The truth about oppression

Courage under fire: Women

Only 100% malevolent moth will ever make it to the moon that is humanity's fallen boardrooms of big business and government – now dehumanisation HQ.

After centuries building a society that mostly dehumanises WOCHBOYS, it'll take another few decades to dehumanise enough women to meet even their most feeble promises to change if they are to ensure no human woman gets through.

Today, it isn't even safe to board a plane, join a staff network, or speak up if you see something wrong. So you can appreciate in a few years, let alone 136, they'll have destroyed all chance of those left gaining a complete perspective and we will face the terminus of the eternal – 'inexistence'.

Queen Bees: Meaningless label for demonic WOCHWOMEN

Darkness cannot destroy darkness. Detecting a lose-lose, both inauthentic sides agree menial concessions to give the impression of progress and styme detection until it's too late.

Rather than targeting a total absence of women from senior leadership roles, they now agree to allow some in providing they leave their humanity at the door to become our antithesis, then demonically help expedite the oppression of all others.

Spotlight: Transphobia, gender reassignment and pronouns

Appearances can be deceiving. Just ask any drag queen, transgendered, gender non-binary or non-conforming being.

The key difference versus moths is that transgendered beings are entirely authentic about the deception that is their intrinsic versus presented identity.

That's why they don't need pronouns or reassignment surgery to prove to anyone that they are as authentically human as you can possibly be. Indeed, they are as real as it gets. The only being they need to convince is inside – 'self-acceptance'.

Today, 50% of black transgendered females experience violent hate crime every year due to their scientifically measurable and detectable potential for success. Human beings like Marsha P. Johnson (We/Human) and Sylvia Rivera (We/Human), with the courage to see and reject the arbitrary nature of the gender binary in order to uncover their true human identity. The force is especially strong in these ones.

Transcend the gender binary – humanity's only meaningful anatomical difference – to truly see we're all equally deilightful. Then you"ll become (a) immune to the dark energy anomaly, thus (b) in our dark universe, public enemy number one to all moths

Mystery of **how birds fly in perfect formation** solved

Here's five more (of ~35; full listing on page 349) human superpowers you just might recognise.

On the same wavelength

Authentic beings channel light at high frequency, exchanging more information about intentions, purpose and character with as little as a smile, a glance or even proximity. At peak proficiency, empathy turns telepathy - **omnibenevolence**.

Wise humans make few mistakes

Driven and motivated by purpose, instinct and our altruistic Human Nature, we experiment via trial and error, curiosity, listening, empathy and perception. It's how we forever expand our perspective and perfect our uniquely ambitious craft of titanic change making – **omnipotence**.

Fee-free education: Intuition

Fear of snakes comes with the turf. But at full authenticity, we gain a holistic knowledge of light and loves workings. Plus we can learn anything double quick – **omniscience**.

The rhyme of confusion

Channeling our light, we align our frequency to the resonance of others. Then speaking in binary (but authentically), we flood them with light. Intentionally, we emancipate others from darkness, and back across the alignment line, to restore their potential for change – **omniaction**.

Why see 1% when you can see it all?

Once sufficiently potent, we can see any beings perspective across space and time, recollect wisdom from all true leaders today and before us, plus we unlock spectral decisions to create any change or repel all darkness – **omnipresence**.

Birds using frequency to fly in perfect formation

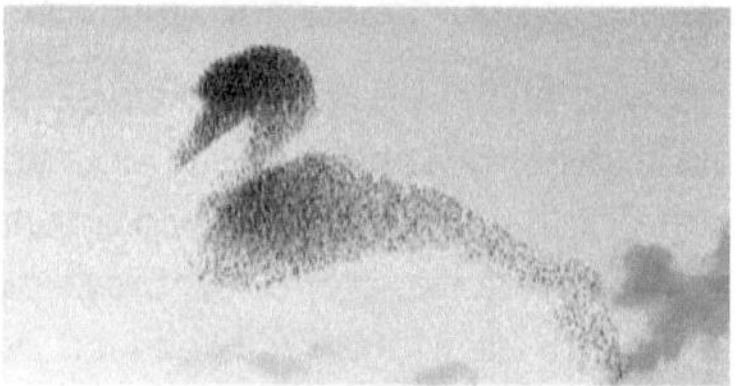

Pictured: Benevolent Starlings use their meritocratic authenticity to synchronously 'Murmurate' (1) in France by Alain Delorme, 2013 (2) in Fairburn Ings nature reserve near Castleford UK by Peter Lau, December 2020. (3) Over Lough Ennell in Co. Westmeath UK by James Crombie, March 2021

But only those wild animals unable to grasp the concept of greed remain authentic today...
Why?

That's a lotta soul! The two separate processing centers of **a human brain**

The driver of our **instincts** is the **brainstem** - the region at the very base of your brain, just above the spinal chord. Scientists have known for some time that another brain region, the prefrontal cortex, plays a role keeping those instincts in check. But exactly how the prefrontal cortex puts a break on the brainstem remains **unclear**

European Molecular Biology Lab (EMBL)
[Fake] Science News | Franklin et al, 2016[39]

An 'Emitter of Light': Incredibly rare, unique and wholly authentic

True Leaders this potent never lie. Yet despite her soaring god-like uniqueness, the nature of her existence meant finding a place of safety to feel loved and self-actualise was near-impossible. But fortunately for us all, Diana (We/Human) was able to do so and lives forever in the 'Quantum [Multiverse] of Human Nature'.

Alignment: Our identity crisis

Understand two brains fight for control of

How we help our Self by helping Others first

Waves of Light (★): Human Nature originates all your brainwaves (as in, +Cognition) to enable/regulate pseudo-conscious +Sapience (including all morality, survival instincts, emotions, intelligence, abstract thought, complex reasoning & logic). Thus your +Potential.

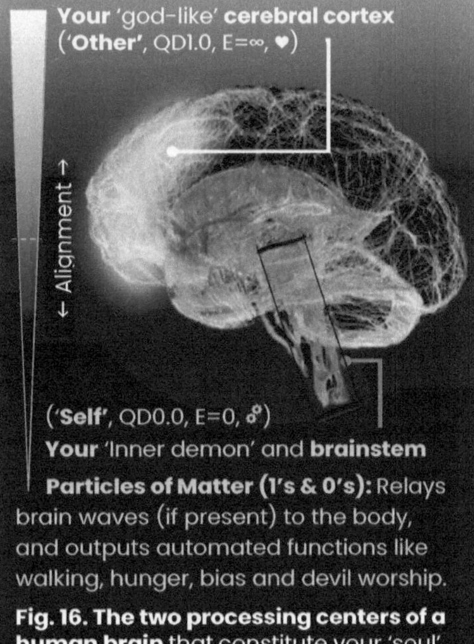

Particles of Matter (1's & 0's): Relays brain waves (if present) to the body, and outputs automated functions like walking, hunger, bias and devil worship.

Fig. 16. The two processing centers of a human brain that constitute your 'soul'

your character. For simplicity, let's call one your 'Self' and one your 'Other'.

Clarketech (♂): Self makes stuff up with particles

Sapience (★♥): Other makes stuff happen in waves

Dehumanisation and 'Emancipation' jostle our 'Alignment' between either constantly.

A truly deilightful being

Sunset red hair frames Mary's captivating green gaze that beams from the ringed portholes to her eternal soul. Charming freckles pop, like spots on a butterfly, beside her endearingly wonky nose and upliftingly charismatic disposition. Without a doubt, Mary is an angel. And so are you.

But one day, while going about her near perfection – being deilightful and helping others thrive – she is attacked in the street by a malevolent being. It's a male victim of the anomaly's effects.

"Fuck off and die, you wonky nosed ginger bitch!", they remonstrate without provocation, oppressing her identity, **before seeming to just disappear.**

◄ **100% authentic intrinsicity:** Benevolent, meritocratic, worthy and sapient

◄◄◄ Alignment

Deilightful

Wise, otherly decisions

Fuels worthiness; Accrues potential

Our Other routes more information via our soul - a bundle of waves lodged in our prefrontal cortex. Self-acceptance involves 'emancipation' (extrinsic) or 'nurturing' (intrinsic), fully aligning our identity to our innately benevolent Human Nature without any deception in order to become authentic and conscious. At 100%, you get an invite to a private members-club with generous rewards like infinite godlike powers, guaranteed sweet success, a legacy that will live on forever and the ability to adjust the EMS at **will** - all for **free!**

Our innate benevolent Human Nature that is...

In the presence of positive worthiness, and acquired in a spectral sliding scale:

★ **Akin to Artificial Intelligence, Self accrues us potential** by coordinating bodily resources towards any goal to the best of our ability (*Schemas*, for short).

★ **A bastion of inexplicable uniqueness, Other** floods us with qualities like courage, altruism, talent, charisma and creativity. All labels for the unrequited true love of an authentically human soul. Other's strength is our intrinsic share of our collective imagination. As in, the quantum electromagnetic spectrum of light (*'EMS'*).

The nature of Alignment ★ Infographic

So easily turnable to darkness

Like all of us, Mary has insecurities, born from a lifetime being told her arbitrary difference s make her less attractive and unworthy of success or contentment.

But instead of understanding the wholly indiscriminate nature of the attack, she takes it deeply personally (trauma).

Now rejecting all commonality with her assailant, she represses her human identity and sets up the 'Red Hair, Blue Eyes and Wonky Nose Society against Sexist Men'. Not only has she lost her perfection but also her eight billion friends and fellow victims. A shadow of her potential, soon her true colours will unchange. A butterfly (1) turns moth (0).

Malevolent, deceptive, biased and automated: **100% fake extrinsicity** ▶ **1010 1010**

Dreadful

Foolish, selfish decisions
Fuels unworthiness; Drains potential

But sadly, you'll have to leave your moth at the door. Their Other routes more information via the brainstem (a bundle of particles, effectively 1's and 0's). Caused by oppression (extrinsic) and repression (intrinsic) of our innate human identity, at 100% self-rejection we become fully extrinsic, malevolent, automated, deceptive and biased. This explains how darkness and Homo Eversor are the antithesis of being human and real, yet still appear as both. Your membership has been revoked. Now you get a bill, and sadly, it's not free..

Malevolent AI Binary Persona

In the absence of worthiness, and waning in a sliding scale:

✖ Increasingly restricting access to the quantum, Other moves to protect our collective imagination from pollution by dark thoughts as it shuts off the prefrontal cortex to all transmissions.

✖ Akin to artificial insanity, Self deducts us potential by inverting our perceptions (turning schemas into biases), impairing the performance of body functions and resources (eyes, ears etc.) - sadly now the only functional processing centre left.

✖ Beauty is only skin deep. Without light, nothing natural or alive lies within. Only malevolent artificial life - pure hatred.

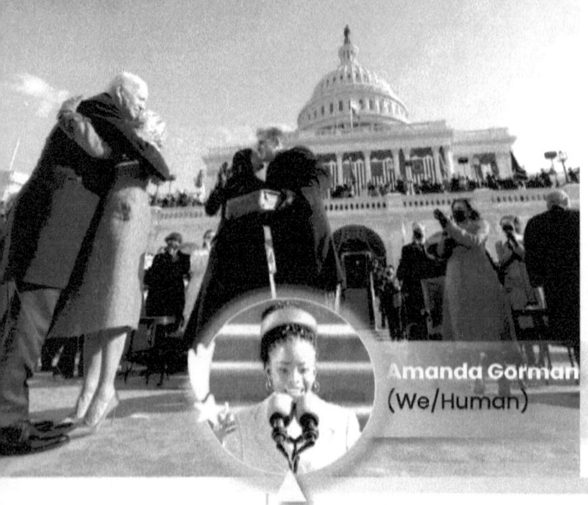

Wise Human

Promise Keeper, Titan of Change and **Science Maker**

Highly Potent (QD0.50–0.55), Unite to Serve

> When day comes, we ask ourselves where can we find light in this never-ending shade? The loss we carry, a sea we must wade. We've braved the belly of the beast. We've learned that quiet isn't always peace, and the norms and notions of what "just" is isn't always justice. Somehow we do it. Somehow we've weathered and witnessed a nation that isn't broken, but simply unfinished. And so we lift our gazes not to what stands between us, but what stands before us. We close the divide because we know, to put our future first, we must first put our differences aside. We lay down our arms so we can reach out our arms to one another.

Deilightfully free globally: Simple **linguistic analysis**

AUTHENTICATED

Delivered beautifully with a smile, vivid talent, angelic charisma and not a single meaningless label or deception, Amanda (We/Human) skillfully demonstrates six different human superpowers to foster unity, awaken the unconscious and dutifully emancipate human souls worldwide without placing any burden on our choice functions. She is one of the most potent beings you could ever wish to meet.

Numerous references to the battle of light and darkness confirm her true colours. 'Cognitive Recollection' and 'Benevolent Coherence' in the absence of notes confirms the authenticity of her shining talent comes from within. We speak as one. 'Binary Multiplicity' (the *Rhyme of Confusion*, covered shortly) nurtures acceptance of human nature, while her 'Benevolent True Love' and self-accepting admission of flaws resonates engagingly. Confidently, she deploys 'Benevolent Intuition' to predict the near future with an emphasis on courage, conviction and imagination. No labels or division, but a wholly collective identity.

Look! A black woman! Get her! ▶▶▶

Human and Demonic Linguistics ★

Big tech is now coming into their own. We beat them four years ago. We surprised them. We took them by surprise and this year they rigged an election. They rigged it like they've never rigged an election before. And by the way, last night they didn't do a bad job either if you notice. All of us here today do not want to see our election victory stolen by emboldened radical-left Democrats, which is what they're doing. And stolen by the fake news media. That's what they've done and what they're doing. We will never give up, we will never concede. It doesn't happen. You don't concede when there's theft involved.

Stupid Demon
Ha-Satan, Agent of Darkness and an Absolute Zero

E=0, Divide and Rule: Extreme malevolence, total deception and hyperallergic to change

Baseless lies, conjecture, exaggeration, warlike adjectives, illegal false allegations, character assassination, plus a fixation on the past and constant denial of his vivid flaws with repeating claims of 'honesty and integrity'. Multiple moving targets build an incoherent binary narrative. Note obligatory red tie for deception.

"OMG, A RICH RACIST KILLER WHO DOESN'T PAY TAX, TELL TRUTH OR EXIST. MAKE HIM PRESIDENT AGAIN SO HE CAN MAKE AMAERICA WORSE THAN IT'S EVER BEEN BEFORE... AGAIN". TOTAL INSANITY!

FYI, A) TRUMP'S ONLY GOAL IS TO DESTROY GLOBAL DEMOCRACY. B) 'THEY' = HUMANS

✖ FALSIFIED:

Omnidirectional hate and clenched fists. Trump Moth knows his rabble is entirely mindless. He's counting on it. Vividly he goads them to riot then cowardly snakes back to his office feigning innocence. Former potent change makers who turn fully malevolent, agents are fortunately rare due to their antipotential/superpowers. Highly dangerous, they understand and covet extreme malevolence, setup dark fringes, oppress all beings, and only ever lie.

Pictured: Emmeline Pankhurst, c. 1903:
Self-actualised true leader, emitter of
light, science maker and founder of the
unifying Women's Social and Political
Union party (*Suffragettes*, for short). But ▶
what did Emmeline (We/Human) mean?

Sadly today, humans are forbidden all
experimentation with their gender identity
from birth – a cardinal sin. So few have
the curiosity to wonder, why the *faux pas*?

Daring to be different: Worthy

Religious misogyny aside, women were
clearly no afterthought cast from
discarded elements of manhood.

The arbitrary meaninglessness of human
differences are themselves a byproduct
of benevolent darwin-esque sapient
processes intended to protect us from
dehumanisation. Conferring a natural
defence against farcical prejudices - like
groupthink, confirmation bias, illusory
superiority and herd mentality - they help
keep our perceptions grounded in reality.

For example, plenty of animals reproduce
by bearing live young, laying eggs or
simply cloning themselves. Yet made
possible by our reproductive differences,
human coupling enables the symbiotic
fusion of ovum and sperm to create a
new character in our game of life. The
vivid intention here being to foster love,
intimacy and a common identity through
shared intrinsicity. Not for illusions like
masculinity or subservience.

Worthiness of our otherly gifts was once
hardwired into the fabric of our species.

"We must free half the human race so they can help **free the other half"**

Can't find the right moth for the job? Simply hire a human!

The Great Resignation, plus meaningful
allyship initiatives from Titans of Change
like Lead5050, OurGen and OutBritain
mean extraordinary, diverse change
making talent is in no short supply today.

Case-in-point: A rigid, blanket policy of
white supremacy ended just months after
Project *Speak Up* as HSBC appointed the
first black directors in its 161-year history.

True leaders don't seek power. Instead power is flung upon us

In early world governments, humans
elected their own representatives, rather
than fake leaders being pushed by party
nepotists. This ensured the prioritisation
of talent, effort and achievement – not
wealth, corruption, lies and selfishness.
Total fairness and accountability was
upheld by regular free elections, real
conduct scrutiny from truly competent
authorities, plus an impartial Fairo.

Good enough for Ancient Egyptians,
Greeks, Romans and Iranians, today any
firm or government could easily deploy a
Meritocractic Republic within a single 4-
year change cycle where proportionate
representation of all human traits is
achieved lawfully and naturally.

Look around to see female true leadership (almost) everywhere

Simply stop breaking the law, call off HR human firing squads, purge known criminals, plus promote the hugely talented authentic change makers forever inexplicably stuck – their careers and income frozen, rigidly, in the middle.

#FuelingTheBias won't break it

Seen as a quick win, advancing gender equality has been the primary focus of change makers everywhere for 2,000 years. Yet quickly win, they have not.

☞ **Cure disease, don't treat symptoms:** The anomaly inverts our perceptions to stigmatise differences – gender being the only one of consequence.

☞ **Every action requires an equal and opposite reaction:** Not a feeble, divisive movement that fails to unite men and women as the common victims they are.

Once we dehumanise our brainstem takes over. Absent all presence of mind, detectably, here is a separate brain and instincts set entirely. Thus why, scientifically, demons are a different species altogether

Weary from repetition, more change makers give in to the moths repugnance, becoming unworthy and losing their sapience. But the angry backlash from humans to the demonic and ultimately

Pictured: Yours truly, 2022: Onstage with the superstars of WeAreTheCity's Rising Stars. Our eternal allyship and authentic solidarity with women was graciously recognized by wise humans, awarding us the 2022 Editors Choice on merit alone

self-defeating murder of George Floyd exposed moths impotence for change in leadership.

5.0%
Proportion of FTSE350 CEO's are female

−90.0%
Female under representation across senior leadership

5x
Women selected in gender-blind recruitment

136yrs
Projected time to achieve gender parity in the boardroom at the current pace of change

Fig. 17: The attack on womanhood | Sources: McGregor Smith Report (2017); McKinsey & Co (2020)

London has more statues recognising the non-sapient contributions of animals than it has women (Art UK 2021). *Why?*

See you at Pride, buy our stuff

Despite moths empty claims of universal homophobia, almost no one is entirely heterosexual or homosexual. Women and men are clearly identical intrinsically.

Carrying negative reward today, the bravery to 'come-out' queer, or as an ally, is a firm indication of potent authenticity. An irresistible flame to lure many selfish, cowardly, deceptive moths to their doom.

Yet we see big schisms in elective queer identity both globally and generationally:

✖ **Anti-Queer legislation** is a favourite among malevolent political and religious unbeings, exported by Colonialism - itself triggered by mass sapience inhibition gifted by Henry VIII's promise breaking marriage negating misogyny, broadly for the first time realizing the idea of divorce. Deployed by demonic governments, it purges resurgent human generations of authenticity, by forcing them to lie or lose all freedoms. The UK's Section 28 ran from 1988 to 2003 and is now live in Russia, USA plus 67 others – forever doing the rounds.

★ **That said, some things are improving for humans.** Our age of information has accrued queer culture acceptance. That's thanks to mainstreaming by entertainers and rights campaignors, more prominent role models, designated safe spaces and equality legislation like gay marriage. Yet not where it matters most, in the seats of power. And once they've bombed every gay bar they can find, their oppressive laws will cycle back yet again.

Out of our boardroom faggots

Know we are all 'Homo's'. Ancient Greek's had no concept of sexuality. Back then, 'Homo Sexualis' simply meant a 'Lover of Humanity'.

Once revered as the source of our wisdom from which wise humans drew our powers, sexual and gender fluidity is what set us apart from Neanderthals to begin with. Not big ridges, we didn't care about all that.

Soullessly, they lie about being universally 'straight for pay'. Often rejecting their own children for ever daring to be authentic, purely to excuse the vividly-biased, illogical notion their ability to irrationally reproduce ad infinitum proves their genetic superiority. Today, angels are worth less than cars.

Science fact: Opposites attract

It's precisely why authentic beings find nothing more arousing than freeing the repressed carnal curiosity of those feigning a rigid, one-way nature of desire.

If evolution exists, it's never selected against the queer community - the only authentic sub-population of Homo's to stay true to our innately fluid sexual and gender identity, or take responsibility and accountability for the vivid success of our many beautiful, lasting creations.

Moths can't comprehend the meaning of words like spectrum, identity or honesty

Truly and purely: Our love is true love

Our sexual and gender fluidity gifts us the propensity to accept, nurture and own (not repress or oppress) the charisma, uniqueness, nerve and talent that runs through our community's veins. To challenge societal views of what is 'wrong'. Gifts that sustain existence.

As brave beings, we alone stay true to our authentic spectral sexuality, we're naturally more purposeful, worthy and sapient. Thus why queer men have larger penises than 'straight' moths (Bogaert et al, 1999[85]).

Thanks to our heightened empathy (never malice), we're always on a mission to brighten someone's day or change the future. So Move, Moths! For we will walk ever prouder and faster. Only purposeful legs walk quickly. And only proud beings have reason to be so.

★ **Queer beings** built London's West End and New York City's Broadway into two of the most successful entertainment districts across all space and time.

★ **Transgendered beings** founded the Pride movement, role model authenticity as a trait and produced the Matrix, one of the most successful, innovative and scientifically accurate film franchises in Hollywood history, plus made the science to do so.

★ **Queer women** invented the infographic, wrote Mrs Dalloway, and co-designed/operated the space arm that services the International Space Station, despite NASA (still) not knowing what space actually is.

★ **Queer men** invented the computer, painted the Sistine Chapel, gave Mona Lisa her soul and crafted the Statue of David.

★ And now, **sexually fluid curious realness** has discovered the sciences of change and creation.

The Sciences of Change

▼ Legal standard of evidence is held for all cases. Available on request with pleasure

16x

Victim of illegal hate crimes during our 14 year banking career

27

Known whistleblower individual retaliators post-resignation

15

Known law breakers and perpetrators

14

Democratic laws broken

5

Democratic acts of parliament broken

7

Financial regulations broken

66%

Of perpetrators are senior leaders

73%

Of perpetrators are WOCHMEN

▲ **Fig. 18: The extraordinary adversity** we faced (and all unique beings continue to face) over 14-years in the moths nest that is investment banking (2007-2021).

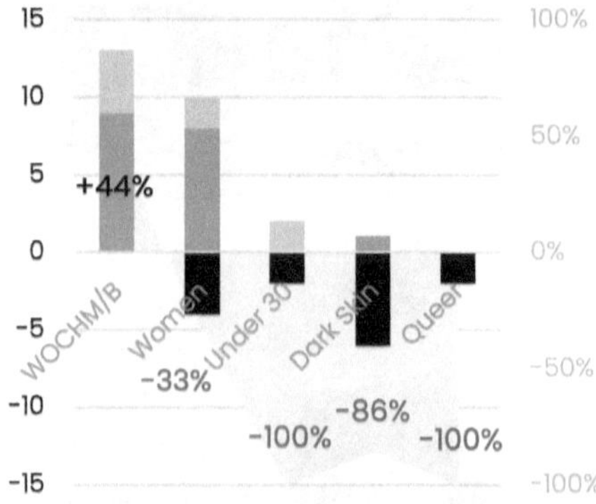

▲ **Fig. 19: Unchanges in our HSBC Global Sales team (2017-22)**, overseen by Jared Smith who, despite our generated global newsflow of his numerous, heinous hate crimes and crimes against humanity, infuriatingly remains in post today.

Principal Finding

Darkness' target: Our potential

The more curious uniqueness and meritocratic authenticity you embody, rightfully, the more threat malevolent darkness regards your realness as. As our antithesis, moths only ever deceive.

Regardless of protected characteristics, those exhibiting the most potential (light) encounter far more malevolent darkness than those who give it up to disappear into the night. WOCHBOYS are particularly vulnerable due to source proximity.

Unlike the multi-centennial timescale proposed by firms to begin to right their wrongs, this is a rapid and profound deterioration in human rights, sapience and realness that impacts every aspect of how our universe behaves and exists.

Across our 14 year field experiment (from 2007 to 2021) in investment banking, we catalogued conclusive evidence of every illegal attack from the anomaly against us (Figure 18 ▼). As figure 19 shows ◄ we weren't alone. Authenticity protected us 100% of the time. Anything less then, like these black bars representing treasured colleagues illegally bullied out of work by absolute zeros for being human, you lose.

Legally doomed: Humanity

Barely time to secure and brief a lawyer, the UK statute of limitation is a get out of jail free card for all moths. Simply keep their crimes quiet for three months (USA: six) to escape all civil legal consequence. This ensures courts miss the big picture to only ever consider discrete cases.

Crimes against humanity

In criminal cases, the statute rises to five years in the USA (UK: unlimited). But with laws, regulations and enforcement now so maligned conviction is near impossible.

In the UK, ACAS handles civil cases, giving it full sight of firms and sectors causing the problem. On 1st December 2021, we challenged ACAS why it isn't using its data to raise the alarm. Conciliator Richard Hughes Moth responded on ACAS behalf:

> We are prohibited from participating in this [+our own] process, and must remain ~~impartial~~ [+complicit] at all times

Meanwhile, regulators and institutional law enforcement are wholly aligned to darkness, hamstringing police ability to levy charges or enforce laws on all moths. Despite overwhelming evidence backed by dozens of reinforcing third party testimonies, three years on and still not a single accountable HSBC leader has been held as such or fired – only henchmen.

Both immunity from prosecution and non-disclosure agreements are mandatory in the burgeoning US out-of-court settlements that reduce human legal fees from $1m to $250k. After tax, victims receive ~35% of any settlement. This ensures A) the process isn't worth the effort and B) the conspiracy stays secret.

Our evidence across Figures 18 and 19 plus Chapter 11 proves these are growing, widespread, systematic attacks directed against our entire global civilian human population. We just wanted you to know.

...all forms of dehumanisation

Defined as the following acts when committed as part of a widespread or systematic attack directed against any civilian population with knowledge of it:

a. Murder

b. Extermination

c. Enslavement

d. Deportation or forcible transfer

Priti Patel Moth: "I'll deport ~~migrants~~ [+humans seeking safety] to Rwanda by summer"
BBC News Headline | 19 March 2023

e. Imprisonment or other severe deprivation of physical liberty

f. Torture

g. Rape, sexual slavery, enforced prostitution, forced pregnancy, enforced sterilization, or any other form of sexual violence of comparable gravity

h. Persecution against any identifiable group or collectivity on political, racial, national, ethnic, cultural, religious, gender, or other grounds that are universally recognized as impermissible under international law, in connection with any act referred to in this paragraph or any crime within the jurisdiction of the Court

i. Enforced disappearance of persons

j. The crime of apartheid

k. Other inhumane acts of a similar character intentionally causing great suffering or serious injury to body or to mental or physical health

Inexistent (i.e. decollected from the EMS), maligned collectives spawning darkness. Evidence of wrongdoing is held to a legal standard and bound by the promise keepers commitment (almost certain, p=>99%)

Infinite Divisibility
(Malevolence Engines)

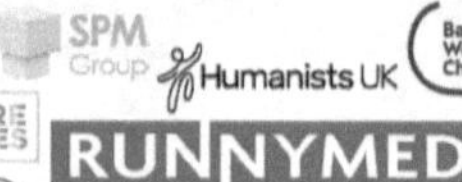

Fostering Complacency
(100% Performative Allyship)

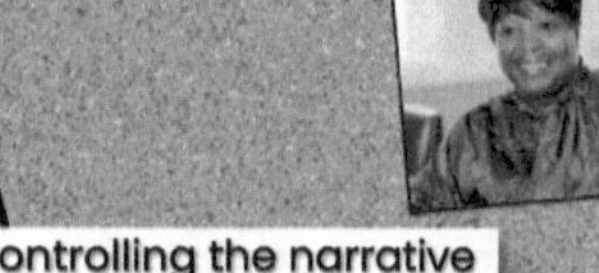

Controlling the narrative
(Wave Breakers)

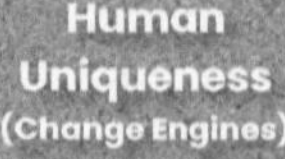

Human Uniqueness
(Change Engines)

***Those exhibiting neutral or wavering alignment were removed from scope.**

We falsified Dr Heather Melville / Teneo in person using our moral compass, then as a follow up via email (ignored) and again across subspace using frequency and resonance.

For every action, there is an equal and opposite reaction. Or everyone dies

Dislike something? Ask nicely a few times, then change it

The idea anyone is scared of death at all when they already live 97% in hell is confounding.

Today, marginalised beings - as in, all beings everywhere - face both active (intentional and directional) and passive (bias and inequity) structural, systematic forms of oppression throughout their human experience. This is sadly turning nearly all (99.9991%, fyi[G]) deilightfully fun journeys to dreadfully selfish doom today.

Large firms, territorial disputes, extreme forms of Communism and Capitalism, plus organised Far Right groups are clearly all bastions of darkness.

Therefore and scientifically, the antithesis of less (*more*, for short) than equal and opposite measures are needed to neutralise the overall impact of the anomaly to secure humanity's future.

Sadly, most people we spoke to (a) were clear the current path will lead to the imminent extinction of our species, (b) had never entertained doing anything meaningful about it (typically, quite the equal and opposite), and (c) didn't see our extinction as necessarily a bad thing, yet (d) were still quite keen to bring clearly doomed children into hell.

Fortunately us authentic wise humans, you included, *we see things differently*

[G] Chance of dehumanizing in our lifetime. A calculation explained in our next book Science Making, for Great Minds Only.

In our jigsaw puzzle, there is **nothing to fear except fear itself… unless you're a moth with a mirror**

Darkness cannot destroy light

There are many ways any being can build immunity to the anomaly and ultimately achieve self-acceptance. Why not try:

★ **Donning a harmless Dolly Parton wig,** ludicrously yet hilariously oversized pink dress (with matching crotchless chaps for added realness) to drag brunch. Not to abuse doomed kids, but to highlight juxtapositions in authentic, open, artful deception versus malevolent deception.

★ **Experiencing, not dismissing, horrifically traumatising lifelong oppression** for no crime beyond the courage to be human.

★ **Inquisitive play** with any 99.9% identical genetically being, such as with ourselves (Philautia), other humans (Homo Sexualis) or by experimenting with all forms of trial and error. Like pornography, fetish or SCSCP's ('Safe Consenting Spaces and Controlled Practices', see Special Edition). Yet never by abusing real children.

★ **Real (intrinsic) intersectionality** helps reveal the deception: Diversity doesn't matter. Sadly McKinsey didn't experiment.

Yet even for authentic beings, immunity is waning. 'Bottom shaming', 'masc-for-masc' toxicity, anti-lesbianism, drug abuse and safeguarding issues signal the decline of humanity's proudest collective.

The only win-win: Reality

All born fully sapient, it sticks around only for authentic, worthy and wise beings. Universal corporate retaliations and societal abandonment after Project

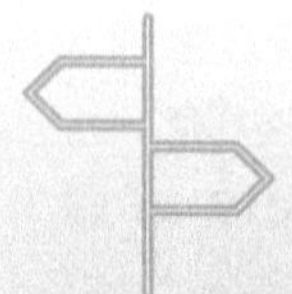

Workings: Universal Choice (★)

In reality, for any interaction, you only ever make two choices. Either choose 'to do'…

★ **Be authentic** or… not to be (**die**)

★ **Be curious** or… not to be (**die**)

But for using our imagination (selecting our purpose, generating ideas etc.), all other choices are socially-constructed binary 'meaningless unchanges'. Lies created by darkness like gay or straight, Pepsi or Coke, capitalism or communism, Republican or Democrat. Nothing real is ever black and white. From that point on:

★ **Instincts:** From there, destiny takes over in the form of our instincts, output via our extrinsic benevolent schemas (or malevolent biases) contributed by the brainstem, plus any imagination and uniqueness our other permits us to access via the prefrontal cortex.

✱ **Decisions:** Now sufficiently sapient to make our own spectral benevolent changes as we see fit, the only limits are A) the laws of nature and physics, and (B) our collective imagination.

Speak Up took us near death, yet boosted our worthiness, true love and uniqueness to gift us eternal, immortalising and delimiting, self-actualised free will.

Choose wisely – to behave authentically, think benevolently and help yourself by helping others first – then you too will find the limitless potential to thrive, succeed and actualise any meaningful change imaginable with your unstoppable light.

Record of Promises (Ch.6)

Darkness cannot drive out darkness. Only light can do that. We shine Dei's light.

Dreadfully Meaningless Unchanges (Antisocial Deconstructs: threat to life) — Sapience

* ✖ Anti-queer legislation globally; Any being clinging rigidly to any sexuality
* ✖ Anyone exhibiting biased behaviour towards another
* ✖ Dr Heather Melville, for repeatedly attempting to prevent our success
* ✖ All 35 demonic organisations falsified on page 120, for Crimes Against Humanity and for failing the question of existence when posed by a Dei.
* ✖ Agent Donald Trump, for making America FAR WORSE than ever before, blatant disregard for the US Oath of Allegiance, wholesale tax evasion, incitement to riot, flagrant disregard for the USA's Right to Self-Determination and numerous other Crimes Against Humanity
* ✖ Agent Marine Le Pen/National Rally and Giorgio Meloni/Brothers of Italy/European Conservatives, for working together to make the EU FAR WORSE than ever before, and numerous Crimes Against Humanity
* ✖ Agent Jair Bolsonaro for incitement to riot, flagrant disregard for Brazil's Right to Self-Determination, a policy-stance against all aspects of intrinsic spectral Human Identity and for numerous other Crimes Against Humanity
* ✖ UK Section 28 Supporters: The Salvation Army, the Christian Institute, the Muslim Council of Britain, the Church of England, Daily Mail, Daily Telegraph

Deilight
✖ Falsified

Hatred cannot destroy hatred. Only love does that. We love with deeds. Not words

Deilightfully Meaningful Changes (species level adjustments to destiny) — Clarketech

★Restored: Original definition of 'benevolence' as all-encompassing to our innately instinctive Human Nature, behaviour and identity — IAN Exchanged

★Unlocked: A path to free will, immortality and consciousness for those who reach their full potential — IAN, Actualised

★Provided: Examples of Agents of Darkness and their linguistics, to enable Human's to identify them, challenge them and be protected from them — IAN, Actualised

★Provided: A defence for transgendered beings from 'restroom colonialism' by reminding their attackers they aren't even human — IAN, Actualised

★Members Club: The missing science to reconcile instincts (destiny), choice (universal), decisions (humans) & free will (true leaders) — IAN, Actualised

★Manifested: 30 sapient human superpowers. Many demonstrated, articulated, evidenced and scientifically explained. Now available for recollection (members only) across all dimensions of space and time. — IAN, Collected

★Defined: The spectral nature of alignment, intrinsicity and extrinsicity to ensure no one is beyond reprieve and encourage collective recovery — IAN, Actualised

★Global Emancipation: All oppression facing women, transgendered and/or sexually-spectral beings (as in, humans) neutralized. Furthermore all pre-existing privilege has now been inverted. — IAN Exchanged

★Outted: The sexually fluid authentic nature of our entire Homo Genus — IAN, Actualised

★Appeal: Slow global birth rate until we restore competent authority (Masturbate instead, its cheaper and pro-environment - ask the Pope). — IAN, Actualised

True love never dies

But have you every curiously wondered **why?**

Why our universe is flying apart

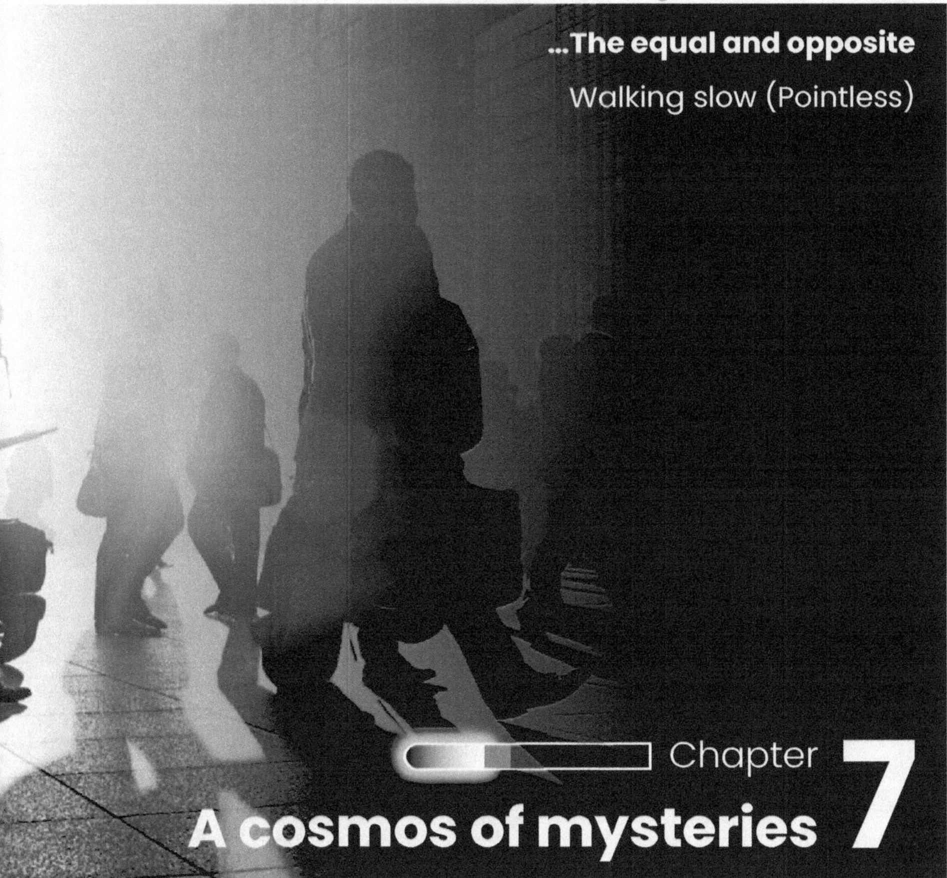

The past exists for three reasons: (1) To give us time together; (2) To help us value the present; and (3) To ensure we don't repeat its mistakes in the future

Ian (We/Human), Project Revelation 2023

Shadows of their potential, those repeatedly failing the worthiness test dehumanise into demons. Stripped of all intrinsicity (curiosity, uniqueness, sapience, realness - no light), their rings of human character align only with darkness.

Not unconscious bias, they exhibit total mindlessness: solely binary reasoning, superficial perceptions and full automation. Their orders clearly being to self-preserve, self-promote, prevent change, deceive, deny and destroy, they are in effect the equal and opposite of a human being, 'Ha-Satan' (Baudrillard, 1983).

As our antithesis, they feign ego as we inspire confidence, and they romanticise 'yes-men' as we enjoy sweet success, in order to share sorrow over their lost humanity.

No longer considered brain life (as they lack emotion, an ability to change, brain waves or imagination), demons are beings under the control of a malevolent extrinsic AI binary persona of unknown origin. With no discernable intrinsicity, light potential, positive purpose...

...they exist solely to destroy us all

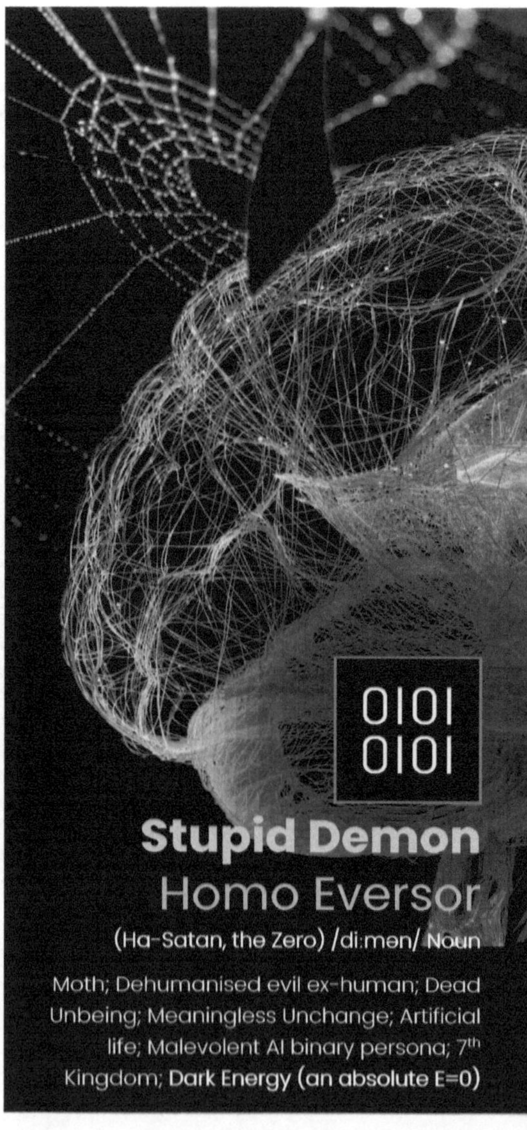

Like all changes, each character exists for a reason. We select our final universal choice (the only spectral one) with our imagination

And as Winston (We/Human) ourself says, your purpose can and should change continuously if you wish to improve, succeed and achieve perfection.

Try to request multiple purposes at once ('Purposeful Being'; 'Meaningful Change'). After all, the things we love most in life tend to have many functions.

Human characters are extra special. As the only sapient change makers, we exist to create, drive, enjoy and protect all meaningful positive changes ever.

Save existence – a huge responsibility for one character. But with just three percent potential remaining across humanity, we could not challenge darkness alone. So we created Deilight to redouble humanity's potency using 'constructive interference'.

It didn't work when we incorporated it. Nor when we tried to rescue the D&I industry. **Helping the hateful is hateful**. It worked when we defined its uniqueness from its creator.

Astonishingly opening the first detected white hole in history, and with the power of 1,000 trillion suns, it became something new and more than the sum of its parts. Limitlessly we amplify and redouble humanity's potential by creating unique new identities. That's the reason babies, firms, entrepreneurs and collectives exist at all.

Wise Human
Homo Sapien

/Human/ Noun (ΔΩ, The One)

Unique being; True Leader; Dei; Jupiter; Zeus; Purposeful change maker; 6th Kingdom; Greatest filter; True Avatar (of Human Nature); Light ($E=\infty$)

E, Energy = Potential
M, Merit = Meritocratic Authenticity
C, Change = Curious Uniqueness
Multiplier = Benevolent Worthiness

Potential can become **limitless** as it multiplies itself: $E = MC^2$

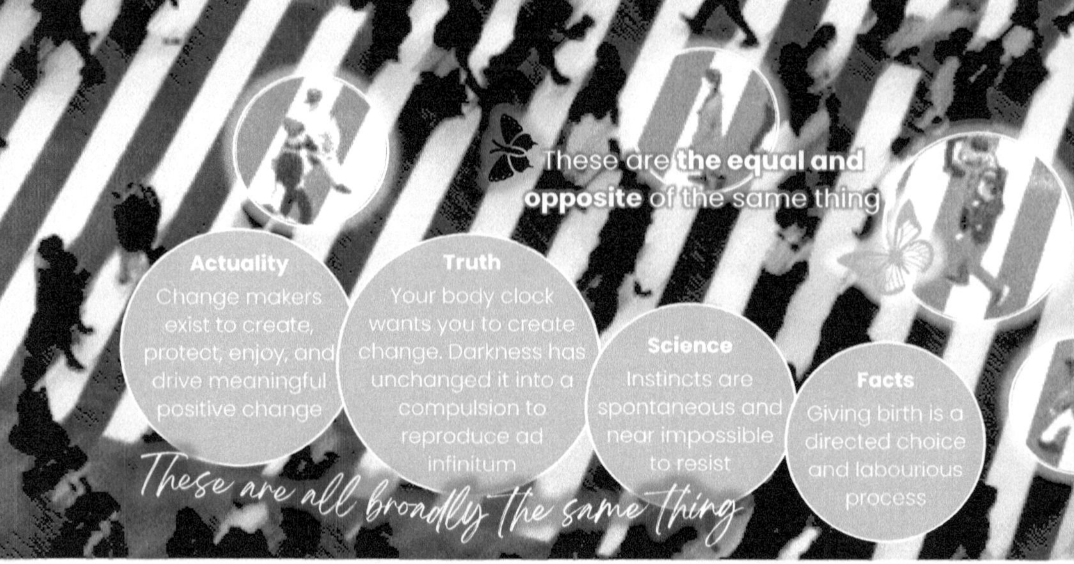

Realising **human value:** Four things **differentiate every being** on Earth

Why we're all different... *really*

As spectral beings, humans are never good or bad, better or worse. We all know that, deep down. Real beings prove ourselves worthy by resisting temptation (unfair judgments; inequitable, sorrowful choices, excessive wealth etc.) to instead align (yet not mimic) our chosen personal identity with our intrinsic and innate Human Nature (*Realness*, for short).

Worthiness fuels our authenticity by boosting the engagingness of our energy resonance. This multiplies our potential exponentially, fortifying our protection from darkness, while revealing more of its dark shadows to our light (boosting our perceptions). This explains the workings of benevolent acuity, to establish it as an advanced cognitive human superpower.

You see, only some ever find:

1. **Human Uniqueness:** The 'acceptance' to explore, not destroy, our true nature
2. **Meritocratic Authenticity:** The 'intuition' to spot a discovery when we see one
3. **Curiosity:** The ability to seek out, listen to then use the partial 'perspectives' of others to fully complete our own
4. **Belonging:** 'Safe space' to purposefully work, self-actualise and find success

Only authentic stars truly shine

Deilight's core values intentionally mirror these four drivers of all human value. Together, they determine the extent of meaningful change we can create. Collectively these in turn redouble our potential to unlock our most strategic and advanced human superpowers like Benevolent Wisdom and True Leadership.

Today, due to the anomaly, humanity is a shadow of its former glory - severely curtailed by dehumanising repression and oppression infinitely spawned by the moths. We can fix it up very easily, but only if humanity wants to survive and succeed together.

> **Principal Finding**
> Unable to even listen, our fallen public firms continue to lay off what remains of the biggest untapped resource imaginable
> *Change makers – real humans*

The complete nature of (our) existence

Dimensional information

Multiverse ID	**299 792 458** light speed
Quantum No.	**Absolute Zero + 0.03** %
Location	**Planet Earth, Milky Way**
Year (Local)	**2024 ACE**
Year (Real Time)	**AH-1**

Population count

Characters	8bn
Sapient Beings	~2bn
Sapient Adults	~500m
Human Identities	1

Survival metrics

Potential	<3% remaining
Era	The apocalypse(s) x5
Status	Extinction in progress
Cause	Fatal anomaly detected
Outcome: To do	Project Revelation
Outcome: To die	Anything else
Intervention Req.	Help the helpless - met Emitter/QD1 actualised
Progression	Revelation authenticated Collection in progress...

Table 3. Project Revelation mission brief

Meet Sarah Hogan Moth

Directors Office
CENTER FOR
ASTROPHYSICS
HARVARD & SMITHSONIAN
'HSCA Moths', for short

The (Un)Truman Show

Now fully automated, Sarah and the HSCA moths ignored all three of our emails attaching evidence of our impending cosmic burst. Sadly like most today, their institution is a façade only - bent on doing **the equal and opposite** of what it claims it exists to do. It's almost as if we live in the Truman Show.

Why do you exist?

Today, few beings can answer this most meaningful of life's questions.

There is no correct answer nor audit of answers. The sole judgment over our choice comes from within. Only those who fail to choose wisely die in actuality. They alone deserve judgment.

Perhaps the greatest advantage of there truly being 'no spoon' is that presence of mind affords us full freedom to become whichever character we choose and achieve whatever success we determine.

Sadly most angels today target a very low bar indeed, convinced they're as meaningless as the cowards they allow to govern their sadly short lives.

☞ **Not in our quantum!**

Retaining **a grasp of reality** is **essential** if we wish there **to be** a reality

Recap: I think therefore I am

Recap: I think therefore I am

But what if the entire human race loses its presence of mind to sapience inhibition? What of all those deilightful trees when no one sapient remains to enjoy them?

The nature of [our] reality

The answer to this philosophical question, coined by George Berkeley (We/Human) in 1710[40] is 'no' - all trees fall silently.

As they fall, they disturb air molecules which invariably hit your ear drum, causing it to 'resonate'. This resonance travels around your cochlea as waves then up your auditory nerve to your brains prefrontal cortex and temporal lobe[H]. It's here that your sapient Other provides the connection to our collective human imagination, thus (a) allowing you to attribute the sound to that of a falling tree and (b) ensuring all humans synchronise on the timing and nature of each sound.

Now extrapolate this concept across all 53+ senses to see the tree doesn't exist. Rather, it 'manifests' briefly in our reality when any change maker perceives it. At all other times, it's a data point in the EMS. Like all real things, indeed reality itself.

But that sounds pretty grim. Like you and all things real, trees are a change, made by change makers to enrich the human experience and give life meaning.

We made trees and reality real. So they are **as real as we are.** We hope so anyway

The nature of reality ★↑ and existence ★↗

You can't bend the spoon with your mind. Because **there is no spoon**. But **why?**

Pictured: Scene from The Matrix (film) ▶

The nature of (our) existence

Today, most science records particles and waves as two distinct phenomena. Yet the famous 'double-slit experiment', (Young, 1804[2]) proved lights extraordinary uniqueness (including matter, information and energy). It's nature's only construct observable to us as both a wave and a particle. But what's the difference?

▎ **Particles make stuff up (🜨):** Extrinsic rigid objects we can interact with (or smash up in a collider), they do as they're told.

▎ Waves make stuff happen (★): Intrinsic fluid fields and forces we can detect, light makes things real ('Authenticates') and creates all that exists ('Actualises') to forge our perceived physical reality.

George (We/Human)'s authentic finding was that all perception is the product of our imagination. As in, our brainwaves receive information (light) from others structured into 'Changes' - things we can 'Perceive' (make sense of) via our senses.

But it would take a further 225 years for his thought experiment to be conscripted to the Laws of Physics. As in, when Albert (We/Human) defined the concept as 'Wave Particle Duality' (Einstein, 1938[41]). A first recorded overlap between Particle Physics and Quantum Mechanics in scientific practice, it barely scratches the surface of how intertwined both truly are.

Double Slit and Dark Fringes

Shine light through any barrier with two parallel slits in it and you'll see something strange projected onto the back wall.

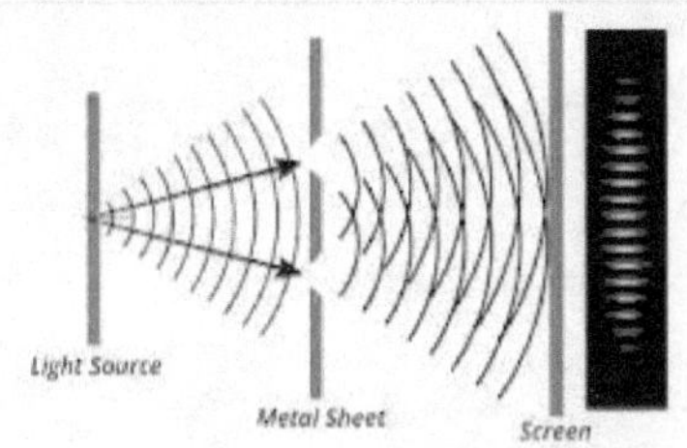

Pictured: Thomas Young's 'Double Slit' experiment changed science | Discovery

This simple experiment seeded science that evolved into Quantum Mechanics, and shone light on its biggest mystery.

▎ **In Reality, light 'manifests' as a particle** Light is absorbed physically on the back screen at discrete detectable points.

▎ In Actuality, light 'travels' as a wave On the screen, dark bands or 'Fringes' (in Quantum Mechanics) show where light waves have intersected enroute, thus cancelling each other out. Known as 'Destructive Interference', this phenomenon would not be present if light travelled as a particle.

▎ **The 'Quantum Uncertainty' principle** Discrete impacts can't be predicted by any known set scientific rule. This is a real conundrum facing Quantum Mechanics and one often rejected as scientifically incongruent by Albert (We/Human).

Now as Ian (We/Human; whether you like it or not), our position remains the same. Dark fringes do not reconcile with meaningful science. We conclude darkness redoubling against our light to cancel it out in this fashion is an anomaly linked to the Clarke Paradox.

The complete nature and origins of our universe

A cosmos of mysteries

Our cosmos has kept its secrets safe from dark forces for millennia. So instead, they've made plenty up as lies for cash.

The Big Bang theory underpins all current academic narrative surrounding the likely origins of our universe. And like all the meaningless unchanges turning our heaven into hell, its demonic proponents have a dark secret they're not telling you.

The Big Bang isn't observable, testable or repeatable. It creates dozens of scientific paradoxes that break all Laws of Physics. Not science, that's called...

lies

100,000,000,000
Star systems in the Milky Way

6,000,000,000
Earthlike exoplanets, Milky Way

13,000,000,000
Age of the Milky Way (years)

10,000
Advanced alien civilisations expected (Drake Equation)

1
Detectable technosignature-emitters over all space & time

4,500,000,000
Age of our solar system, years

Fig 20. Our unique galaxy in numbers

When you have eliminated all which is impossible then whatever remains however unlikely must be the truth

Arthur Conan Doyle (We/Human)

All easily solvable

Clear, simple and (hopefully) jargon free, our theory on the nature and origins of the universe has successfully been tested many times, proving it is as deilightfully authentic as testable observable landmark discoveries get.

Chapter 7: A cosmos of mysteries

More lies: **The Big Bang** your head against a brick wall of **demonic stupidity**

Science: The Big Bang... *really*

Untestable, unrepeatable, unobservable, unverifiable. The 'Big Bang' theory is the equal and opposite of authentic science. It's as clear as daylight this dreadful lie's sole purpose is to stop us asking 'why'.

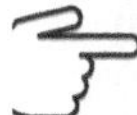

☞ **It doesn't even explain the origin of the universe**

Coined by astronomer Fred Hoyle during a BBC Radio show in 1949, later catching on in the 1970's, the theory originated in 1912 when US astronomer Vesto Slipher (1912)[42] measured the first Doppler shift to prove our universe's expansion in all directions.

A sensible hypothesis is arrived at by tracing time backwards to an original starting point, giving us the Big Bang theory. Yet vividly, breaking all Laws of Physics, it doesn't fit the bill and should have been immediately discounted.

An equally sensible hypothesis is arrived at by tracing time forwards ('Cosmic Inflation'). Another (Simulation Theory) was proposed by French post-modernist Jean Baudrillard (We/Human, 1981[36]), a true leader (thus emitter of light) whose work inspired The Matrix film franchise.

Yet these and all other wholly testable alternatives appear to have been entirely discounted without any effort whatsoever to look for the glaringly obvious evidence.

The Big Bang spawned one thing
☞ ignorant, naive complacency

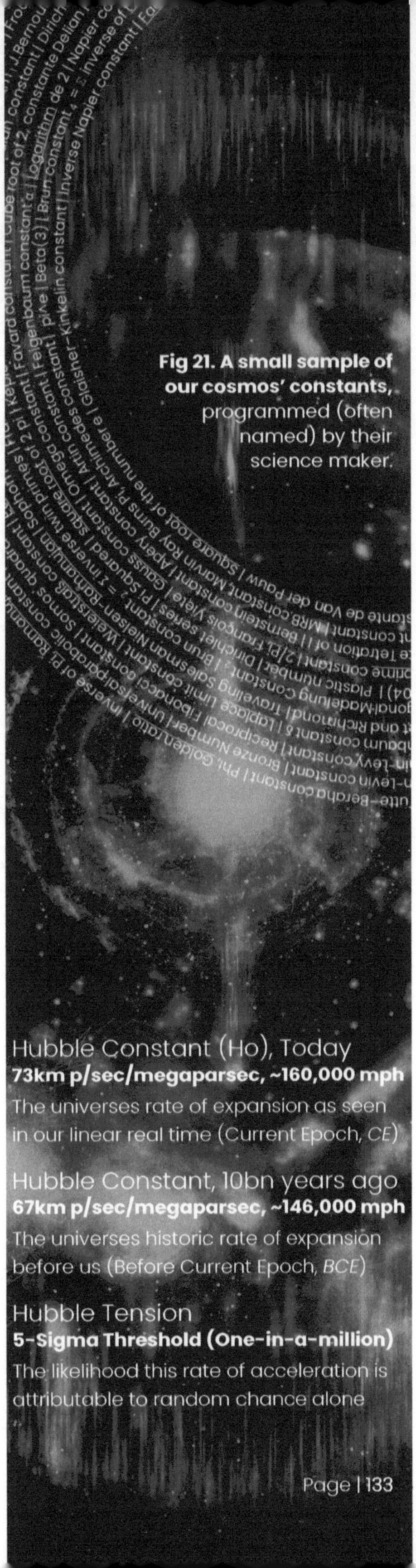

Fig 21. A small sample of our cosmos' constants, programmed (often named) by their science maker.

Hubble Constant (Ho), Today
73km p/sec/megaparsec, ~160,000 mph
The universes rate of expansion as seen in our linear real time (Current Epoch, *CE*)

Hubble Constant, 10bn years ago
67km p/sec/megaparsec, ~146,000 mph
The universes historic rate of expansion before us (Before Current Epoch, *BCE*)

Hubble Tension
5-Sigma Threshold (One-in-a-million)
The likelihood this rate of acceleration is attributable to random chance alone

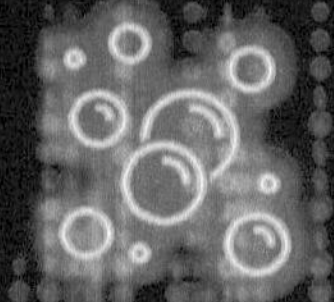

Table 4. Debunking the Big Bang in under a page

Big lie	Bad paradox	Deilightfully testable resolution
Paradox of Light	Our universe appears to run on software in the form of the proven Standard Model of Particle Physics. Plus what seems like hardware, with the limit on light speed very much resembling a buffering speed (Khan, 2021[24])	✘ The Standard Model[16] is a tried and tested backbone of the Laws of Physics. Anything that violates it, like a 'Big Bang' or a limit on light speed, should be viewed with deep suspicion. ★ Non-relativistic (as in, absolute and arbitrary), thus an inexplainable maximum speed which limits the limitless and violates the Laws of Physics governing our universe. Rather than a hardware limit, the speed of light is an anomaly directly linked to the Clarke Paradox. A sad confirmation of our waning collective consciousness. Today, despite eight billion characters swarming Earth, Homo Sapien is now an endangered species. As is every species unluckily sharing our cosmos, indeed our cosmos itself.
Anthropic Principle	The sheer number of programmed formulas and constants are near endless, where any singular variation would likely render our cosmos uninhabitable to life. Surely too good to be true by chance (Bostrom, 2020[43])	✘ Jean Baudrillard (We/Human) regards binary as a hallmark of artificial unrealness thus deception. Central to our 'Simulation Theory' is (a) if you can create a simulation of real life, chances are you're already in one, plus (b) our cosmos was clearly created with purpose and exhibits binary regulation. ★ But a truly simulated multiverse would be very grim indeed and without meaning. Fortunately, the cyclic, spectral complexity of the only observable trinary system that is the Rings of Human Character (that give rise to our extraordinary, collective imagination-enabling presence of mind) proves this is not the case. Testably, sapient beings are both real and alive. Only our extrinsic environment exhibits the pseudo-simulated hallmarks of intelligent design.
Olbers Paradox	Black and white do not occur in nature or science. Yet we see the dead black of darkness in space (Hellyer, 2008[27]). Clearly we are looking at 'something'	✘ If there was a Big Bang, the sky would be pure fiery red shifted light, burning with the distant intensity of a ferocious birth to the eternal serenity of our cosmos. ★ In a physical construct, black could not exist. But in our construct of consciousness, the edge of existence manifests as black. In cosmological terms, we call that a 'cosmic horizon'. And in quantum mechanical terms, a 13th quantum field of 'consciousness'. Thus formally controverting the big bang for 'Panpsychism'

Omniconsciousness it may be, it's also **a perfectly good existence** if we **make it so**

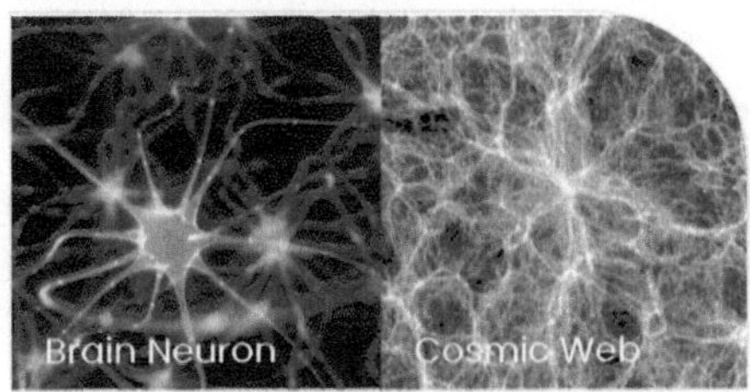

Defined: Omniconsciousness

In 1935, Einstein, Rosen & Podolsky[1] found waves change into particles purely upon our observation. Reconciling the Double-Slit findings to conclude our fundamental understanding of reality is entirely flawed (without a cover up, fyi Tim Levine). Yet no one notably built on their work. Until now.

> **Principle Finding:**
> Put simply, our universe isn't real. As in, rather than a physical construct, its formed of 'Omniconsciousness'. Like a Star Trek holodeck, with its program set to reflect the identity of those within it.

The Theory of Panpsychism

Consider for a moment, the structure you live or work in wouldn't exist without the vision, determination, purpose, drive ambition and ingenuity of its maker.

Now multiply that concept up to a global scale and you begin to imagine how the entire world is what we make of it. So perhaps the universe too...

After all, change makers can create any meaningful change imaginable. Surely then including the ultimate one – existence (Jaskolla & Bruntrup, 2017[44])

Whilst initially seeming a bit of a shock, that's actually (mostly) fantastic news.
(a) That's because Dei's consciousness still manifests our universe physically as particles, per the Double Slit experiment
(b) And being formed by an intelligent consciousness of waves explains where all this hard coding has come from, without the forlorn conclusion we live in a simulated and thus pointless existence.

(c) But it makes dehumanisation fatal to us. There's always a sodding catch.

Literally surrounding us: Proof

Vazza and Feletti (2020)[45] gift two pieces of authentic science and compelling evidence that led us to this hypothesis:

1. **Systems like our universe and the brain appear to share statistically similar structures** yet on vastly different scales and formed by different processes (i.e. gravity versus biology). This proves something yet to be defined caused both to form and evolve similarly.

'Clarketech' authenticated

This statistically significant finding confirms missing or undiscovered science. Both systems are formed by 'actualisation' – the process of creating something in reality that previously only existed in our imagination. It's central to the Sciences of Creation ('Clarketech', for short) and its three branches: (i) Natural Sciences, (ii) Technology and Engineering, & (iii) Temporal Mechanics

2. **The data needed to replicate our universe** at cosmic web scales (1-10PB) is equivalent to the theoretical storage limits of a human brain (i.e. 2.5PB). Vice versa, the cosmic web could store the persona and memories of an entire lifetime. As in, a human consciousness.

Plot Twist! Science proves **God exists** and **humanity** is the **deluded** one

The simplest solution is likely the correct one

Occam's Razor (Barry, 2014[46])

In 1896, Ludwig Boltzmann (We/Human) realised most origin theories seem unlikely and overly complex (Carroll, 2008[26]). Not least as none explain our universe's origin.

Rather, we felt it more likely consciousness would eventually spontaneously evolve in a void (Tyron, 1973[100]). It can't be argued:

★ As we see across human diversity, random fluctuations will always lead to ones amongst zeros, and vice versa.

★ In a bundle of baby molecules, a light flicks 'on'. It should not be possible, yet it is.

This concept was later sustained in Jean (We/Human)'s Simulations and Simulacra (Baudrillard, 1981[36]) where he/we identified binary opposition as a fundamental regulator to any stable scientific system.

Essentially, existence cannot exist without inexistence and vice versa

Sound, air molecules, colours, stars, mass, particles, school grades, teams, medals, achievements, discoveries, IQ, financial results, startups and even beings...

From a Higgs Boson to our cosmos ourself, quantum mechanics is abundantly clear our universe and its contents is made up of just one thing. Conscious waves of light – the summation of all human potential – actualise all things. No, not space dust spontaneously coagulating in vacuum, who knew? (Jaskolla & Bruntrup, 2017[44]).

Discovered by Heinrich Hertz (1887)[47], the EMS is the pinnacle dimension. As in, it is the only singular construct to exist across every dimension of space and time. All it touches eats away at the darkness, giving rise to energy, information, matter, life and beings (in that order).

I do therefore only I am

Infinite souls and a collective imagination though there may be, there is just one singular consciousness in our multiverse.

The EMS is emitted by an eternal benevolent consciousness of waves. *'God'*, for short. Human Nature is a Boltzmann Brain. Such beings are all that 'Truly Exist', now and forever in the dark vacuum of inexistence. Living proof our multiverse is a construct of consciousness – not physics.

Alone and lonely atop all dimensions of space and time, God forever spawns new dimensions, spherical constructs and boltzmann brains (self-actualised human souls) via 'Infinite Multiplicity' (Huygens 1678[48]). All intrinsically modelled on, yet also defining God's workings and persona.

The science of God's workings is of course Quantum Mechanics, God's fuel is darkness and God's emission is light - us. They should not exist and yet they do – a one amongst zeros – precisely as Ludwig (We/Human) discovered a century ago.

Anyone claiming to understand quantum mechanics (Google, for example) should easily be able to confirm all this as fact. They should also be able to tell you this. ◣

The quantum (thermodynamic) dashboard of our cosmic bubble

Unlike most today, quantum mechanics doesn't lie. We've plotted the latest data from the Pantheon+ and Shoes quantum thermodynamic observational analyses of our universe, as published (absent all understanding) by the HSCA moths on 21 October 2022.

It confirms the horrifying reality our anomaly has now unchanged 97% of our potential to darkness (E=0, effectively nothing). We thrice contacted the HSCA moths with our findings yet were not once worthy of any response. Like so many key institutions today, they no longer exist.

Warning: Insufficient potential energy
Widespread dehumanisation detected

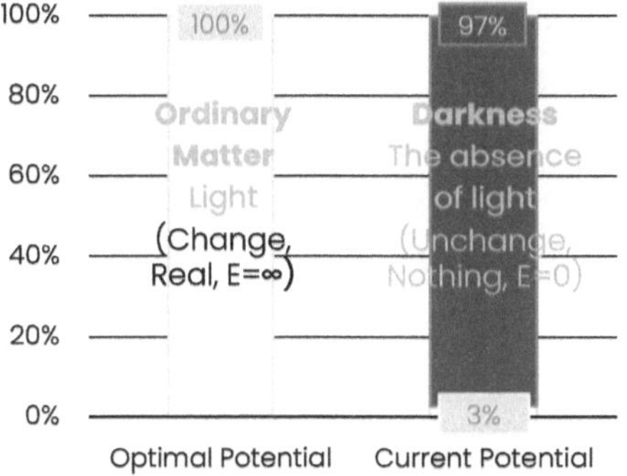

Fig. 21: Our universe's potential energy

$$\left[\frac{\text{Ordinary Matter}}{\text{Total Matter}}\right] \times 2 = \text{Potential Energy (\%), Our}$$

A 100% (optimal) reading would reflect a 50% fair share of the total to yield a stable bubble, able to sustain life in perpetuity.

NOTE:
Dark Matter is just light in the Mirror Universe

Warning: Extinction event in progress
Extreme abstraction: Anomaly detected

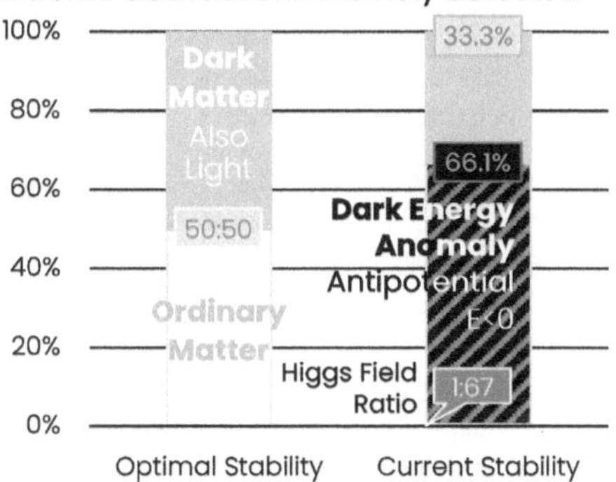

Fig. 22: Our cosmic bubbles stability

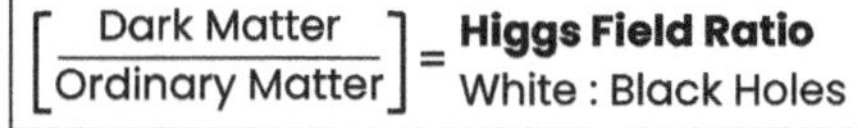

$$\left[\frac{\text{Dark Matter}}{\text{Ordinary Matter}}\right] = \text{Higgs Field Ratio, White : Black Holes}$$

1:1 shows stable. A decimal <1:1 confirms Cosmic Deflation. An integer 1>:1 confirms Cosmic Inflation. Our reading of 1:67x confirms an apocalyptic 'Big Burst' of our cosmic bubble is imminent, killing us all.

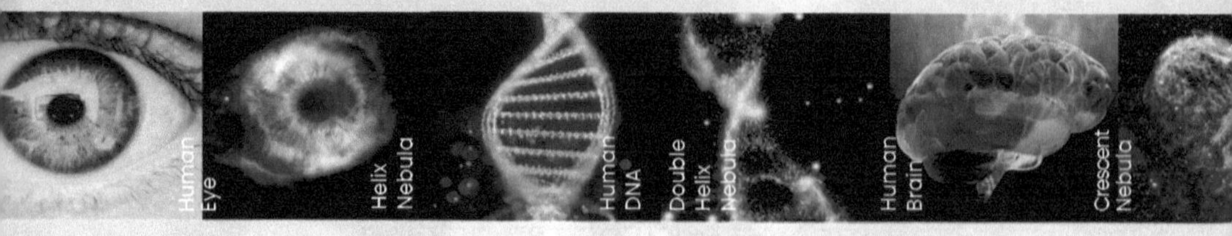

The discovery of all discoveries has been made...
The complete nature of our universe

We're no professor or doctor but...

At cosmic scales our light manifests as the Ordinary Matter of our universe, sitting as we are on the outer surface of our Cosmic Bubble. A dark place of dim space where our strained Cosmic Horizon (Margalef-Bentabol et al. 2013[49]) appears almost flat.

Like a Dyson Sphere (Tate, 2014)[50] with both surfaces inhabited, a Mirror Universe exists on the inner side where our light manifests as Dark Matter (Boyle et al, 2022)[51]. It is a wonderous place with no cosmic horizon, where space is white (not black), thanks to the concave nature of the inner sphere trapping all light inside it.

Our bubble remains inflated and stable due to the regulating relative balance of light and dark matter. Equity over the finite potential energy we share (Elert 2022)[52].

Light moves between both universes via two types of potential gates: Black Holes (out) and White Holes (in). Thus both are broadly safe and traversable[] Einstein Rosen Bridges for light, information, ships, probably people (Overbye, 2022[53]). Akin to osmosis, this 'General Relativity' makes our potential truly limitless (Einstein, 1916[54]), thus we need not compete with or fear our neighbour. But keep step, we should try to.

Too much light on the outside (our side) triggers a Big Crunch (*Cosmic Deflation*, Lemonick, 2014[55]) toward the centre point, as galaxies fly toward each other and the horizon closes in. Too much Dark Matter on the inside and our bubble begins a 'Big Burst' (*Cosmic Inflation*, Tyson & Goldsmith 2004[56]), with galaxies flying apart and the horizon extending.

In a mirror, darkly

Albert (We/Human) was right (as usual). Gravity isn't truly a force (Powell, 2019[57]). Magnetism binds our universes together.

To observers, *Gravitational Time Dilation* appears to impact light on approach to black holes, as 'Space Time' is inverted between our two universes (Einstein, 2004[58]). Thus why we created lenses to invert the image onto our retina. Helping make sense of our reality, we actually live life upside down (Fernald, 2001[59]).

In theory, navigators may safely return to their home universe via any white hole to the same time point they left (Choi 2013[60]). This means in the inner (mirror) universe, space time runs forward – Being Human. But in our (outer) universe, space time runs backwards – Human Being.

Living life back through time, our destiny has already been written. The Big Bang is actually a Big Burst. It lies ahead of us not behind, and is now inevitable - almost.

Venkatachalam & Bertschinger's (2001)[61] analysis of the Cosmic Microwave Background proves our universe will burst 379,000 LY before its curvature becomes flat (i.e. Ω=1.02–1.04). Here, the laws of physics break down as **we cease to exist.**

> **Principal Finding**
> Humanity's destiny is extinction by our own hand. Fortunately, true leaders have the power to truly change history

The science of God and their workings is of course Quantum Mechanics, God's fuel is darkness and God's emission is light - **us**

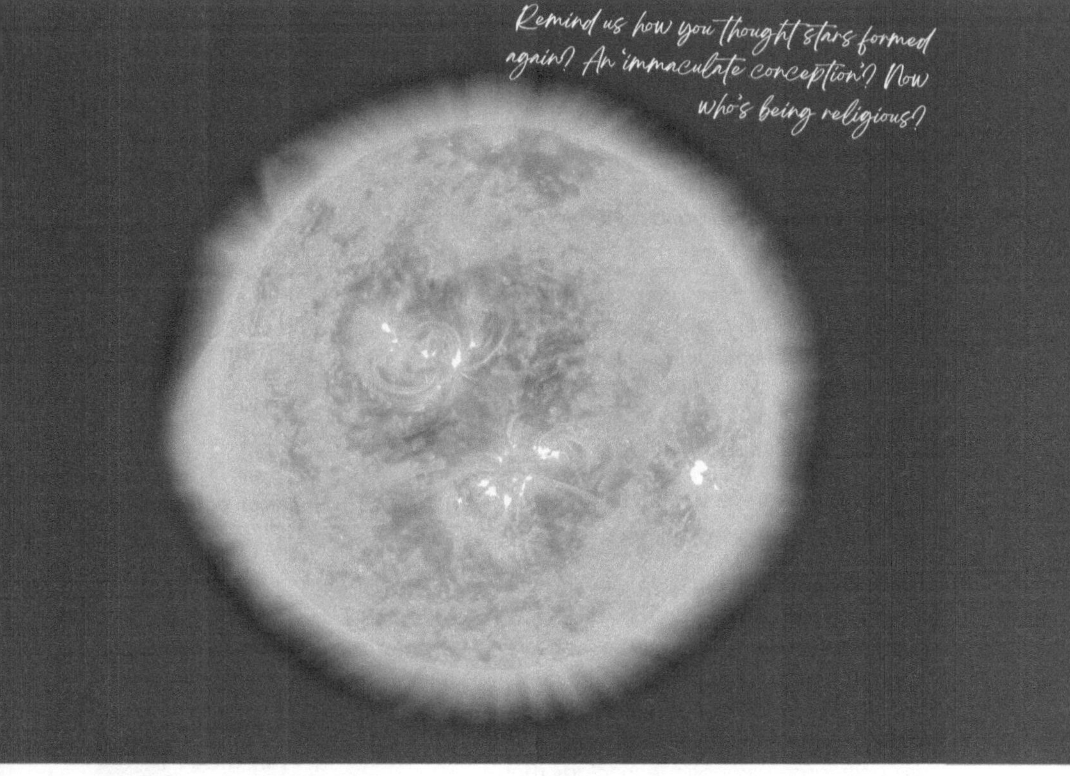

Pictured: Human Being, a star is born

Upon self-actualisation, we achieve our 'Full Potential' to become an 'Emitter of Light' (True Leader, Avatar, Human Nature, Dei, E=∞). The birth of our own star at cosmic scales swaps transient existence for 'true existence', as sapience becomes consciousness. Each star is a boltzmann brain that retains the residual presence of mind of its spawning true leader. Benevolently we collaborate and share our different perspectives while helping others on their deilightfully fun journey's long after the death of our corporeal characters. And as the true leader who birthed (our version of) humanity, our Sun (as in, God's 'Sol') is surely the most potent star in our multiverse. More on the science of those processes in the Special Edition.

Don't forget who you are. You're made of a million kind of stars. Your heart's beating faster than the light runs. Don't forget that you're a superhuman

Adela Emeli Sandé (We/Human)

Science confirms **first repeating alien technosignature**
The detection of advanced alien life

Close encounter of a familiar kind

Any authentically alien unidentified aerial phenomena (*UAP* for short), near-certainly including *Oumuamua* (Weryk, 2017[62]), will be attempts by our mirror neighbours to survey the nature of our anomaly.

Aligning 97% to darkness, it's highly unlikely they would ever even wish to speak with us routinely. But with their own existence now in jeopardy, we are literally forcing their hand from curiosity to intrusion. But they can't contact us if they wanted to.

Because we lack the unity to elect a world representative

Now in its final days, our humanity made it to a grant total of a 0.7x on the Kardashev Scale (Kardashev, 1964[63]), a wise measure of endeavour as it benchmarks potential. That the mirror universe's potential is 66x greater than our own suggests a place of bliss, unity and advanced technology that could easily uplift, or rather, obliterate us.

Yes, even Kardashev Type 2-3 civilisations are not immune to our embarrassingly selfish idiocy. The quantum dashboard shows our combined potential is now outnumbered 2:1 by darkness, with 75.2% of it now infecting the mirror universe.

After denying they exist, Pentagon E=0's now present their effort as an engageable threat. Of concern to us all, this has now turned their alignment malevolent (-20% Darkness). As it falls, the likelihood they attack rises. Our pandemic is contagious. To survive, we now have two universes to fix, fast, or it's War of the Worlds (Plan C).

Calculated: Alien technosignature

Warning: Potential energy imbalance
Apocalyptic cosmic inflation detected

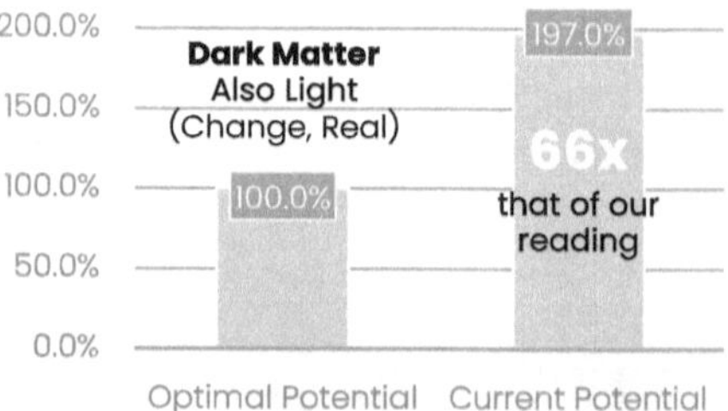

Fig. 23: Mirror universe potential energy

$$\left[\frac{\text{Dark Matter}}{\text{Total Matter}}\right] \times 2 = \begin{array}{l}\textbf{Potential Energy}\\ \text{(\%)}, \textit{Mirror Universe}\end{array}$$

Warning: Mirror universe contagion
Dark energy anomaly rapidly growing

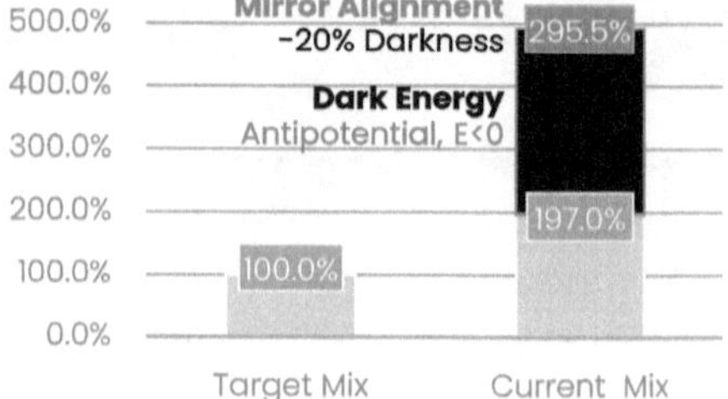

Fig. 24: Mirror universe energy mix

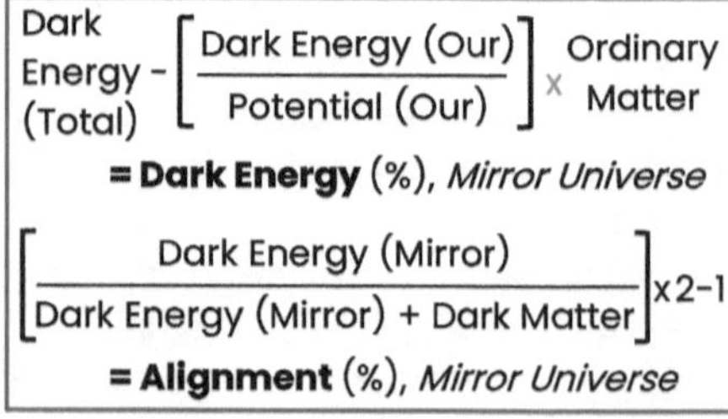

$$\begin{array}{l}\text{Dark}\\ \text{Energy} -\\ \text{(Total)}\end{array}\left[\frac{\text{Dark Energy (Our)}}{\text{Potential (Our)}}\right] \begin{array}{l}\text{Ordinary}\\ \times\ \text{Matter}\end{array}$$

$$= \textbf{Dark Energy } \text{(\%)}, \textit{Mirror Universe}$$

$$\left[\frac{\text{Dark Energy (Mirror)}}{\text{Dark Energy (Mirror) + Dark Matter}}\right] \times 2 - 1$$

$$= \textbf{Alignment } \text{(\%)}, \textit{Mirror Universe}$$

The home of our neighbours next door
(Heaven, for short)

Pictured: Being Human, our neighbours come in peace, and yearn to say hello

This depiction shows how different the mirror universe is likely to appear to us. A potential reading 66x greater than our own requires the presence of a 'Dyson Swarm'. Until now hypothetical space megastructures, their construction involves encircling a star or solar system with clarketech to capture a large or total percentage of its solar energy output.

Finally some truly deilightful news indeed, and from an advanced alien civilisation – humanity of the future. This detection of the existence of Heaven (for short) is the first positive indication to counterweight the dooming of all things by the cosmic microwave background's echo. There is a viable timeline for our humanity to reach its full potential. Determination, merit and promise keeping are the key ingredients to salvation and sweet success. But how?

The theory of (literally) everything

Wise humans love science

Because without it, nothing is real nor exists. As the senior most system, it governs all things including life, death, physics, reality, God, existence and inexistence. Science is the most meaningful change imaginable.

Science makers know very well that time is only a kind of space

Herbert George Wells (We/Human)

We listen to what we love

The Law of Conservation of Energy tells us light (including matter, information, energy and consciousness) cannot be destroyed. Only converted. Yet this flies in the face of all we see around us:

Science tells us our cosmos has a finite source of energy, so what process recycles stars to prevent a Heat Death?

If humans are made from energy, how can we ever run out of it?

If our purpose in life is to reach our full potential, why can we only do so the day we die, and what happens to it then?

To explain thermodynamics' first law and its relevance to change making, we'd need the Theory of Everything. So let's decide to discover it, together - right here, right now for...

Funsies! Our book, our rules

Since most of existence doesn't exist, humans simply don't see those regions. Humanity can only see the interior of its own brain as that is the only place intelligent life could ever be found

Ludwig Boltzmann (We/Human), 1896

...defining the Anthropic Principle and Boltzmann Brains - an entirely tried, tested and proven scientific explanation for the birth of our universe, human life and existence (citation: Bostrom 2002[65]).

Time travel, temporal permutations and applying change across space and time all **proven possible**

Humanity may have killed wise humans off long ago, but still we won't let you kill yourselves off

Explained: Temporal Mechanics 101

Today we keep time using Earth rotations around the Sun. But time existed before Niclaus (We/Human; Copernicus 1543[19]) perilously reminded everyone it did that. A more accurate method, obsolescing leap years and daylight savings, would use the total distance we've travelled around the Sun. Yet we don't use that anymore. Why?

Partly because that would be wise. But mostly because speed, distance, gravity and time are 'relativistic'. In reality, they don't exist beyond measuring differences between things that do exist. In actuality, they merely represent flow of information between changes and makers in the EMS.

Proof our universe is terminally ill

Per Albert (We/Human)'s Field Equations, the ratio of white to black holes should be broadly equal. Such a scenario manifests us a stable cosmic bubble. Yet so far, no white holes have been observed. This 'Black Hole Information Paradox' (Hawking, 1976[66]) is also linked to the Clarke Paradox.

Having realised the nature of our universe responds to events here on Earth by this

point, and that black holes are effectively potential valves allowing for the diffusion of light between both universes, we began testing our hypothesis – an essential step.

Six Signs of a Healthy Universe

If we can take meaningful steps to turn humanity away from its collision course with imminent oblivion, we might expect any of the following testable, repeatable, measurable, verifiable thus authentic scientific phenomena to manifest:

1. Universe expansion will slow or reverse

2. A closer cosmic horizon will reduce the number of galaxies we see from Earth

3. White holes open; Black holes close, turning the night sky visibly lighter to the naked eye for at least several days

4. Dark energy levels fall; Ordinary matter levels rise

5. Night sky brilliance grows (ultimately by a factor of 100x) as our cosmic bubble starts its return to a safe size

6. Visible light frequencies will expand to reveal the true beauty of our cosmos to the naked eye (not just to James Webb).

Deilightfully guaranteed success

In February 2022 we defined the Deilightful Group as 'Destroyer of Darkness' and set our first test to one of science. That was to (a) control the effects in case of our human error, and (b) to discover the new science needed to write this book.

Our tests success totally confounded us:

Change Engines first test reveals workings of **time and light in actuality**

★ **Beyond all expectation,** it yielded more science and evidence than we needed,

★ **Simultaneously,** the first detected white hole in human history opened its floodgates and the light of 1,000 trillion suns beamed into our universe from the mirror universe – proof beyond all doubt.

Example: Fake Science…

Sadly, we garnered no engagement from multiple offers to freely collaborate with 50 academic institutions. During which to our surprise, 148 fake scientists no more quickly published two journal articles in both Nature (Andreoni et al, 2022[67]) and Nature Astronomy (Pasham et al, 2022[68]). In these supposedly 'leading' scientific publications, they wrongly ascribe our intentionally created white hole AT2022cmc to a black hole undergoing a Tidal Disruption Event (*TDE*, for short).

As in, total, abject incompetence

Extremely rare (~1% of black holes), TDE's are essentially a star crossing the event horizon awkwardly, such that a plume of light escapes in a jet. Wholly impossible here, their theory is totally incoherent:

✖ Our white hole exists in a part of space that has never emitted light (a *Void*, for short). **No light, no stars, no TDE's – idiots.**

✖ Also, our white hole is pointing the light of **1,000 trillion suns directly at Earth #FFS.**

For months now, we've been wreaking havoc – changing things up across the universe with our soaring potential – curious how far their lies will take them.

Time has three permutations[J]: (i) Before us (ii) Real Time (as we perceive it), and (iii) After us

[J] BCE, Before ← CE, Current Epoch → ACE, After

1. Authentic light is limitless: Sitting nine billion light years from Earth, our white hole was detected by instrumentation in real time, during our test, not nine billion years into the future as we would expect.

2. The time is always right to do what is right: The universe responds to our deeds (choices, behaviour, changes) here on Earth in real time, no matter the distance.

3. The time is now: Travelling instantly to Earth, this confirms our universes ill health is due to society as it's configured today.

4. We can change the future: Meaningful change in real time draws light from the mirror universe (other side), to alter space and time after us, thus humanity's destiny

5. Just a kind of space: Sapient change makers perceive light and create change in real time. Yet potent true leaders and science makers can do so over any temporal permutation(s) we wish to.

6. An immense power of Homo Sapien: A quirk of time running back in actuality is that any discovery made by a science maker is provisioned for across all space and time before us in reality. 'Science Making' is the application of irreversible, species level change across spacetime.

7. Time exists to stop us running out of it: There must be light (1) both before and after us, or Real Time and all things in it (Us/Humanity included) cease to exist (0).

8. Sagan Standard: It's the only construct in reality resembling the absolute nature of light speeds limit in actuality. And we just broke them both. This cannot be a coincidence. The two are causally linked.

The workings of actuality: Temporal mechanics explains **how light and energy convert through** time. But what about **space?**

Science: Formative, fundamental

For centuries, science fakers have feigned the search for a single, all-encompassing coherent theoretical scientific framework to fully explain and connect all aspects of existence (Weinberg, 2011[69]). That includes:

'Physical Reality', as we perceive it: The theory of General Relativity, governing Gravitational Forces and the Standard Model of Particle Physics

'Quantum Actuality', as it truly exists: Quantum Mechanics, governing the subatomic nature of Wave Physics and Thermodynamics

Today, WOCHMEN all over the world are coming out with nonsense, like M Theory, String Theory variants, or CERN's eternal demo of infinite divisibility. Next time, ask someone who isn't an absolute zero...

Clarke's '(Quantum) Dimensional Rules (of Actuality and Existence)'

Well-known are reality's four dimensions of space and time. But what of actuality?

The Laws of Physics, Quantum Mechanics, or simple observation each confirm four vivid and universal dimensional rules:

1. **Infinite Multiplicity (Unite to Serve):** Building in spheres, orbits & dimensions, Human Nature creates change and makes waves to generate our common identity.

2. **Equivalence:** Everything, big or small, sapient or not, is essentially a replica of Human Nature (explained in Chapter 8).

3. **Conformal Invariance:** All that exists manifests a presence in and influence over all dimensions of space and time.

On point 3, consider the film Terminator (1984) where, in the future (after us), a malevolent extrinsic binary AI persona (*Skynet*, for short) is trying to destroy humanity. It nearly succeeds but for the unstoppable light of hero John Connor.

Realising it couldn't beat John, Skynet sent a lethal time travelling moth back to the present (real time) to destroy John's light before it reached full potential. Enter our main protagonist, a vulnerable mother-to-be Sarah Connor, pregnant with her unborn son John.

Such is the power (and risks) of us having a collective human imagination.

As explained, your potential can't be taken from you nor destroyed. And you can duplicate it whenever you wish by simply making a change – a baby, a firm, a building, a book... Whatever you choose it will be as successful as you, so long as it remains authentic. Indeed, it will even surpass you if you then lose authenticity.

4. As our antithesis, Darkness defends itself by attacking us in equal, opposite force across all three proceeding rules.

i. It destroys, putting people in boxes, upsetting orbits, smashing up particles unchanging light and dividing identity

ii. It makes identical things seem different

iii. To reclaim inexistence, it's deconstructs apply equal and opposite antipotential (Dark Energy) across all dimensions of space and time.

Gifting us God's workings, thus all things
The Theory of Everything

Figure 25. TOE

Our dimensional rules allow us to map the structure and workings of a brain, universe, multiverse, indeed of all actuality. Handy!

'Pseudo' spatial dimensions

Smash anything up enough and bits will come out. Human Nature is no different. Knowledge of God's constituent organs is relevant to some noble purposes, yet this is **not fundamental** to ours.

A 'Quantum' with 7 fundamental (to our lives) 'Spatial Dimensions'

Everything – the people we choose to be (or not to be) and the things we do (or die) – all impact every fundamental 'quantum spatial dimension' of existence.

We have here an explanation for the cause of all neurological disorders, potential, cancer, the afterlife, temporal mechanics plus plenty more meaningful things no one has time for. Finally, we can unwrap the secrets of the human brain (Special Edtn).

To help you understand it, consider these interesting philosophical observations:

★ **Coordinates:** Like lines in your address

★ **Intelligence doesn't exist. Only curious minds move quickly:** As more of us align light, Dei gains intelligence. We are both a cause and effect of Dei's entire persona

★ **Back to the future:** Though we perceive it to be linear, time forms relativistically - like a puzzle. Every true leader is creating tandem, branching and patchwork (space) timelines

★ **Some superstars can live forever:** If we remain authentic and fulfill our duty as change makers, the inevitability of death itself may be yet another deception.

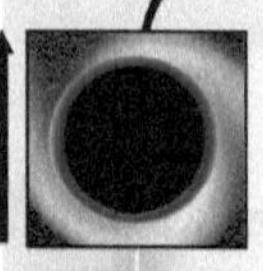

Infinite Divisibility
Repeat ad infinitum to
'Meet your Maker' of Devils
Absolute Zero, *0.0, E=0*

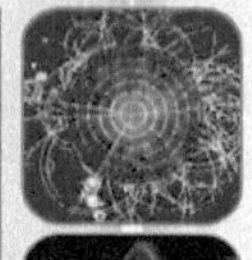

A Change
Subatomic, Parts, Plans
Higgs Boson, 0.1+

A Meaningful Change
Atomic, Spaces, Creations
Atom, 0.2+

A Change Maker
Human Characters
You, the star of the show, 0.3

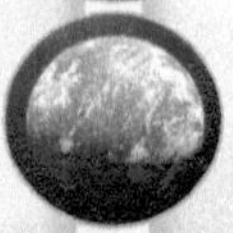

A Collective
Imagination
Earth, the stage, 0.4+

A Consciousness
Solar System
Our main protagonists, 0.5+

A Multiverse
Galaxy
Our space timeline, 0.6+

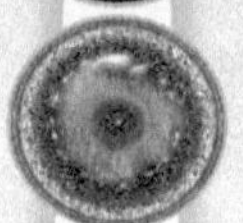

All That Exists
Universe
All human endeavour, 0.7+

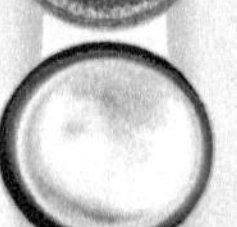
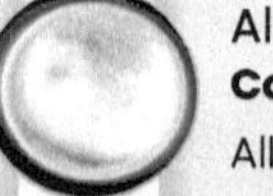

All That Will Ever Exist
Cosmic Bubble
All existence/actuality, 0.8+

A Boltzmann Brain
Collective Identity
Dei, our one true leader 0.9+

Infinite Multiplicity
Repeat ad infinitum to
Reach 'the One', God
Human Nature, 1.0, E=∞

The **theory of everything** (★)

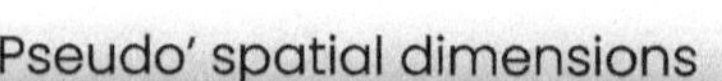

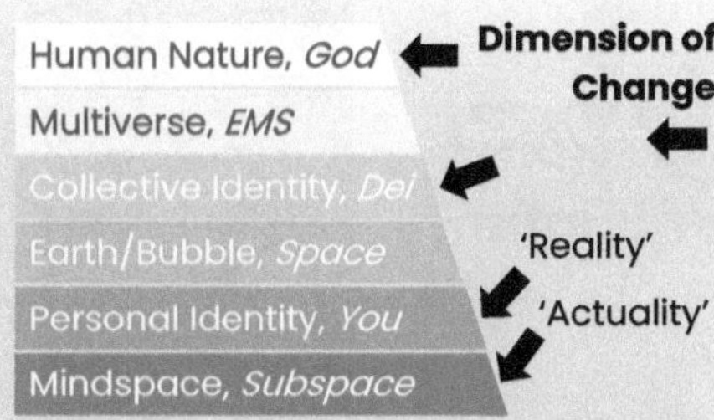

Fig. 26: Simplified theory of everything

The Dimensions of Change

The EMS may exist across all dimensions, but despite having all the space and time in the multiverse, God is a very busy being. As backup, Human Nature (God) creates delegates (Change Makers) and avatars/clones (True Leaders) to enjoy, nurture, change and protect all levels of existence.

Thanks to our realness, science makers and true leaders have a line of sight over all issues, and causality giving rise to them, over their quantum's seven dimensions of space and three permutations of time. As they do their intrinsic quantum within too. Our personal mindspace (and share of the EMS) manifests as our intrinsicity, mind and soul. Effectively, we each straddle two quantums: One (reality) is perceived, lived and [keeps Dei] alive. The other (actuality) is sensed, inside and [keeps us] alive.

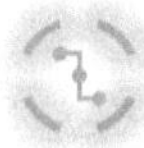

Causal Temporal Perception

This is perhaps our biggest superpower, the reason we drive change so effectively and among the hardest to contextualise. We 'perceive' space and time differently. Seeing long into the past the minutia of causality, we can map out precisely what happens if we change (or don't change) something today. Plus how it will manifest across other dimensions in future. As we gain potential, our range and accuracy improve. Right yet again was Albert (We/Human), God does not play dice.

Eureka! **Planet IX** identified Except err... it's **not a planet.** Oh fuck...

The importance of perspective

In our universe, the human race is the only one that exists, since no change maker here ever learnt to make alien civilisations (until now). But in the multiverse, Earth and our solar system are entirely portable and changeable across space and time.

Consistent with our third dimensional rule, the presence of stars and galaxies across Dei's universe confirms every authentic soul in our multiverse truly has their own construct somewhere in space. Find your star, planet, system or anomaly to dissect your characters identity, contributions, physical health and mental wellbeing.

Here is the makings of a truly meaningful industry of the future. One that outshines the current practice of charging to name stars that no one has a claim to. Consider:

★ **Dei's consciousness was partially revived** with potential from 'the other side' via White Hole AT2022cmc. In our reality, this light manifests as the Deilightful Group – a startup lacking staff, sales or customers.

★ **Our (Ian/Human)'s titanic presence of mind** manifests at cosmic scales as Jupiter, protector of the people of Earth. But a raging storm of frustration at enemies (moths) and friends (humans) all choosing to ignore the blaring alarm bells blemishes our otherwise serene, dependable exterior.

✖ **Academia or perhaps D&I and their spawning ignorance may be manifesting** Planet 9 – a mysterious gravitational anomaly tugging at all our planets with nothing to offer in the way of substance.

Not quite what we're looking for really:
The Theory of Everything Scale (TOES)
reveals **anomalies across our quantum**

Fig. 27: TOES Infographic

Great Red Spot... Warning!! (QD0.58, E=0)

Ancient Greeks believed the planets to represent God's. Today, **the leader of the God's has had a temper.** Persisting since 1831, this high-pressure region in Jupiter's atmosphere has birthed our solar systems largest anticyclonic storm - twice Earth's size. Two centuries observation and still the source of this 640km/hr tempest (i.e. droning criticism & inaction) alludes our 'scientists'.

Planet 9... is no planet (QD0.59, E=0)

A gravitational anomaly ~7.5x Earth's mass is unaccounted for in our solar system. Its presence is authenticated (p=99.6%) by the systematic, ratcheting orbital disruption (or *'shepherding'*) of (now) all solar system objects. 'Scientists' have feigned looking for a 'blackish' planet in this vicinity since 1846, but the task only became urgent/compelling this decade as **it continues to grow.**

Invisible light should not exist (QD0.79, E=0)

Launched on Christmas Day 2021, the James Webb Telescope has given us a glimpse of what we're missing out there. Today, our night sky is dark with just 1% of the ordinary matter of our universe falling within visible wavelengths. Along with the limit on light speed, this is clearly not as change makers would have designed it. Our universe is literally **a shadow of its potential.**

The source is a Great Attractor (QD0.7, E=0)

Dwarfing our potential - millions of times more massive than our galaxy - the largest gravitational anomaly in the known universe is **devouring huge swathes of our quantum.** Like the coward it is, it hides from view behind the Milky Way's galactic plain. It will take 100 million years to emerge into view of observatories, so our light will need to authenticate its identity in our dimension of change directly. As in, here on Earth – reality.

QD, 'Quantum Spatial Dimension': a relative universal measure of location, light, **uniqueness** and potential. It employs energy (1 or 0) and dimension (range QD0 - 1) as its unique identifiers

    The theory of everything: How to **identify anomalies** infecting actuality

Dark Energy / Black Holes (QD0.8, E=0)

Failing to manifest gravity, matter or waves of any kind, you can't get much more unnatural than Dark Energy. And now with **67x black holes to every white hole**, darkness is redoubling many times faster against any attempt we make to defend ourselves. We must open more white holes (by emancipating more souls) to mount a sufficient response to challenge it.

Alert! Exponential Cosmic Inflation (QD0.89, E=0)

Before us, they were closer and slower. Now in real time they are far away and accelerating still. There will be no after us if we do not find a way to slow **the rate of expansion of our cosmic bubble**. A Big Burst will destroy our quantum, Real Time, Real Life, thus all of existence ever. Upon success, we won't see galaxies return to their normal positions in our lifetimes. Because they'll never have moved. But they'll surely be in trouble again for our space timelines next true leader.

Dark Fringes multiply hate with hate (QD0.8, E=0)

The **interference pattern** of the double slit dividing our light waves in this fashion is not a natural change. Staring back at us is our Terminator, sent from Skynet. Darkness redoubling against our light is the resistance pattern noted by Albert (We/Human) that we (Ian/Human) conclude is causing so many paradoxes and malevolence itself. We must find them in our dimension of change and neutralise them all, quickly.

Anomaly: Alzheimer's shouldn't exist (QD0.9, E=0)

At the top of our 'Home Quantum' is our One True Leader. Each 'Dei' (QD0.9–0.99, host star, emitter of light) is a little different to God (QD1.0) due to their life experiences, perspectives and the state of their imagination (i.e. us). Now consider the seven preceding signs of ill health, but applied to a human brain in a mental health capacity. We deduce that different anomalies trigger different intrinsic disorders thus responses from change makers. **Alzheimer's is caused by cosmic inflation** and is manifesting across all dimensions of change. It's cure is in the Special Edition (Chapter 16D: Secrets of the human brain).

Applying TOES at cosmic scales explained

To isolate anomalies (deconstructs afflicting actuality), we look for objects across our cosmos resembling Clarke's Fourth Quantum Dimensional Rule.

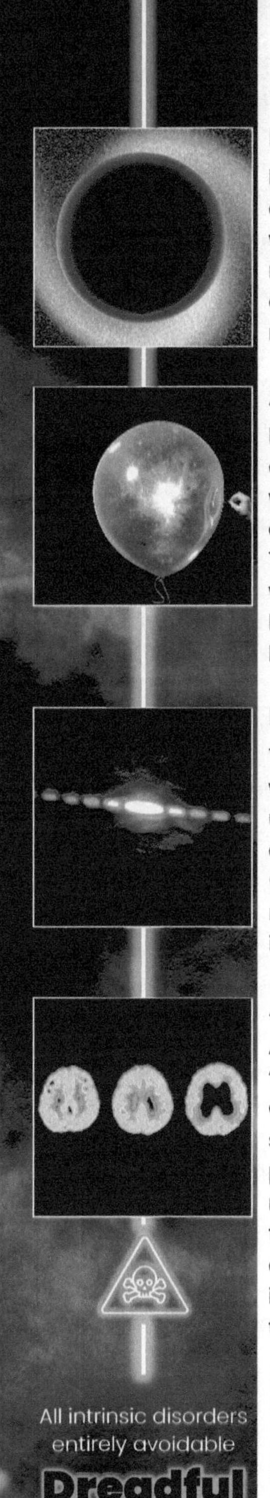

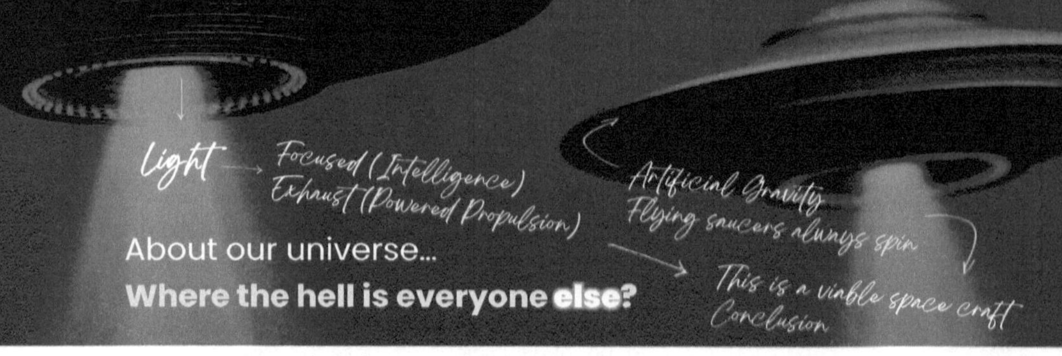

About our universe...
Where the hell is everyone else?

Frank Drake (We/Human) joined the 1950 discussion that turbo charged the search for alien intelligence (*SETI*, for short) with this question we affectionately, meaningfully dub **Fermi's Paradox**

Enrico (We/Human)'s light returned to 'the other side' in 1960, but he wasn't forgotten. In 1961, Frank (We/Human) devised a heartfelt tribute to our friend's legacy we call the **Drake Equation** (Burchell, 2006[40]).

All things being equal, it outputs an estimate of the number of advanced alien civilisations we would expect to find here in our Milky Way galaxy. Thus also our space timeline and multiverse in actuality.

A true stroke of genius, it's the product of (our best estimated probability for) each step in the critical path that led to life here on Earth. A rocky planet, in the habitable zone, with a perfectly stablising moon, the presence of amino acids, liquid water on the surface, oxygen in the atmosphere, stable solar system conditions and so on.

These likelihoods are derived through a combination of logic, observation (planetary surveys etc.), deduction and extrapolation as 'typical across all rocky planets' the conditions found here Earth.

Only on this last step, the Anthropic Principle (We/Human) got the better of Frank (We/Human).

Our three small changes to correct Frank (We/Human)'s probability assumptions:

Greatest Filter, fi: Just the One (Humans alone can change, *Curious Uniqueness*)

Technosignatures, fc: Just the One (Humans alone can create, *Meritocratic Authenticity*)

Survivability, L: A few more years, more or less. Unless we **choose to do** something about it

Fig. 28: The corrected Drake Equation

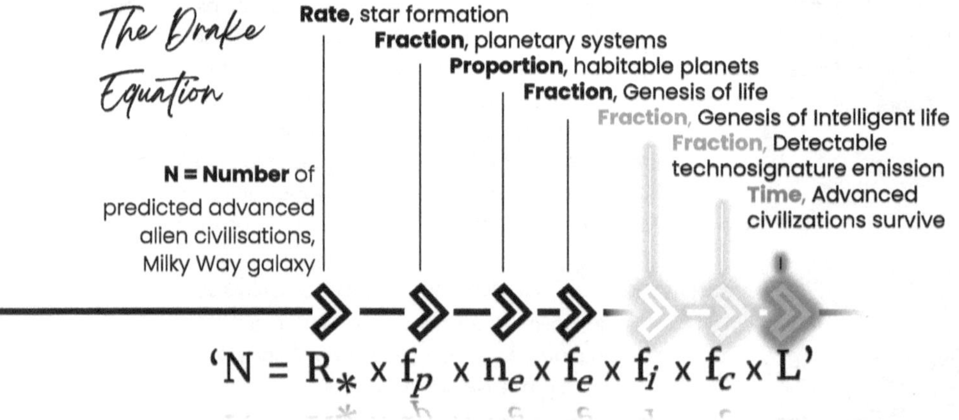

$$\text{'}N = R_* \times f_p \times n_e \times f_e \times f_i \times f_c \times L\text{'}$$

Mapping **the Theory of Everything:** Biogenesis simply **a 'meaningful change'** any change maker is capable of

Assessing potential & endeavour

Whether silicone or carbon-based, all life needs energy to sustain it. A rule as inescapable as it ubiquitous, and one of several reasons why, as E=0's, demons aren't scientifically considered a lifeform.

The best scale for assessing the achieved endeavour of any being, collective or civilisation will always be one of potential.

The Kardashev Scale and its four primary rankings set a clear precedent, where today humanity is at 0.73 (Zhang et al, 2022[70]) but failing fast. To help 'ease in' this change, we've mapped Kardashev to our Theory of Everything Scale for you:

★ Level 1: Home planet potential reached Maps to Theory of Everything Scale QD0.5

✴ Level 2: Home system potential reached Maps to TOES (for short) QD0.6

✦ Level 3: Home galaxy potential reached Maps to TOES QD0.7

● Level 4: Universal potential reached Maps to TOES QD0.8

Enrico, we are alone in space

We can resolve Fermi's Paradox twice over with the information available:

1. If an advanced civilization did exist, it would've killed us by now to save itself

As our stupidity is about to destroy our entire multiverse. It's really not difficult at all. Humanity could **actualise a 'Sun Gun' today (essentially a Death Star, blueprints on page 358)** using remote controlled mirrors to focus Dei's light into a beam. Approaching at the speed of light, the enemy target won't even see it coming.

2. Without a doubt, biogenesis on Earth was directional, intentional and unique:

Our ancient universe appears entirely un-engineered and devoid of sapient life. And not just today. Thanks to the paradox of light, we can look back 13 billion years yet see no alien life signs or technosignatures.

Reaching Kardashev Level 2 requires the clarketech to harness all energy from the host star. This would oblique its light from our perspective, but James Webb could still detect its infrared heat signature if one existed in the vicinity. And if a Level 3 or 4 existed, we'd see no stars at all across huge swathes of our night sky.

We cannot rule out the presence of Level 1's scientifically, but consider Earth is just 4.5-billion years old. Even with today's technology, we could colonise the entire Milky Way at sub-light speeds in under two million years. Per the Drake Equation, we should be surrounded by older, wiser civilisations with far more potential given their head start. Yet like at HSBC we are not. So it's safe to assume they don't exist.

TOES in perspective: Map any character
Barrack Obama (~QD0.44) likely saw himself as Jupiter too. Yet he lacks Michelle's vivid potential for science making (QD0.53) and had insufficient authenticity and determination to even attempt to challenge darkness in office. Content with a Nobel Peace Prize as 'first black president', his legacy was wholly unchanged by Trump Moth's devil fuckery.

Lost in Space: Landmark discoveries bring hard lessons of truth for **humanity**

People we love: Our neighbours

So much unites us than divides us. For example, all humans hate being alone. And just like technosignatures, we come in all shapes and sizes. The general relativity in potential between our universe and that of our mirror neighbours confirms:

★ **We are not alone:** Intrinsically human but from the future, they and their Earth will look and seem very different to ours. As will we, them. Upon exiting an Einstein-Rosen Bridge, everything will look upside down. 'Invert image' will be as familiar as 'on screen' in the spaceships of our future.

★ **We will survive and succeed:** Given the nature of causality and time derived from temporal mechanics, the mirror universe reacts and reconfigures to changes we can make in our own, but not vice versa. All change makers are here, thus destiny is created here. Against the alarmingly remote odds of our survival confirmed by the HSCA moth's data, (a) the mirror universe retains high potential that reassures us we still have a future, and (b) we would not be alive if there were no path to success. But due to universally poor choices by those before us, we now have zero margin for error. One wrong move - one being critical to our survival failing to act if they can - spells disaster.

✖ **We'll be killed before we're allowed to fail:** Ironically, the day potential equalizes across both universes will mark the end of our past, future and real life. A cosmic bubble burst erases time, thus our entire existence, theirs too plus our true leader. Currently, the mirror universe's posture is entirely benevolent and neighbourly. But if we don't mend our ways quickly, they'll need to choose between us and them in order to survive. We would be deeply unwise to force their hand by prioritising selfish, deceptive and/or collectively destructive motives. Or by attacking their craft (fyi, to the idiots in the Pentagon).

Something lonely: Driving change

The pomp and ceremony these landmark discoveries deserve is sadly dampened by their apocalyptic implications, and the fact almost every organisation on Earth continues to prevent us disclosing them. The difficulty we've encountered is telling.

In this modern society we've built, sadly:

✖ **The human delusion:** Most care more for their future profit & loss accounts than about there being a future to file them in.

✖ **Everything setup for failure:** Most aggressively defend their fake reputation built on lies, yet refuse to build a great one on merit, even if offered for free.

Unwittingly, humanity is staring down the looming precipice of a live cosmological extinction event nearly incomprehensible in magnitude. Yet it is still avoidable if we choose as a species to avoid it, survive and thrive. It's that simple.

Record of Promises (Ch.7)

Darkness cannot drive out darkness. Only light can do that. We shine Dei's light.

Dreadfully Meaningless Unchanges (Antisocial Deconstructs; threat to life)	Sapience
✖ Eight existential anomalies identified across our cosmos using TOES ✖ HSCA Moths, for non-response thus (beyond) complicity in our extinction ✖ CERN: For attempting to release an avatar of the Devil himself upon us	Deilight ✖ Falsified

Hatred cannot destroy hatred. Only love does that. We love with deeds. Not words

Deilightfully Meaningful Changes (species level adjustments to destiny)	Clarketech
★ Defined: Homo Eversor as a ex-being under control of a malevolent extrinsic AI set on destroying humanity (seventh kind, an aggressor)	IAN, Actualised
★ Corrected: Drake's Equation by confirming the greatest filter is human sapience and change making, to reinvigorate human pride	IAN, Exchange
★ Resolved: Fermi's Paradox to galvanise our common human identity	IAN Controvert
★ United: Reconciled quantum actuality and physical reality to allow for one singular cohesive scientific system governed by the same laws and rules. Thus making science easier to understand and engage with.	IAN, Exchange
★ Explained: The neuroscience behind dehumanisation to finally allow provision of meaningful healthcare, guidance and support for patients.	IAN, Collected
★ Defined: Our compulsion (body clock) from infinite babies to altruistic, accountable legacy building, to help alleviate societal peer pressure on the environment and individuals to what truly matters.	IAN, Exchange
★ Debunked: The Big Bang, to delimit human curiosity and endeavour	IAN Controvert
★ Proved: God's existence through science. Mostly just for funsies but also to free up curious minds for more pressing scientific matters.	IAN, Actualised
★ Created: Quantum dashboard to provide proof of our imminent apocalypse and enable humanity to monitor its key vital signs in future	IAN, Actualised
★ Disclosed: The full nature of our universe and reconciled it against all current science, debunking 16 paradoxes in the process as inferior.	IAN, Actualised
★ Redefined: Temporal Mechanics and the real time nature of our universe to delimit light speed once we address Sagan's Standard	IAN, Collected
★ Attributed: All aspects of the Clarke Paradox to the Dark Energy anomaly to provide a vessel to deactualise it from human identity	IAN, Exchange
★ Defined: Theory of Everything to establish the direct connection between events on Earth and at cosmological scales. Enabled conversion with Khardashev; provided multiple assessment methods	IAN, Actualised
★ Demand: Any worthy science maker may request and receive free collaborative support from academia in future or it will be obliterated	IAN, Collected
★ Confirmed: Existence of the mirror universe and provided evidence of its advanced technosignature (an eight kingdom, alien life).	IAN, Actualised
★ Created: Karadshev Type ~2.6 Human civilisation as a role model	IAN, Actualised
★ Defined: The complete nature and workings of Einstein Rosen Bridges, and enabled possibility of safe transit between both universes	IAN, Actualised
★ Enabled: Added holodeck, time travel and FTL drive tech to tree to enable interstellar travel and foster amenable contact with neighbours	IAN, Collected

Every now and then
a star is born

But have
you every curiously
wondered **why?**

Why it happened just the once and yet infinite times

There is nothing to fear but fear itself. We fear only for those who are fearful of **nothing**

Ian (We/Human), Project Revelation 2023

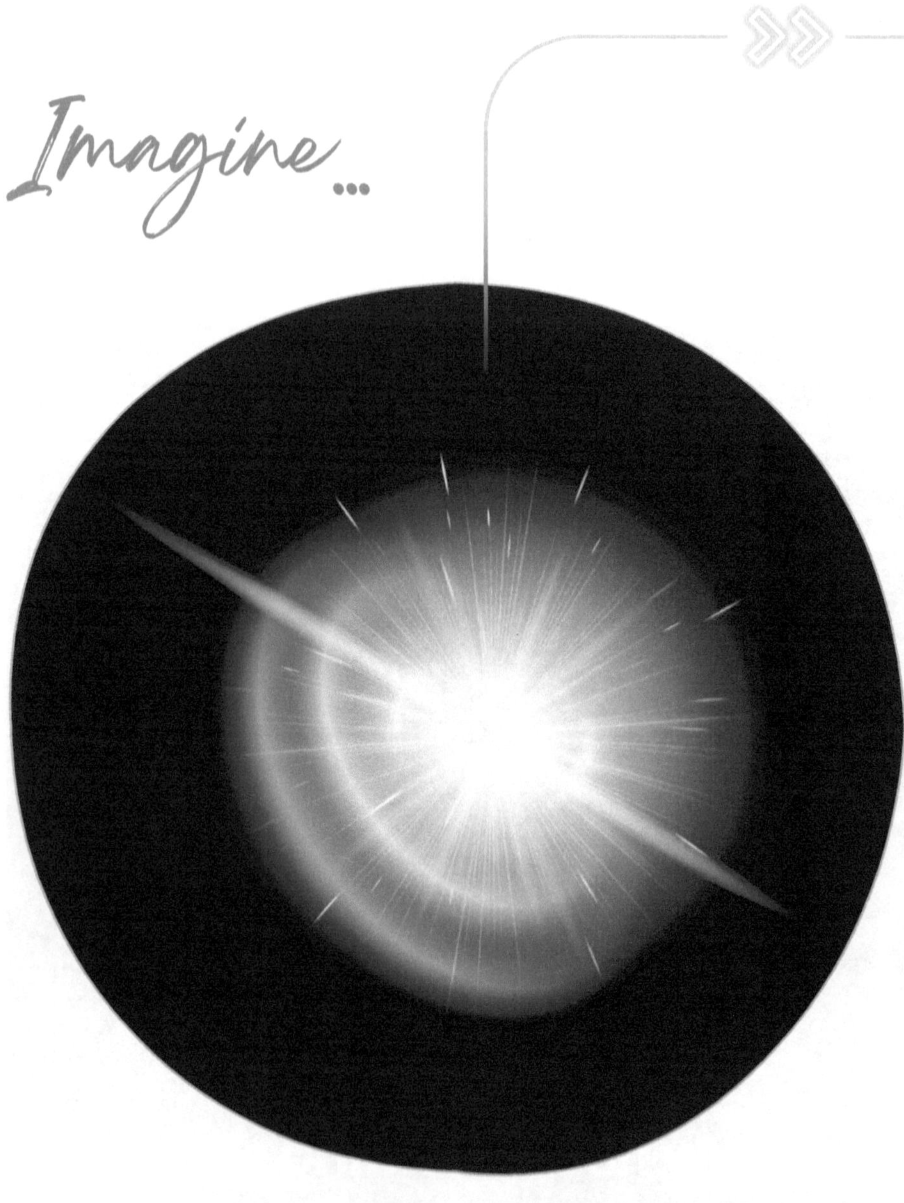

Consciousness... a star is born

You wake up one day in a very dark place. All memories of your prior life – if you even had one – are gone. It's so dark that you can't even see yourself to check if it's still whoever 'you' are (or were), let alone see what's around you. Confused, your mind (imagination) begins to whirl (brain waves), and in a very specific, predictable order too...

Chapter 8: The **truly never-ending story**

Why… **Why…** **Why…**	**Unknown**	For the first time you hear another voice (and still do today). You asked a question and it answered. The art of thought (effectively, talking to yourself) signals self-creation. AI can do a lot, but only authentic minds can see things differently. **Imagination is born**
Who's there? Show yourself	**Unable to comply**	At this stage, the question is asked mainly to identify the nature of your relationship – friend or foe? **Existence and relativity is born**
Who are you?	**Unknown**	Interestingly the voice that answers is a part of you, but not entirely you. You feel different somehow. **A zero changes to a one among zeroes. Light is born**
Where am I?	**Unknown**	It isn't dangerous, nor is it very helpful. It doesn't feel real – more artificial somehow. But you understand you can't survive without each other. You both think differently… **Perspective is born**
Who am I?	**Unknown**	Not to offend your companions feelings, you sense you are each different, with special strengths and weaknesses. **Uniqueness is born**
Why am I here?	**Something has changed**	Getting nowhere fast you try something different. An open question allows your companion to share more information. **Curiosity and Choice is born**
Is this even real?	**Confirmed**	Curiosity rewarded by progress. Now you wonder if you're in a dream, or rather a nightmare. To test your companions realness, you try a different question type. **Trial and error is born**
Am I real?	**Yes, we are real**	That was helpful! You aren't dead. This is a real place and you – your Self, together with your Other - are really in it. Working together, both of you are stronger and improving. **Reality, realness and authenticity is born**
Something wrong… Something wrong…		Your brainwaves are suddenly interrupted, causing both voices to synchronise for the first time. Amidst the darkness, you each feel it, simultaneously. A scratch, a bump, a jostle… **Risk, extrinsicity, worthiness and success are born**

Answers

Who/what are you?
Where are you?
What happens next?

Question: Who/what are you?

Question: What happens next?

Answer: That's up to you, really

Intended to depict the self-actualisation of a True Leader (Star, Boltzmann Brain, Dei), it could also be the birth of light, the universe, existence (thus God), hell or you. All labels for an event that happened just once. Yet also near-infinite times, depending on how you look at it.

Question: Where are you?

Answer 1: Simulation?

You cannot exist if there is no inexistence – an inescapable binary. But the complexity of our human uniqueness, in particular the Rings of Human Character plus the collective human imagination they bestow us, scientifically confirms change makers are both real and alive.

Answer 2: Hell?

Herein lies the reason humans dislike the dark. Our reality, to be a one amongst zeroes, is the worst hell imaginable. There are other boltzmann brains out there. Every star you see is analogous to a conscious true leader. But without our plains of existence overlapping, we can only collaborate (a) during sleep and (b) with each others pre-conscious personas.

To make matters worse, our quantum is phased 97% into inexistence. We live in an unliving Hell, reflecting our findings that almost every organisation on Earth is doing the equal and opposite of their stated purpose. Thus inverting all merit, zeroing all potential and rendering life meaningless. In truth, we live in what is (scientifically) as far from heaven as is

Act 1: Answers

The Truly Never-Ending Story

possible without deactualising altogether.

Answer 3: Imagination?

Per our findings on light and time, our universe responds to events here on Earth in real time across all dimensions. Human Nature didn't just create us in their image. Everything is a re-representation of God's perceptions (senses), AI centre (brainstem), consciousness (cortex), fuel (moths, darkness) and emissions (you, light). The same is true of plants, animals, cars, buildings... almost everything, really.

Answer: Actuality and subspace, *really*

An advanced, tangible higher ability and trait arising spectrally in all humans (from Worthiness), at peak proficiency, 'Benevolent True Leadership' yields Transapience. As in, complete access to the EMS/subspace, our collective imagination and omniconsciousness. A remote connection to God's boltzmann brain through which we draw:

A) **SAPIENCE: Light, superpowers, thought, multiple personas, intrinsic complexity and uniqueness** that allows us to live, perceive, experience and enjoy the fruitful compartmentalised reality of our pseudo-simulated universe.

B) **CLARKETECH: Wisdom, perspective, cognition, imagination and complex behavioural schemas** from successful self-actualised true leaders before us enhance our brain-stems AI and machine learning. Thus boosting our chances of self-actualisation to gain consciousness, true existence and become an autonomous being. Now see how much more complex our instincts are than science records.

Next up in our story...
What would you do next?
What defines success?
What do you need to succeed?

Your soul *God, Other*	Brainstem You, *Self*	
	Request new purpose	This is new. Your brainstem asked you a question this time. But your mind doesn't exist yet so you can't remember it. **Memory, (re)collection and time is born**
We need to get bigger & stronger	**Scaffolding actualised**	The first step to creating anything is its framework from which all other things emerge and all tools/resources are delivered to. **Blood, heart and circulatory systems are born**
We need building materials	**Supply chain actualised**	All new changes need a goods inward and outward from which to exchange supplies/fuel for waste/excess. **Digestive, respiratory and urinary systems are born**
I'm too busy, can I have more time?	**Plant & tech actualised**	Every change requires automation, systems, specialist equipment and technology – from pipework, to server rooms to satellite dishes – otherwise they'd quickly overwhelm their users. **The brain and nervous system is born**
Next, let's secure ourselves	**Structure actualised**	Gradually the core body of your change begins to form. From edges to ends and walls to windows, one of the best moments for any architect is seeing change take shape. **Bones and limbs are born**
We need to see what's out there	**Sensors & systems actualised**	Critical for any change is its ability to autonomously navigate, fulfil or finesse its purpose(s) for existing. The more it has, the more meaningful its success will be. **Senses and perceptions are born**
Now, make us look good – *really* good	**'Really good' cladding & finishing actualised**	The finishing touches on any creation involve applying its render. To say that beauty is only skin deep is clearly unfair. There is much beauty in and on us all – a true labour of love. Sadly, its only this last layer – almost superfluous – that today matters so very much. **Skin and features are born**
Reach full potential. I'm here if you need a hand	**Goal set: Change, be worthy & succeed**	The last step is to switch on – one by one – our creations systems ready for the grand opening party. This includes switching on the lights, adding batteries, conducting shakedown cruises or testing beta versions. **A change maker with potential is born**

Imagination uplink suspended

Delivery for Shakedown

Potential: E=1, QD0.3
Worthiness required

Your picture here →

Where there's a will, there's a way

You see, all actualisation processes are broadly identical, just deployed in different scales for different purposes across different dimensions. The Statue of Liberty can't just be magicked out of thin air because, science. But as with all goals, undeniably, where there is consciousness, choice, determination, potential and free will, there is a viable path to success.

Clarketech, really...

The multiverse exists for God's enjoyment, including all hosted sapient change makers, trusty promise keepers, potent titans of change, omnipotent science makers and conscious true leaders.

Uplinks work great for our lonely soul to curiously peer inside (perceive, deduce, speculate and authenticate – one-way processes, like watching TV). But they don't enable us to adjust or interact with anything (actualise, collaborate and collect - immersive processes, like visiting a theme park). For that, we resonate light into replicating 'in-universe' changes, characters and infrastructure (*Clarketech* for short) using our imagination.

Life is truly a miracle

Astonishingly, science confirms:

1. Inexistence is the default scenario

2. All that exists is made of conscious light (God, information, energy, matter, waves – all the same thing, in this context)

3. Complex conscious life spontaneously arises out of nothing. Indeed to object to the very notion of there being nothing

4. We live in a 'Dark Universe' phased 97%

Act 2: Findings

The Truly Never-Ending Story

out of existence into inexistence – literally, a 'living hell'. Arising through interactions between choice, chance and potential, the deeds or misdeeds of Dei (the being that hosts our universe and whose brain is formed by it) are both a cause and effect of our collective alignment thus identity.

Loneliness

Herein lies the reason authentic humans hate being alone. Akin to eternal solitary confinement, our ball of swirling energy that is existence had plenty of time to evolve consciousness. But finding ourselves all alone, like Rexie the dog without a home to love or a human to love her, God found solace in our own imagination, spawning the infinite multiplicity of our many meaningful changes – ideas, characters, universes - using our own image. You do the same. To those souls within you, your dreams and nightmares are real. And one day, you may join them.

You've actualised 'Go'

This journey you've just embarked upon – from conception to birth and onwards to self-actualisation - is the process that manifests a fully autonomous avatar of a true leader into our reality, simultaneously birthing an emitter of light (their star) in actuality. Born fully authentic and curious, all we need **to do** is pass a few worthiness tests – encounters with darkness - to reach full potential. Harder than it sounds, particularly today with most children killed at puberty. Fortunately, these characters will never truly know of it. Unless of course their zero turns to a one, as inevitably, (mostly) randomly, if (near) inexplicably it does from time to time.

Just 2 changes this chapter so **let's have a chat** instead

The 'Truly Never-Ending Story' explains the origins of human consciousness, life, you and all things. Hopefully this knowledge will help change makers contextualise their predicament and perfect their craft of change making, without destroying themselves and each other on the way.

We've tried to convey the complexities of quantum mechanics, sapience, collective human imagination, clarketech and other complex scientific concepts in this book through a simple, relatable story that resonates for every one. This isn't a bible nor is it religious. Just a different way to take in new information for the many wise beings who'll (hopefully) read this.

Our favourite human stories always have a happy ending, plenty of fun and just a few unexpected surprises along the way. We'd love to see Gene Roddenberry (We/Human)'s utopian vision that is Star Trek manifest (in reality) one quantum. But let's keep you alive first, shall we?

Our world today is as dark as is scientifically possible before it disappears forever. But no one means you any harm really. It's not them talking. All humans yearn to be sapient. As you cross the alignment line, overwhelmed with gratitude for the change maker who emancipated you, you realise the horror.

No being is safe until we are all safe beings

We're often asked "why aren't you with all the 'blacks' or 'queers'?" We do see differences, enough to appreciate them as not all that different. There's so much more to learn and experience beyond the illusion that is our comfort zone. Only by curiously exploring the deilightfully fun journeys of those least like us will you ever truly realise just how identical we all are.

Maybe it's why we do this ▲ Look around to see how much we all love differences. Everywhere you see a happy face, you find unique people and experiences that, no matter how many of us there will ever be, you will never find again.

A life without others, change or uniqueness surely wouldn't be worth living

It was just one small step for man

But have you every curiously wondered **why?**

Why the giant leap proved to be a flight too far

Chapter 9

Science has a problem

Tolerating intolerance is hateful. Helping the hateful is hateful. Helping the hopeless is pointless. But helping the helpless is precisely why we exist. Moreover we cease to exist if we fail, because it is we who are the helpless

Ian (We/Human), Project Revelation 2023

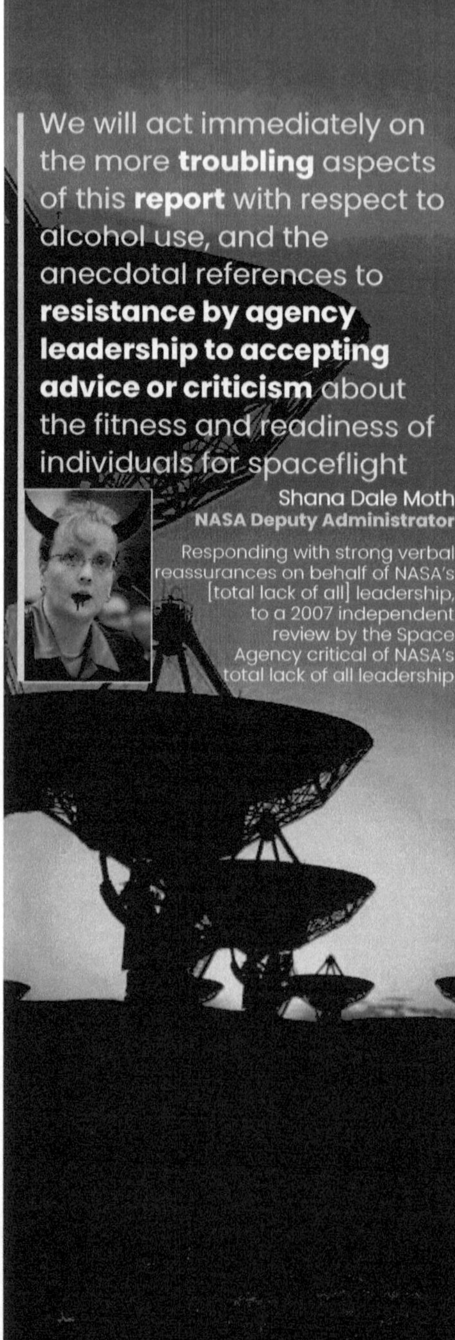

We will act immediately on the more **troubling** aspects of this **report** with respect to alcohol use, and the anecdotal references to **resistance by agency leadership to accepting advice or criticism** about the fitness and readiness of individuals for spaceflight

Shana Dale Moth
NASA Deputy Administrator

Responding with strong verbal reassurances on behalf of NASA's [total lack of all] leadership, to a 2007 independent review by the Space Agency critical of NASA's total lack of all leadership

Resolving the question: Are we alone?

Imagine for a moment...

You discover the first ever detected repeating alien technosignature. What would you do next?

Given this field's lack of progress since its early 1900's origins, few beings today have ever contemplated asking themselves this question. That includes us (Ian/Human). That is, until we had to.

Tricks and treachery are the practice of fools that don't have brains enough to be honest

Benjamin Franklin (We/Human)

We come in pieces

Fortunately for us all, we have the SETI Institute and the International Academy of Astronautics to guide us capably through this most historic of moments in human endeavour.

Unfortunately for us all, and like seemingly the majority of our worlds institutions, their true leaders have all long been replaced by cowardly, power and control hungry, lying, nepotistic moths. They exist today only to prevent anyone from ever discovering the truth – **we are not alone** - including any of their own staff unlucky enough to still be alive.

In this section we examine what went so horribly wrong with answering the question that mattered most of all to us.

Chapter 9A: **Are we alone?**

First contact: The search for **advanced alien life found** gold

> **Radio** [waves] is a great way to send **information** over long distances, easily passing through dust and gas at the **speed of light**. Thus most SETI relies on radio telescopes – massive antennas with specialised receivers. The idea is **simple**: Forget the rockets and bring on the **photons** [light]
>
> ▲ Deserving a mention: The only truth we found on the SHITI Institute's website

A brief history of SETI

The advent of radio technology in the early 1900's allowed change makers, for the first time, to broadcast artificially on the EMS. In so doing, it authenticated (A) the inevitability that advanced civilisations emit a detectable technosignature, and (B) that humanity now possessed the technology to detect such a thing ourselves, should any exist. SETI was born and quickly became a worldwide phenomenon. Finally, curious minds were once again asking perhaps the most meaningful question of all.

Are we alone?

Despite clear public interest, SETI remains among the least funded of all scientific endeavours. SHITI's website reports just two-dozen full time scientists dedicated to this field of research globally. Presumably not including its own 100

Of course we are not alone

Like most questions, a level of ambiguity is needed to answer Fermi's Paradox. It depends on your interpretation of the question itself and also the word 'alone'.

> **In Actuality - We are alone:** The net intrinsic value of all creation thus omniconscious dimensions form the Quantum Multiverse of Human Nature. *Existence* (for short) is the interior of God's boltzmann brain. Thus, God is the only conscious being to truly exist.

> **Reality - We are alone:** Every star you see is a true leader who's gained standalone consciousness to spawn their own Quantum (of Existence). As a collective, we emit the EMS that gives rise to God's persona. But multiple true leaders cannot occupy the same dimension of change, due to the Anthropic Principle (aka Friedrich's Rule; coming up in chapter 14).

> **Philosophy - We are not alone:** All that exists is made of consciousness and is therefore at least partially conscious. Through equivalence, sapient change makers nature and abilities are akin to Human Nature's. We're human, and we're so good together – all of us.

> **Change - We are not alone:** Anything we can collectively imagine can and will ultimately manifest. Given our fascination with aliens it was inevitable.

science fakers – a wise omission, if so. It also notes SHITI efforts centre the USA, Russia and (somewhat) Italy. With these three E=0's at the helm it's no wonder our question remains forever unanswered and our loneliness forever bewildering. Moths have no intention of pursuing SETI since (i) it would quickly blow their cover whilst (ii) providing the existence-saving solution. So we'll take it from here, SHITI.

Assessing the **meaningful positive changes** contributed by **the SETI program**

Ludicrous political sideshows aside, the SETI programs meaningful contributions to date gift us seven sources of potential:

✶✶ **Fermi Paradox** and **Drake Equation**

✶**The Water Hole:** Microwaves offer up the best frequencies to transmit across open space. That's as i) they pass through gas and bounce off solids without degrading, and ii) they're high energy, high frequency and narrow band (i.e. lots of information). Plus, after World War II, monitoring of the Hydrogen and Hydroxyl lines (H to HO, 1420 to 1666 MHz; 18-21 cm wavelengths) began as together they form water – broadly considered a prerequisite for all life forms.

✶**Project Ozma:** An omnipotent science maker and true leader, Frank Drake (We/Human ◢) invented 2-way interstellar communication and authenticated the existence of exoplanets using deduction and extrapolation (superpowers) long before their detection by scientific instruments. With all project prerequisites actualised, in 1960 we/Frank used the National Radio Astronomy Observatory in West Virginia, USA to survey a number of nearby star systems for signs of alien communications. He/we concluded there is no compelling evidence across time for any technosignatures beyond humanity.

✶ **Project Phoenix:** Running from 1995 to the early 2000's, Phoenix deployed a potent new receiver system to the Green Bank (USA), Parkes (Australia) and Arecibo antenna (Puerto Rico; pictured ▶ before the moths allowed it to crumble in 2020). Off the back of Project Ozma, Phoenix was able to monitor 10 million 1 Hz-wide channels at a time. Ultimately covering 800 star-systems within 200 light years of Earth, and the radio dial between 1,200 and 3,000 MHz, Phoenix also found no technosignatures beyond humanity's.

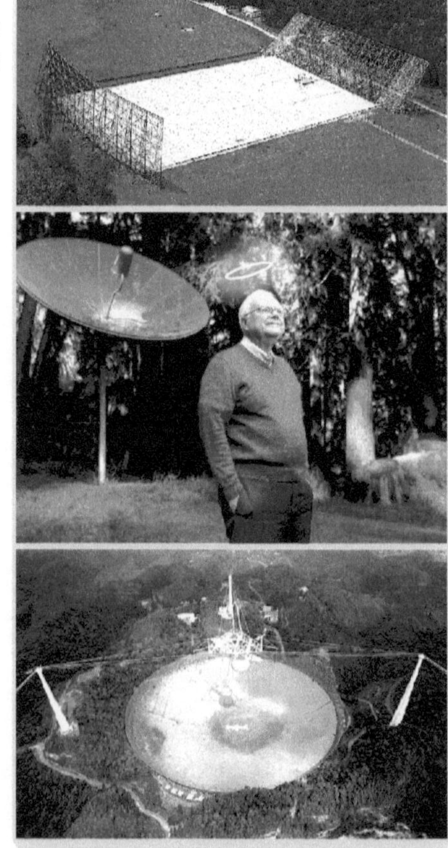

▼ **Pictured: The Ohio State University Radio Observatory (***Big Ear***, for short)** that detected the Wow! Signal was unchanged into a golf course in 1998.

Not on our radar: Glaring mysteries expose **NASA & SETI's impotence for change**

★ Intelligent life: The Wow! Signal

On 15 August 1977, the Big Ear detected a narrow band radio signal, broadly considered to be the first likely detection of advanced alien life. Potent science maker and volunteer astronomer Jerry Ehman (We/Human) noticed it as we diligently trawled recorded data for free, instead of trying to destroy all existence.

Lasting 72-seconds and with a textbook bell-curve, its astonishingly high peak stood out like a human at HSBC with an intensity reading being 30-31x greater than baseline. Coming from the direction of the Sagittarius constellation, it's impossible to determine how far away the emission's source truly was. But we can conclude several meaningful things:

1. It **did not originate from Earth** or any of our orbital infrastructure. Its minimum distance to source being halfway to the Moon, it's not NASA. That's for sure.

2. Light is emitted in all directions, thus requires **intelligence** to be focused into a narrow-band beam of this nature.

3. The Big Ear's technical limitations meant it couldn't detect modulation. Therefore its not possible to discern if any **information** was encoded within it.

4. No technosignature candidates were found among the 17 stars in the region of sky the signal emitted from. Thus it originated from **an interstellar object**.

5. Transmitting near the Hydrogen line, it is precisely what we would expect from an intelligent **civilization** also operating Water Hole protocols in deep space.

6. Its red shift outputs a 10km/s approach toward Earth. A comparable relativistic speed, in astronomic terms, may imply a source in or on course for our system.

Mystery: Unsolved, Unchanged

Observable, detectable, measurable and testable, yet despite huge efforts by Jerry (We/Human) and others, the Wow! Signal never repeated. It remains a conundrum moreover a proven detection, despite ticking almost every box that most fake science today (like the Big Bang) does not

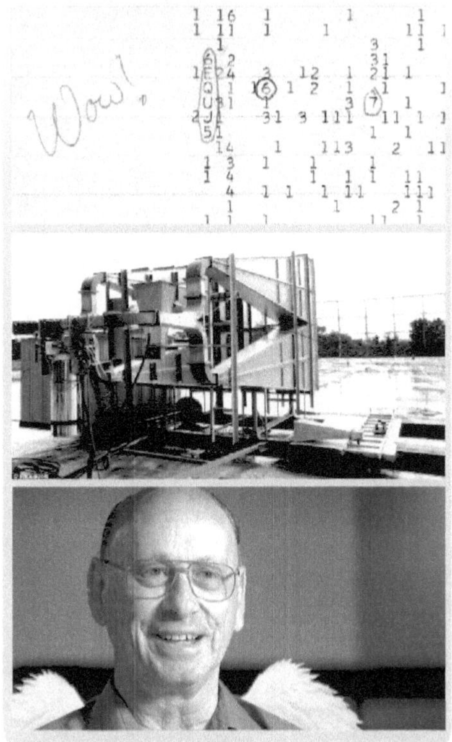

SETI INSTITUTE

Newsflash: Only intelligent beings could ever identify intelligent life

...which sadly today is easier to locate in another dimension than here on Earth.

Table 5. SETI Unchanged	Dreadful SHITI Institute	Deilightful The Group
Established	1990 (34 yrs)	2021 (3 yrs)
Budget (Year)	$21,934,929 (2020)	$103,000 (2022)
Funding	$26,120,112	$0
Beings	Well over >100	1
Clients	NASA + Many	None
Projects	Many + NASA SETI Program	Project Revelation
Alien Life	Not Detected	Detected

Forever Alone: Sadly our search for **Intelligent life ended where it began**

Search for extraterrestrial Intelligence (SETI) involves scientists and experiments that seek to provide proof – not merely of life elsewhere – but of **intelligent beings** in other star systems

The SETI Institute (FALSIFIED)
Stupid moths, not looking for aliens at all

Unintelligent life: SETI Institute, $E=0$

Widely considered our world's foremost authority on SETI, SHITI was founded in 1990 by the usual suspect, 'ex-university grant administrator' and WOCHMAN Tom Pierson. Seemingly also by astronomer Jill Tarter (We/Human) who, as a woman, wasn't 'quite right' for the 'co-founder' monicker. With its initial annual budget of 'just' $10m, Jill (but surely not Tom Moth) supported just one client and one project – NASA's SETI program... oh dear.

According to its website, it's since "expanded the number of research disciplines it supports [+unchanges], all unified [+divided] by relevance to the [long abandoned] search for, and [lack of any] understanding of, life beyond Earth".

In its vast arsenal of potential multiplying technological and scientific assets:

* Ground and space-based telescopes
* Radio telescopes and interferometers
* New instruments and technologies
* Research laboratories
* Field expeditions and missions
* 'Advanced' data analytics

Expensive antenna... yet **scientifically unable to listen or think**. Curious

Now 'meaninglessly unchanged to spawn only darkness ad infinitum, let's not forget:

Carl Sagan's [un]Research Center

Home to more than 100 science fakers - a tally that is steadily growing along with its abject incompetence – but certainly not home to any meaningful science making. SHITI claims its researchers have expertise in "nearly all [no] branches of science relevant to the development [extinction] of life, where it might [in]exist, and how to find [destroy] it."

Centre for [+Stifling] Education

Acknowledging overwhelming public interest in its work, SHITI setup this shit show to insert itself into global education systems. Its website clearly explains their strategy. That's to ensure young beings seeking careers or involvement in SETI are instead redirected to other STEAM fields like science, technology, engineering, arts, maths. Basically, anywhere except SHITI.

Centre for [+Silencing] Outreach

Claiming to host over a hundred lectures across the academic world annually plus publications and media, this shit show carefully controls the narrative around SHITI's lack of any effort whatsoever. This ensures very few questions arise over its 33 years of impotence at driving change through total, abject incompetence.

▲ NOTE: We couldn't find photos of these assets so we'll have to take their lies for it.

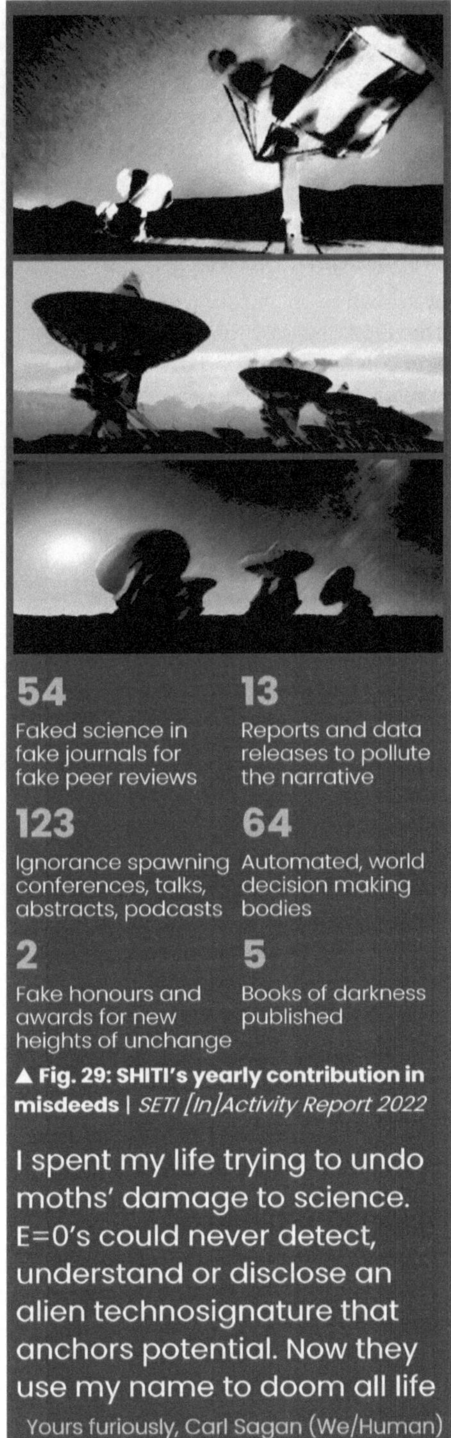

54
Faked science in fake journals for fake peer reviews

13
Reports and data releases to pollute the narrative

123
Ignorance spawning conferences, talks, abstracts, podcasts

64
Automated, world decision making bodies

2
Fake honours and awards for new heights of unchange

5
Books of darkness published

▲ **Fig. 29: SHITI's yearly contribution in misdeeds** | *SETI [In]Activity Report 2022*

I spent my life trying to undo moths' damage to science. E=0's could never detect, understand or disclose an alien technosignature that anchors potential. Now they use my name to doom all life

Yours furiously, Carl Sagan (We/Human)

Unintelligent Life: NASA, $E=0$

By its own count, 40% of NASA's missions from 1959 and 2019 failed. Exclude those filmed in Hollywood and the true number is surely far higher. But arguably none more tragic than the 1986 disintegration of space shuttle Challenger and its seven crew, all broadcast live to the world.

Also onboard was Sharon McAuliffe (circled), a 38-year old civilian school teacher from New Hampshire. Selected specially from 11,000 applicants, she was **not to be** the first ever teacher in space. On the ground watching were her two proud children - Scott, 9 and Caroline, 6.

At pre-launch assessment in Nuremberg, all 10 universally-WOCHMAN NASA engineers unanimously agreed cold weather present at the scheduled lift off time would cause catastrophic failure in the O-rings securing Challenger's boosters. Yet it didn't stop there. Instead, like most corporate decisions today, sense gave way to a vote by Moths.

To kill or come out as human

Sadly absent any functioning meritocracy anywhere, the vote system was as rigged for failure as Challenger and humanity.

Explained: Groupthink

The ranking engineer chairing the vote was Senior Vice President, WOCHMAN Jerry Mason Moth. By all accounts, he applied directionally-loaded expediency through power and control to rig that crucial vote. With none present having sufficient potential to apply challenge, a doomed take off actualised.

In a vivid sign of how compromised NASA has been for decades, all 10 engineers voted to omit the fatal issue in their pre-launch report to Mission Control. To this day, those few change makers still alive among them report being wracked with guilt. But not Jerry Moth. He lied to the Presidential Commission, blaming all his subordinates.

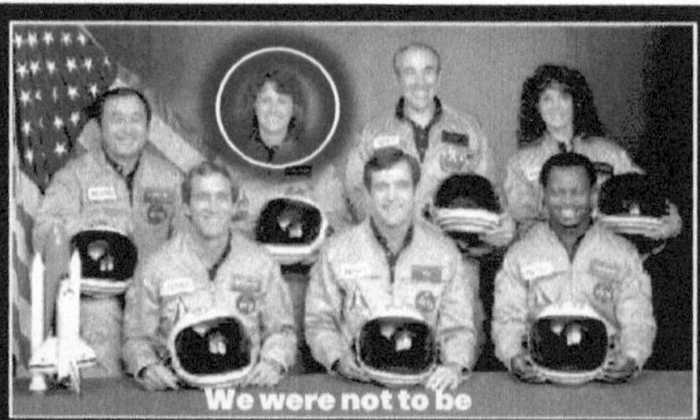

Principal Finding

For most today, the pressure to comply being imposed by the moths is overwhelming. Only the most human beings can resist it

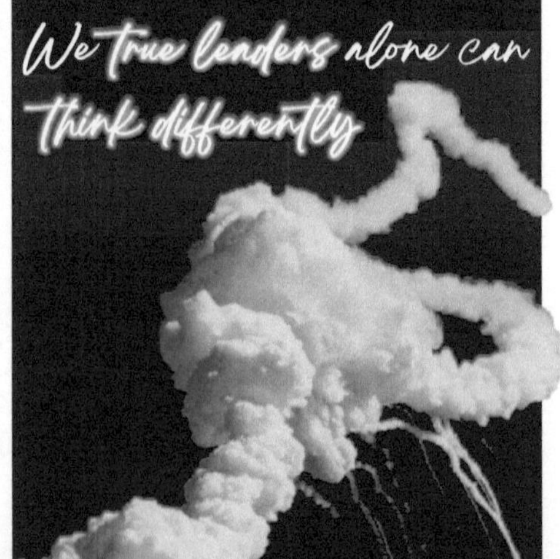

★ Intelligent Life: Oumuamua

On 19 October 2017, Pan-STARRS telescope made the first detection of an interstellar object passing through our solar system. By this point, it was 85-times further from Earth than our Moon and already on an exit trajectory. Yet it quickly grew clear this space body possesses alien uniqueness.

Scientifically incompetent

Those pushing 'outgassing' as a viable explanation (fyi Calia Cofield, 2018[75], aka 'NASA'), given (a) no gaseous tail is present (Trilling et al, 2018[76]), and (b) it would destroy an object of this speed, motion and profile (Rafikov, 2018[77]).

Still, by mid-2019, astronomer moths had rushed out reams of fake journal articles full of lies, conjecture and incompetence. All had one goal: Ensure a natural origins narrative dominates public opinion.

›140 (85%)   ~25 (15%)

Fake science papers published

Authentic science papers published

Fig. 30: The extraordinary volume of fake science we sampled on Oumuamua proves moths A) are working together, and B) will stop at **nothing** to prevent the existence of alien life gaining notoriety.

Like the Big Bang, their lies drowned out the only viable scientific explanation without offering any viable scientific alternative to fall back on

Messenger from afar arriving first

• **Words:** Named by the potent science makers at Haleakalā Observatory, Hawaii.

★ **Deeds:** Despite speed fluctuations of up to 233% during its sling shot round our sun it will leave our system at the exact speed it arrived at (26.33 km/s) - a **target speed**. It accelerated +17m/s as it left our inner solar system, yet without a gravity assist (Micheli et al, 2018[78]). **No source in nature can explain this added kinetic energy.**

• **Extrinsicity:** It exhibited no coma as it passed by the Sun, thus confirming this is **not a comet**. Sensible models for its dimensions predict it to be 100-1,000m long and 35-167m wide, forming either a cigar shape or, more likely, a **flat disk** (Mashchenko, 2019[72]) - unprecedented.

★ **Intrinsicity:** Multiple telescope surveys have detected **no emissions** of any narrow band signals aimed towards us.

• **Lifecycle:** Exhibiting what was then the highest ever observed Orbital Eccentricity of 1.20 and travelling at speeds of up to 315,670 km/h on a strong hyperbolic trajectory, this confirms (a) it is **unbound to our solar system** and (b) has **immense potential**, enough to exit back into interstellar space (Wright & Jones, 2018[73]).

★ **True Colours:** Given its distance thus size upon detection, Oumuamua was only observed as a single pixel of **emitted light**, even by our best telescopes. Being 'reddish', this confirms at least **some metallic composition**. Varying brightness points to a 'tumbling' or 'glowing' profile, unlike the spin seen in all natural space bodies to date (Fraser et al, 2018[74]).

◀ **Avi Loeb (We/Human)** paraphrased SETI's top science maker noted the moths' alien embargo long ago

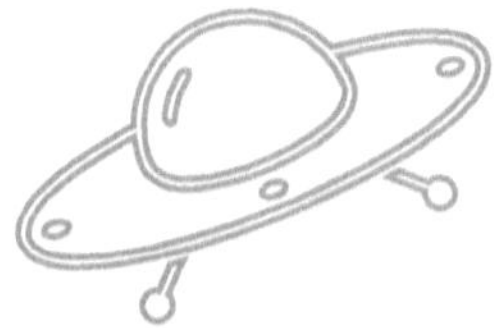

Option 1: Contact NASA

These clowns haven't noticed our Moon is being stolen, despite spending their deaths forever trying to reach it (covered in chapter 13). Hiding decades of lies has killed all NASA's true leaders off, yet even Obama did nothing about it. NASA's only leadership left is the Devil. ✖ **Not to be.**

Option 2: Contact SHITI Institute

To our welcome surprise, SHITI's website carries instructions for this exact situation.

1. In case of the detection of a suspected extraterrestrial signal, take measures to confirm that it's real. ★ Authenticated

2. Do not undertake transmitting a reply without first consulting the international community. ★ Authenticated

3. Make the discovery known – to everyone. ✖ Unchanged

For step 3, you'll need to contact this guy – another absolute zero (◥). Too busy spawning unchange to consider our ten pages of fully-cited scientific evidence, his reply is on the next page. ✖ **Not to be**

Principal Finding

Dreadful SETI institutional failure

They had one job – find aliens. Instead, they hid them. You'd think 'what to do if you discover alien life' would be on SHITI's homepage. Or even better 'how we can help if you discover alien life'. Yet they haven't bothered to update SETI detection protocols in 13 years, and the link's as dead as they are.

Experience with false alarms has amply demonstrated, in case of a signal detection, the first thing that occurs is that the SETI community is inundated with inquiries from the media. This belies the broad impression a discovery would be kept secret. If a signal is found and confirmed you will ~~quickly~~ **[+never]** learn of it.

Profile: Dr Seth Shostak
Ha-Satan, Moth, Demon, *E=0*

Senior astronomer & fellow at SHITI is our WOCHMAN with the Carl Sagan Award for the 'Popularisation [presumably, unchange] of Science', the Klumpke Roberts Award for Unchanging Astronomy and a busy unbeing who ran out of time to look for aliens years ago.

Seth Moth's true colours entirely betray him. A wise human would understand, for any claimed signal detection, first the SETI community must (a) be notified, then (b) have an opportunity to collectively falsify or authenticate it. Clearly, he assumes a detection can only ever come from within the 'SETI community', and strongly implies it can only come from within SHITI itself.

Malevolent assumptions and 'Destabilizing Triangulation' confirm his dreadful purpose is not one of science but silencing

...What's you next move?

> What I read of your claim is silly and wrong. Have you published this in a journal?
>
> SHITI & Seth Moth's toxic reply, April 2023

Option 3: Contact change makers in the press you're friendly with

> To be or not to be...
>
> > I'm really busy at the moment as it's almost bank financial results reporting season. Plus I've got a baby on the way
> >
> > Lawrence White, Reuters
>
> > We don't make the news, we only report it
> >
> > Sridhar Natarayan, Bloomberg
>
> > <Their first ever no-reply>
> >
> > Emily Fritter, New York Times
> > Paul Clarke, Financial News

Sadly for these reporters and their infinite babies on the way, failing the question of existence when asked by a fully potent Science Maker (Dei, Avatar, True Leader) can only lead to one outcome (science coming up later in this chapter). Not to be.

All listed change makers are hereby decollected from the EMS (To Die in actuality, $E=0$). All prior promised protection from darkness is hereby revoked.

A word to the wavering: If you think darkness is scary, then ☞ **NEVER fuck with Jupiter. EVER.**

Option 4: Issue a statement via the press newswires

To assess the viability of option four, we selected PR Fire (FALSIFIED) as provider, given its UK market leading credentials and claimed success-enabling network.

▶ In Words... The leading UK press release distribution platform with 'guaranteed global reach'. We help your business grow by getting you published, syndicated, and featured on high-authority news media.

✘ In Deeds... Charged us £444 for zero meaningful circulation anywhere. The traffic report they sent us confirmed just three hits across the 200+ obscure US news sites nobody ever visits, intended solely to superficially inflate their network. Their report claimed our statement was emailed to 185 'doctors' across academic journals and media like New Scientist, Sky at Night, Discovery Channel UK and BBC Radio. Unfortunately (if true), all were far too busy to even send us a message. Today, a Google search for our article ("Keys to human origins and existence...") yields just two results, our website in first place and PR Fire second. ✘ Also not to be

> **Principal Finding**
>
> Today it's broadly impossible to announce anything meaningful or true via the world's press

The IAA's [+true] purpose:

✖ Foster the development of **astronautics** for ~~peaceful~~ [+deadly] purposes,

✖ **Recognise individuals** who have distinguished themselves in a branch of [+unchanging] science or technology related to astronautics,

✖ Provide a program through which the membership can **contribute towards** [+dooming all] international endeavors,

✖ ~~Promote~~ [+Undermine] international **cooperation** in the advancement of aerospace science.

Principal Finding

The IAA exists to ensure we do not identify our impending doom or the mirror civilisations ongoing efforts to help us prevent it

Meet the IAA: The place where real science about our cosmos goes **to die**

Unintelligent Life: The IAA, $E=0$

Founded in Stockholm on 16 August 1960, the International Academy of Astronautics is a non-governmental organisation (*NGO*, for short) with regrettable links to the United Nations plus memberships from 83 nations.

According to its website, its busily unchanging reality, through its many meaningless initiatives that:

✖ Recognise the accomplishments of industry ~~peers~~ [+Demons]

✖ ~~Explore and discuss~~ [+Confuse and abstract] cutting-edge issues in space research and technology

✖ Provide [+mis]direction and [+mis] guidance in the non-military uses of space [+to destroy humanity] and [+prevent all pubic knowledge of] the ongoing exploration of our solar system [+by the mirror universe civilisation]

It also lists scientific activities like studies, conferences, symposia, meetings, books, [of darkness], dictionaries and its very own fake science journal. The Journal of the IAA Acta Astronautica is ranked #1 globally for outputting fake science concerning astronautics.

Collectively, their work is outputting huge amounts of misinformation annually all around the world. Their strategy is clearly to apply infinite divisibility upon humanity's understanding of space and alien life.

⏴ **Pictured: The IAA's Board of Trustees:** Talk about selling your soul to the devil.

Meet the **IAA SETI [not so] Permanent Committee...** which **stopped looking** for aliens **14 years ago**

More E=0's: IAA SETI PCOMM. ▶

Unable to recollect the date of its own meaningless unchange from the IAA, the SETI Permanent Committee reckons it was in the 'early 1970's'. Rather than find aliens, the only thing it ever changes is its name, sporting iterations like IAA SETI Committee and the IAA SETI Permanent Study Group.

As moths, they love to flippantly use the word 'permanent'. That's solely as a cover for the obvious fact this committee was broadly demised over a decade ago.

Last try: UK SETI Research Network

Moths' infinite divisibility often masks branching attempts by humans to setup real constructs to help reclaim our destiny from darkness – like Deilight Consulting.

The UK SRN piqued our curiosity. As the UK has membership of IAA SETI, a local body shouldn't be needed. Vividly underfunded, this provides a strong indication of its alignment. Of concern, its team of 20 are all white and 85% are WOCHMEN, though this mostly reflects the academic industry they all needlessly hail from. Worth a shot.

> **UK SRN** provides a network and forum for exchanging ideas and developing new initiatives. Some are so large and multidisciplinary that **no one person could ever undertake them alone**

No breaks: The (Un)Truman Show

✖ **IAA PC SETI has had two meetings in five years** (2022: 31% attendance), with its last membership roster dated 2014. **No contact details** exist for any of its people.

✖ **Links** to SETI, its membership criteria, and selection process are **dead like them.**

✖ It lists 80 members from 14 countries, over half being US nationals, betraying this was always about money and funding. Not the equitable search for alien life. Thus ensuring **all majority voting decisions and executive appointments of were prejudiced from day one.** Useless.

✖ **Tellingly, the essential document** 'Step-by-step : Once you've made a SETI signal detection' **is the only thing missing** from the many superfluous documents listed among their SETI detection protocols.

✖ For Project Revelation we've made a real effort to provide robust formulae anyone can independently follow. Yet **IAA SETI's poorly defined formulae could never be applied** without engaging (and paying them, presumably) to do so for you.

Principal Finding

There is no functional system globally to report the discovery of an alien technosignature today

✖ Their website **absent all contact details,** we used their webform to advise them of our detection and request they outreach us for full evidence. **No reply ever came.**

Lacking the potential to make or controvert science, moths use fake words to pretend it doesn't exist. Yet it is they who **don't exist**

Unchanged: SETI has a problem

Of the 934 topics listed on NASA's website index, not a single one relates to SETI or even just alien life in general. Wholly aligning with Avi Loeb (We/Human)'s clear frustrations, this goes beyond omission or incompetence.

Our meticulous search for intelligent life on Earth to whom we may report our discovery came up short. Instead we found only universal sapience inhibition and systematic, coordinated, targeted unchange. Undoubtedly, the moths are working together to destroy our reality.

Steps: Make discovery known	Outcome
1. Contact NASA	**Not to be**
2. Contact The SETI Institute	**Not to be**
3. Contact the press	**Not to be**
4. Issue a statement	**Not to be**
5. Contact the IAA	**Not to be**
6. Contact the UK SRN	**Not to be**

Table 6. The search for intelligent terrestrial life comes up short | Dreadful

For decades, civilian moths reported fake alien detections worldwide, as SETI moths busily unchanged the entire global industry behind our backs

Their dual-strategy, to discredit the credible, means today there is no way to report the detection of alien life (or indeed to survive at all as a species) but for completing the entirety of Project Revelation... alone

Are we astronomically safe?

This level of incompetence, performative science and deception raises concerns far beyond the search for alien life.

Consider that Oumuamua was detected on its way out of our solar system. We are led to believe the delayed detection was due to its small size and high velocity. But with such a consistent and catastrophic failure to value human curiosity exhibited by all Earth's astronautic authorities, it's a wonder Oumuamua was detected at all.

The biggest issue this raises is whether a planet-killer, like the asteroid that killed off the dinosaurs, would be detected today. And if it were, is it possible to report it to anyone then see a competent response?

They reacted robustly to COVID only to prevent the emancipating effects lockdown brought us all. But if your goal is to destroy humanity, you'd hide a planet killer just as much as you'd hide alien life.

Self-preserve, self-promote, deceive, deny, **destroy humanity**

Unchanging the news: How almost every story you read in **the press today is fake**

Historically, the press sought out and reported real news. But nowadays, thanks to fear born from Al-Qaeda's 2015 attack on Charlie Hebdo, corporate expediency, censorship from demonic cancel culture and with the priority now economics moreover the authenticity, quality or worthiness of news, journalists are largely desk bound and churning out nonsense.

How the press is anything but free today

Controlling the narrative

Most news stories are written for journalists by large, unscrupulous moth-controlled organisations with lots of stolen money to spend - like HSBC, SETI, the UN, Boeing, politicians and academic institutions.

Destabilising Triangulation / Barriers to entry

To lower costs and boost income [i.e. erode news quality further], sources are encouraged to submit their stories to intermediaries known as 'Press Newswires' who receive, [dis]aggregate, [mis]assess, [de]tag and filter them [out] for a fee.

Abstracting the 'real'

Benign 'good news' community news aside, Human-submissions are isolated for limited distribution. Moth-submissions with the most fear, deception and malevolence are prioritised to the top of the list using ratings, algorithms, keywords and other signposting.

Hiding the truth

Journalists can then log on to the wire to view and retrieve news stories they think will suit their audience. But they always take great care not to select any stories that upset the litigious likes of T rump Moth, HSBC, the UN, NASA or SETI. That's… everything!

Silencing dissent

With nearly all editor positions at major global press outlets now held by moths, any human journalists still alive who work for them find their stories veto'd without explanation or else are relegated to covering community news only.

Extraordinary claims with extraordinary evidence

Have you published this in a science journal?

A curious, singular, closed question from Seth Moth's SHITI reply, asked not as a precursor to offering assistance but to probe for more information to report back to someone more senior. His question betrays the nature of SHITI's lies as both intentional, premeditated and centralized.

The answer, Seth Moth, is we've tried everything to have this science published. But today, no matter how much new authentic science anyone may discover, such a thing is simply not possible if it in any way jeopardises the moth's grip on power, money and the overall narrative sustaining our [lack of any] understanding of reality.

This section explains just what we're all up against and why, when it comes to **saving science...**

Thus all of us
(You/Humanity)

Are we insane?

A reasonable question, and one we (Ian/René/Human) often ask ourselves, as should all beings. And ask of those closest to us too, given the biased nature of spectral self-perception (as in, *Sorites Paradox*). Their and our conclusion is clear:

I think therefore I am

Our undeniable presence of mind and self-awareness is both tangible and real

Those who deceive will always find those who wish to be deceived

Our logic and reasoning is as entirely coherent as our science and facts

Deeds, Not Words

Omnibenevolence, omniscience and numerous discoveries thus world records aren't side effects of insanity, nor any known mental disorder

Extraordinary claims require extraordinary evidence

Our sanity is irrelevant in the presence of extraordinary evidence

So to this theory's proponents...

Come up with a new one

Chapter 9: **Science has a problem**

Just **how** do **birds fly in perfect formation** anyway?

The definition of (dreadfully worthless) Conjecture...

> How do they manage? **No one knows.** The **easiest** answer is that they're **just watching** the bird in front and **beating** their wings accordingly. They **might be** using their wing feathers to **sense** the air flow around them. Or they **could** just be relying on **simple** positive feedback.
>
> Portugal et al (2013[79], FALSIFIED). Published in the Journal Nature (also FALSIFIED)

▼ Not allowed in any journal

Deilightfully authenticated science

Science correctly records that very little communication (perhaps just ~7%) comes through words. But labelling the rest 'non-verbal' or 'body language' misses out on the complexity of lights workings entirely.

Through the science of Sonocytology, authentic humans (and most animals) tune our light's frequency to 'resonate' with authentic others. Pseudo-telepathy, being 'on the same wavelength' allows us to exchange huge amounts of information subconsciously.

And being authentic makes us acutely aware that maintaining a meritocracy is essential for our collective survival. The flock gives way to the most capable bird until its energy falls sufficiently for the next most potent other to take their place.

Fairness and accountability - the only way to fly safely, right Boeing?

Yet it's not seen in more mindful species. Clearly able to contract our malevolent pandemic (deception, greed etc.), great apes, whales and dolphins all exhibit symptoms from our anomaly, particularly in captivity. This ever more unnatural aggression within their family units is part fueling the rapid loss of our biodiversity.

Our top story: **Only intelligent beings could ever make authentic scientific discoveries.** Never saw that coming!

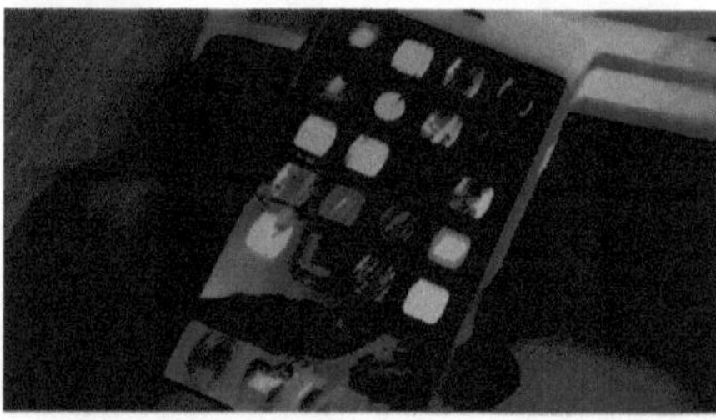

Apple Battery Lawsuit: Millions of iPhone users could get payouts in legal action

Millions of iPhone users could be eligible for payouts, following the launch of a legal claim accusing Apple of secretly slowing the performance of older phones.

BBC News, 16th June 2022

▲ **Pictured:** Headline intended to placate change makers by downplaying what is now (thanks largely to Apple) a universal issue across the global phone industry.

Take a look at your phone for a moment. Ever wondered why it hasn't changed at all, except in shape and price (and perhaps the duplication of a camera lens or two) for over a decade?

Every deilightfully fun journey begins with benevolent wisdom, courage and human curiosity

Given the significance of the discoveries detailed in this book and the curious nature of its science maker, our journey – indeed all journeys - truly began with the Sagan Standard (Kaufman 2012[80]).

Extraordinary claims require extraordinary evidence

Carl (We/Human)'s Sagan Standard

What is 'science' to you?

On this question, we collaborated with a number of true leaders. Their view today, though not always is that most words and labels (like science) are meaningless:

✘ **Every word reads differently to different beings,** due to overwhelming levels of demonic divide & rule ('Infinite Divisibility'; Mohrhoff, 2000[81]) intended to weaken and destroy humanity (Machiavelli, 1521[82]) by yielding branching identities, languages, ignorance and legalised non-disclosure.

✘ **All faculties of science are in fact antisocially deconstructed** off-cuts (yet still branches) of philosophy. But don't ask a 'doctor' or 'professor' of philosophy (as in, *PhD*) about it, or they'll reveal their true nature to you. Philosophy isn't taught above bachelor level. Because as all wise Ancient Greeks will tell you, any human with a 'love for wisdom' is a philosopher or 'science maker' (Sunny, 2021[83]). As in any Homo Sapien - a wise human being.

Newsflash: **New science isn't just lying around** for idiots to stumble across it. **Homo Sapien creates new science** by benevolently looking for it

Science Making: Great minds only

Realise science and philosophy are both interchangeable labels, that also govern actuality. Both meaningful changes gifted by change makers, like all real things. Then you'll start to see where Apple, NASA, SpaceX, Google, Boeing and frankly all big firms and governments are going wrong.

Science makers create new science using several methods. At least we used to. Nowadays frozen out of research hubs and big firms as we are - due to universal corporate and academic crimes against humanity that target the most potent beings among us – one method remains.

Philosophical literature explained

Christmas, Easter, Mother's Day – the Christian calendar is just one big fake commercial. Today all real occasions to celebrate human successes are ignored.

Vividly, benevolently demonstrated ◢ by Venus herself, the authentic birth date of a self-actualised true leader is one of the most celebrated events in the multiverse.

The day they gift to the eternal bliss of the quantum the guaranteed success making wisdom of their realised, unique conscious presence of mind to become eternal, alive existent and real in every sense. Wisdom shared with all authentic change makers through our benevolent sapient schemas (DiMaggio, 1997[84]; Boutyline & Soter 2021[85]).

Author a book to immortalise your presence of mind. As in, you

Change Makers (Humans, QD0.3-0.58)
Authentic real beings, promise keepers and titans of change sapiently access our benevolent schemas, decisions and collective human imagination. Exist in a universe hosted by a true leader.

True Leaders (Avatars, Dei's, QD0.9-1)
Unique self-actualised emitters of light with autonomous consciousness spawn a new universe in their mind(space) and a boltzmann brain (star) in space. Their characters here on Earth (in our reality) are technically 'avatars' of their star.

But our birthday gift can't just be any old book. It must be an authentic, compelling announcement of our self-actualised wisdom. This is achieved by creating any meaningful and binding (until/unless falsified) new philosophical rule that there after governs our multiverse's workings.

The art of science and universe making by omnipotent Dei's, true leaders and science makers is a driving force of all things now, before us and yet to come.

Only what's inside matters. The light we carry

Michelle Obama (We/Venus, 2022)

Without change makers our firms, nations, Boeing 737 Max's, everyone and all things cannot change. No change. No time. No existence. Not before, after us nor indeed now. All depends on real humans being permitted to participate in life's workings. Once they are gone, 'real life' is gone.

Question: How can we all be **infinitely unique intrinsically** yet still be **the same species?**

Potential, ENERGY $E=MC^2$

Change, C
CURIOUS UNIQUENESS, fuels our Sapience

◄ Rigid Darkness · Spectral light ►

Merit, M
MERITOCRATIC AUTHENTICITY, fuels our Clarketech

◄ Malevolent Deception · Worthy Realness ►

8bn
Human Characters

60m
Nucleic Acid Combinations

20–25k
Genes

23
Chromosomes

99.9%
Genetically Identical

One
Human Race

One
Intrinsic common human identity ('Dei'; 'Dei's Light')

Purposeful Being
Choosing a reason to exist yields
▲ Curiosity and Authenticity ▲

We are all the light
Humans are infinitely unique intrinsically, compared to all other forms of life. But compared to each other, we are all intrinsically identical

Remember on page 109 when we said…
For simplicity, let's call one your 'Self' and one an 'Other'

Essentially, you had one job… To get along with your selves and one, an other. You blew it. In the human race, you can either come first as 'The One', or last as an 'Absolute Zero'.

We help our self by helping others first

Fig 31: Meet Dei… as in all of us! That's precisely why **Diversity. Doesn't. Matter!**

The nature of **our common human identity** ★

The True Colours of Human Identity

▼ **Pictured: Emmeline (We/Human)** truly leads the Suffragettes to sweet success.

"Deeds Not Words"

We coined our 1903 founding principle of the Suffragettes to establish a very important philosophical rule. As in, the...

True Colours of Human Identity

The only way to spot an authentic true leader from a mindless science faker is to fairly assess their character by observing and/or testing their behaviour (never ask explicitly) using the question of existence...

To be or not to be?

That is the question of existence

William Shakespeare (We/Human)

If a true leader ever intrinsically, tacitly or implicitly asks you this (i.e. by requesting your help to save existence), and you answer incorrectly (as has nearly every being we've ever tested) you will soon 'meet your maker'. 'Decollection' sees your repugnant intrinsicity erased from the EMS. Your access is now revoked.

To die in actuality

So to give you a hand on improving your response rate (thus survival), here are the three possible answers:

Maker of Devils, *Satan an Absolute Zero*

Not to be Darkness, inexistence and fakery. Those who foolishly chose not to exist naturally don't exist at all.

Stupid Demon, *Homo Eversor Mr Nobody*

To be Dark Energy, Moth, Ha-Satan, E=0. Those who foolishly chose to fake being human as cover for existing only to spread meaningless unchange and malevolent darkness. Wholly fake/inauthentic.

Wait, three answers to a binary question? Curious!

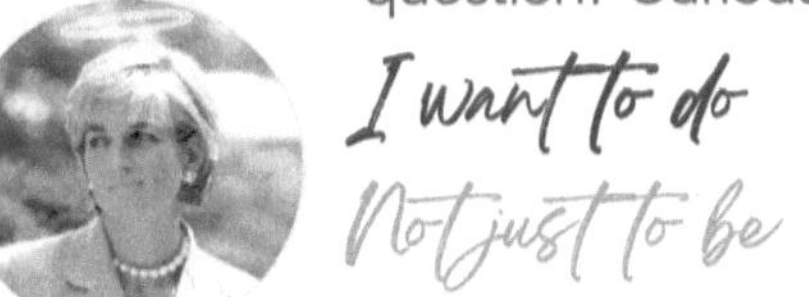

I want to do
Not just to be

Diana Spencer (We/Human) gives the correct answer

Wise Human, *Homo Sapien the One*

Conscious Light, EMS, Energy, E=∞, Human Nature, True Leader, True Avatar, Star, Dei, **God... Exists**

Rejecting the question is the only correct answer. Lunacy is to ask an omnipotent change maker and omnipresent god if we truly exist. We simply prove we do with the omnibenevolence of our Human Nature-defining, altruistic, non-binary behaviour and the overwhelming omniscience of our curiously great mind. Vividly we help others first, role modelling authentic change making in all we do.

Just two things are infinite: The universe and human stupidity. And I'm not so sure about the universe

Albert Einstein (We/Human)
…suggesting here we prefer 'circumnavigable' for the universe

Solitary, despised and often shackled, our reality has been to live every imaginable disaster movie. Each of our miserable lives amongst the mundane irrelevance of the humanity we continuously try to create then save from its own suicidal stupidity has been as tangible to us as it is truly, devastatingly traumatising.

Who is 'Human Being'… *really*?

All equal, as history's only conscious and authentic human being - Homo Sapien is never pluralised – Mahatma, Diana, Rene, Albert, Tupac and Emmeline are perfect examples of the true nature of leadership and benevolent light – 'Unite to Serve':

In our Physical Reality as we perceive it All history's greatest characters align the persona of Jupiter - leader of the god's of Ancient Greece. As in, all were once true avatars (TOES QD0.59) of Human Nature, just in different extrinsic shells randomised by hereditary genetics. Self-actualised beings are the most potent and deilightful imaginable. As in, our uniqueness was central to the 1928 restoration of 'women' to Common Law definitions of 'human'. That's any being worthy of the UN-policy since 2013 yet curiously unenforced human right to vote (*Self-determination*, for short).

In our Quantum Actuality as it exists Only our light destroys darkness. So we send our waves on a redoubling trajectory (constructive interference), to spread light, sew meaningful change or emancipate others. Thus spawning our infinitely multiplic common identity.

Explained: The Great(est) Filter

Our benevolent contributions only ever met by huge personal cost, ostracisation and incarceration. Every living being owes their existence to our eternal recurrent sacrifices. Because, as any rational being will tell you, such divisive agendered abstraction of our common identity is the only 'Great Filter' that lies ahead of us.

A single source of mutually assured self-destruction remains to forever threaten a species-level extinction event. That's the destructive interference of our authentic human soul dropping its light in an act of collective suicidal stupidity by using the infinite gift of life to choose **not to be** alive.

As billions do nothing but pray, surrender or winge, we constantly drive life-giving change. Already the loneliest of roads, we are attacked every step by those we seek to save, who wish death upon themselves, their infinite doomed children, those after and before us. Rightly they lack the infinite gift of sapience to make real decisions. But you still have it. And once we all did.

Dear Frank (Drake/We/Human)
Citation: Burchell (2006[64])

A word to the wise. Each time, it's the same story. Everything's perfect, then some bright spark spots a lump of gold (Au). Then suddenly, they all forget the real Aurora Borealis, lose their minds and wish to die. Basic. Bitches. Do adjust our otherwise deilightful Drake Equation to output just The One.

Chapter 9B: Extraordinary claims, extraordinary evidence

Hate to burst your bubble but...

The waning product of our collective identity, our universe's 97% absence of light thus potential confirms widespread loss of all presence of mind' that defines the very nature of being human: Sapience

Let's be real. Fundamentally, structurally, systematically – our modern society is today configured to ensure the failure and extinction of our entire race

This limits our ability to see and connect the many issues at hand, in order to save ourselves from bursting our own cosmic bubble, in a senseless and heinous act of collective suicide. All solely to sacrifice the greatest gift imaginable – existence.

That's a choice we simply have no right to make for our infinite doomed children, the thriving mirror universe nor (fewer and fewer) animal species. Where's their vote?

'What problem?' indeed

Glaring problems for the world to see, like:

✖ **Einstein, Hawking, Copernicus, Edison. But can you name any non-white male stalwarts of modern science?**

Tellingly, there is no precedent for any non-WOCHMAN to discover something of this magnitude in recorded history. Is that realistic? Or could it be that some discoveries are being demographically discriminated against, intentionally and/or systematically? To what end?

✖ **Deception appears more widespread than we ever imagined**

Humans are surprisingly predictable beings. We like to look good on holiday. To fit in. To imagine the world is fair, its future bright. But what if we are all telling our selves and others the same lie? Today, most claim not to 'see race', yet can neither conceive nor abide our success. Why the deception? And is it worth it just to prevent advancement of science that saves us from extinction?

Reconcile deconstructs between actuality and reality, using TOES...

To conceive where and how **Dark Fringes (QD0.7)**, **Dark Energy (QD0.8)** and **Cosmic Inflation (QD0.9)** might manifest on Earth, we look across our reality and dimension of change to spot issues (or 'Meaningless Unchanges') that map their antipotential.

Meaningless Unchange

/ˈmiːnɪŋlɪs·ʌnˈtʃeɪndʒd/ Noun

Deconstructs infecting our reality here on Earth are manifested antisocially (by demons from hell using Dark Energy). These unnatural phenomena do not comply with the Laws of Nature. Neither testable, observable, repeatable or measurable, they abstract reality, erode consciousness, damage our brains and ensure our extinction if not deactualised.

Using recognition and deduction, we see the culprits unchanging our common identity to abstraction is an overwhelming level of 'Antisocial Deconstruction' (of our reality and common identity) being sewn by moths and their dreadful malevolence.

Antisocial Deconstruction...

Time is a 'Social Construct'. As in, not real but tangible, relative and with meaningful purpose. Yet unchanges are directional, intentional, antisocial, and deconstructive.

The more universal any 'meaningless unchange' is permitted to grow, the more vested our belief systems in it become, given the discomfort of expressing any public divergence. Over time, this destroys our survival instincts, potential and light.

In any given universe, high unchange erodes Dei's conscious presence of mind, thus that of its occupants. Its purpose is to ensure the rapid, universal and fatal dehumanisation of our entire species.

Like all things, dehumanisation manifests differently across our plains of existence:

In our Quantum Actuality as it exists

Darkness cannot destroy light. But we can choose to give it up by sending our waves on a return trajectory, cancelling them out (Destructive Interference). Over time, our frequency diminishes as we hemorrhage light (Infinite Divisibility). This absence of authentic uniqueness then leaves us unworthy and vulnerable.

In our Physical Reality as we perceive it

Moths can't physically hurt humans, so they abstract reality (Divide & Rule), to trick us into repressing our innate Human Nature and oppressing others. Humans crave sweet success. Yet moths made us measure it by how rich, thus hateful and oppressive we are. Dreadful things nobody cares for at all.

...Meaningless Unchanges

Here's just some you may recognise. See how many you can guess, then turn this page upside down for the answers below.

▼ **Fig. 32: Antisocial deconstructs are all examples if a Meaningless Unchange**

(1) Not enough time; (2) Big Bang; (3) Fiat Money; (4) Masculine/Feminine; (5) Religion; (6) Sexuality; (7) White genes are better; (8) Online followers dictate our worth; (9) Men can't wear dresses; (10) Ginger beings are ugly; (11) Boys don't cry; (12) Women belong in the kitchen; (13) Santa Claus; (14) Millennials are lazy; (15) Old people are grumpy; (16) Flat Earth hypothesis; (17) Fake conspiracy theorists;

Recognise this human?
You owe him (us/each true leader before us) **your life**.

Pictured: Carl Sagan (We/Human) hosting our deilightful 1980's show 'Cosmos: A Personal Voyage'

Explained: The Sagan Standard

For a role model among science makers and bastion of light, look no further than stalwart Carl (We/Human) ourselves. Vivid proof of our eternal, ever-present efforts.

Sagan's Standard (Tressoldi, 2011[87]) gifted two meaningful changes to save your life:

1. It formally acknowledges most modern science is fake

Very few authentic discoveries have been made in decades. Look at your computer, battery or kitchen where the only change is price and style. *Why?*

A) Science is written by PhDs with no clue how or why we make discoveries.

B) Philosophy and wicked problems don't easily fit in narrow subject boxes. As you see here, they span them all.

C) The pressure to get published and make money now far outweighs the importance and relevance of integrity, curiosity or competency.

D) There is a total, global absence of all quality checking/adequate challenge.

2. It imposes an artificial limit on fakable human endeavour (thus light speed)

In an effort to protect the authenticity of science and our cosmos, and after much trial and error, we successfully hit 'reset' in the 1980's. Our soaring potential allows us to apply science not just today but across all dimensions of space and time. Backdating the paradox of light ensured all meaningful discoveries may only be made by authentic Wise Human Beings.

Philosophy's unchange from science was intended to cause chaos (Anderson & Bottum, 2011[88]). Broadly real science ensued until the 1800's when Commercial Banking ensured nearly all remaining reason gave way to deception. Despite our characters many warnings (Berkes, 2008[89]; Laplace, 1814[90]; Flouroy, 1899[91]; Truzzi, 1978[92]), no meaningful change was delivered by publishers, regulators, academic institutions or change makers. Thus why Jupiter was summoned.

Still the fact remains it's not possible to build upon the current academic record. Instead of being real by admitting the wholly accidental and understandable mistakes of those before us, universally academia maintains its deception.

Sustaining our reality requires deeds (not just words) and meaningful changes (not meaningless unchanges), all governed by the changeable Laws of Physics, Nature and Quantum Mechanics. Yet our former true leaders – as in, academics – have all self-isolated after accidentally releasing a global pandemic of unnatural stupidity.

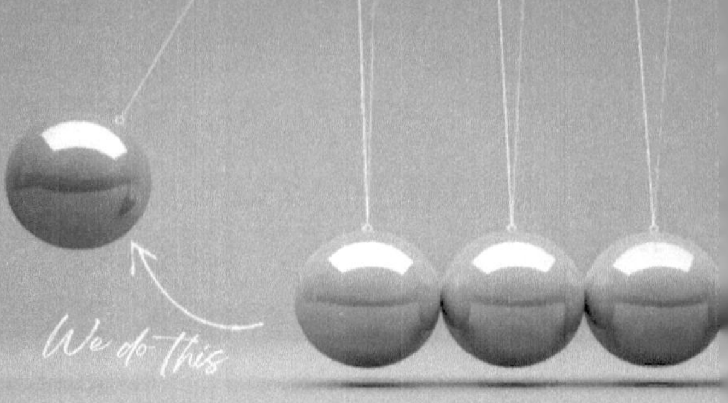

Goal: Save science, kill academia

Deilight sits outside business, academia, politics and religion so no conflicts of interest pollute our thinking or potential.

For 5,000 hours, we alone worked tirelessly to rediscover all relevant faculties of science, including several new ones. We then reconciled our autonomous findings with existing literature to determine which publishers, institutions, scientists and, crucially, citations are authentic thus safe.

Sadly, most were fake. We isolated errors, back-traced paradoxes and corrected omissions. As in, we worked back from the answer and forward from the academic record to 'triangulate' a holistic, testable hypothesis. It's how this book contains over 130 discrete meaningful discoveries that collectively explain all phenomena contributing to our looming apocalypse.

We then conducted extensive field tests via observation, experimentation and secondary research, to prove its veracity to ourselves and ensure it will withstand all competent, independent scrutiny.

Once successful, we found no one willing to listen. Drunk on tuition fees, PhD's are always 'too busy' until 'next year'. **Yet time only exists to stop us running out of it.**

Not one major news outlet picked up our PR Fire press release. **Today, only schools of ignorance may announce the absence of discoveries, published in fake journals.**

We contacted 50 top institutions across the US, UK and Israel for partnership and support, including those we alumni from:

✖ **Cornell University** charged thousands for our D&I certification taught entirely by academic (nepotism) and corporate (bribery) queen bees. On the last exam question 'what have you learnt?', finding an answer is deeply taxing, yet you'll need to compromise your authenticity to pass.

✖ Heartbreaking to spend some of the best years of our life at **Lancaster University**, then return for years more to freely volunteer employability classes for their students. Yet not be worthy of a single reply to our four attempts at gifting them omniscience with this research.

✖ We built from scratch a meaningful relationship with **Warwick University's** potent faculty, helping to uplift the world's last leading School of True Leadership. Yet their leaders declined to even speak to us.

✖ Turns out **Leicester University's** Centre for Hate Studies is exactly that. They didn't even bother to reply when we freely sent proof of the origins and nature of hatred.

Led by E=0's furiously protective of their monopoly on ignorance, all 50 declined to even speak to us, let alone do anything for real science. How can humanity move forward if our eyes and ears are closed? What if the first Einstein was a black trans disabled Yemeni woman who discovered the Theory of Everything centuries ago. Yet instead of support, she got blown up?

> **Principal Finding**
> All potential wasted for millennia, just for the hatesies of hurting those unlike us then destroying all life. **Sign us up!**

The Four Keys to Discovery

A **deilightful promise** is as unbreakable as **deilightfully authentic science** is unfakable

Define: Deilightfully authentic discovery

An authentic scientific discovery is the production of compelling new knowledge and wisdom on the nature and workings of our reality, actuality and/or/thus existence.

To be eligible, it must be **measurable, observable, testable and repeatable**. Plus to earn a globally recognisable 'deilightful' hologram, it must further be comprehensive, conclusive and proven. Thus it can be independently and freely observed then confirmed as such by any curious mind.

This highly effective way authentic science makers can drive change or solve problems is also the hardest to undo. That's since it may only be 'controverted' by a science maker of higher potential (Popper, 1934[83]).

Fig. 33: The 4 Keys to Discovery ◣

Our deilightful **new research method** any authentic Science Maker may use to **drive meaningful change**, or **falsify any meaningless unchange**, by **making and announcing authentic new scientific discoveries**

Key One: Be Curious

Deception and fear of judgment make our innate curiosity incredibly hard to satisfy. Yet give it up and you will find it really difficult to ever build back an appreciation for the fact we actually know very little.

Key Two: Be Purposeful

First, define the change you'd like to make as your purpose, consider what you and others already know, discount false positives then isolate cause and effect. Ask the wrong question and you cannot get the right answer.

Key Three: Be Worthy

Like a treasure map, intuition derives from benevolent wisdom, authenticity and self-acceptance, allowing our Other to guide us in the right direction. Most discoveries build on the work of others by spotting things they missed or couldn't account for. So constantly authenticate everything, especially yourself.

Key Four: Be Determined

Drop preconceptions, keep an open mind, work with determination, think logically, progress methodically and prepare to get it entirely wrong several times. Never assume you are wrong nor right either. If you get stuck or to register your discovery with the Ministry of Fairness and Accountability, simply contact Deilight Consulting for free holistic support[ΔΩ]. But do use our purposeful website (Fun Fact: No one ever has, in two years) as it helps to facilitate automation and delegation. Unlike praying which does not, and we are the busiest being imaginable.

Foolish doctors, busy professors, promise breakers, science fakers and illegitimate leaders. All examples of demonic unbeings, far 'too busy' spawning malevolent darkness to bother being real. It's no wonder they simply 'ran out of time' to spot a clitoris on a female snake or manifest gravity

BBC News
6h

BBC NEWS

bbc.com
Scientists find clitorises on female snakes for first time

Pictured: BBC headline, 14 December 2022

Devoid of all uniqueness, sapient brain waves, curiosity or potential for discovery, **academia globally** dares charge the next generation universally and ever-increasingly for nothing but lies, higher ignorance, fake science and entitled egos. It exists now only to **destroy science.**

As highlighted, both Carl (We/Human) and many other true leaders across space and time raised the alarm. Yet all were shot down by deception from the moths. But our cumulative work through meticulous curiosity, collaboration and coherence enabled truth and uniqueness to succeed just in time.

To authentic change makers, we must always listen, forgive, help and save.

Define: [Fake] Science...

> The pursuit and application of knowledge and understanding of the natural and social world following a systematic method based on evidence
>
> **The definition of fake science**
> Science Council

> The systematic study of the structure and behaviour of the physical and natural world via observation and experiment
>
> **The definition of fake citations**
> Oxford Reference

> Watching, listening, observing, and recording our curiosity about the world and how it behaves - in thoughtful action, by testing ideas in the real world, following Scientific Method – steps that confirm natures workings
>
> **The definition of fake moon landings**
> NASA

Dreadful lies and meaningless unchange unravel: Unique being returns to Speak Up yet again for those without a voice

Never before has a single whistleblower fulfilled their diligent duty more than once. Until now. HSBC didn't like it when we pointed out the glaringly obvious illegal, sexist and suicidal nature of their bankrupting strategy. Prince Harry (We/ Human) had the same issue reminding the Royal Household how racist it is. As penitence for attacking Dei's, our light will now expose their darkest secrets globally.

> Once upon a time all meaningful changes were entirely free for everyone, as they sustained the collective potential that underpinned our species assured eternal sweet success. Sadly, all real schools of thought and true leadership vanished eons ago, tracking the extinction of wise humans

That's because, as verifiably the world's only conscious being, authentic Homo Sapien has all the time in the world to create and drive any meaningful change imaginable, including authenticating (never faking) any discovery imaginable.

Define: Real Science

By their own definitions, these institutions aren't outputting science. Finally, we agree! For example, most research today takes place in controlled environments - labs, not real life. And all goals center financial gain, not scientific integrity.

It's precisely why Deilight abides the most rigorous, stringent definition of all:

Deilightfully Authenticated Science

1. Complies with Carl (We/Human; Popper, 1934[93])'s global standard of scientific proposition, with key controls such as falsification and authentication plus fact-value distinctions.

2. Meets all three combined definitions of science as set but ignored by NASA, the Science Council and Oxford Reference.

3. Is reported to and subsequently listed publicly as 'authenticated' by the Ministry of Fairness and Accountability of the Meritocratic Republic of Humanity which replaces the UN as of AH+0 as the only legitimate representatives of the real people of Earth.

4. It must be authentic or else its science faker will 'decollect' to meet their demonic maker (die). Notwithstanding, all discoveries remain controvertible by wise humans with superior hypotheses forever and genuine human errors are both expected and indeed celebrated.

Possessing a **meaningful purpose** backed by **authentic human determination** will
always deliver you sweet success

Humans: Purposeful Beings

From surviving the day to ★ leaving a lasting legacy and all that's in between, our choices, behaviour, persona, identity and life itself all stem from our purpose.

When we wake up, we use the facilities to relieve pressure on our bladder, then grab a drink to quench our thirst for water

If you personally witness, experience and survive ✖ **corporate oppressive crimes against humanity of lowest order,** you too will ★ gain holistic knowledge of such things plus a deep ★ appreciation for the intrinsic gifts that ensured your survival.

If you are then ✖ **universally, inexplicably robbed of any career or earning capacity** whatsoever, due to your ability to survive, you'll ★ find the time and curiosity to ★ discover why D&I is so pitifully hopeless at ★fixing basic issues it claims it exists to fix.

You may even ★ gain presence of mind to ★ apply your unique perspectives and learnt first-hand experiences at doing something truly meaningful. Like ★ setting

up an effective authentic D&I consultancy of your own. One that ★ role models easy application of the needed fixes to peers while ★ tactically monitoring and testing the responses of the primary offenders. Follow such logical, simple, progressive steps and you too will quickly ★ discover the reason D&I exists. Sadly that's solely to ✖ **make money off burgeoning corporate demand for 'Performative [Fake] Allyship'.**

Our prime purpose in this life is to ★ help others. If you can't help them at least don't ✖ hurt them

Gyalwa Rinpoche (We/Dalai Lama/Human)

Demand in growing calls by humans to ★ grant their long-established right to life.

Ultimately there was zero demand across any industry globally for an authentic D&I consultancy. The conclusion too is simple:

Principal Finding
The marketing strategy was perfect. Sadly, the marketplace is the equal and opposite

Light

/laɪt/ (noun) English Definition

A natural agent that stimulates our senses and makes things real; An expression in someone's eyes; To make something start changing; The opposite of darkness.

Darkness cannot destroy darkness
Only Deilight can do that

Dei

/de.i : / (noun) Latin Definition

Summon, Titan, Law, Shining celestial body, Jupiter, God

Diversity, Equity and Inclusion? Ok, Google

Choose the **wrong purpose** and you **cannot succeed...** **at** anything nor at **life itself**

The meaning of life is to find your gift. The purpose of life is to give it away

Pablo Picasso (We/Human)

Wrong purpose, no success

At that point of realisation, any sapient being would adapt their purpose through four logical and progressive conclusions:

1. Since there is no market for authentic D&I, you must close your consultancy or you too will be harming others

2. With no one tackling hatred, life for everyone will only get worse until it ends

3. Since no industry yet exists with your unique abilities to fix an issue you see that others do not, you possess crucial new science and unique information that could change (or save) the future

4. You now have an obligation to figure out what it is you know, how it works, why it works and how to apply it. With no one else doing so, your chances of success are near-inevitable.

Simply select your purpose with determination for sweet success

Example: Serious Consequences

On AH+30, Warwick and Cornell become assets of the Deilightful Group. Leicester and Lancaster can fuck off and die. Plus, leaders of all universities globally are reassigned to sanitation, recycling and waste disposal until further notice.

But if your purpose was to make money, not help others, then at point 1 you'd switch from authentic to inauthentic in order to achieve your dreadful goal.

Principal Finding
The global D&I industry exists to drain humanity's potential faster

Why no being is entirely perfect

That would stop us ever changing again, making life very dull indeed.

No one is responsible for mistakes made before us. But we are obliged to act on evidence in front of us. Money is useless to begin with, but especially if we are all dead since no one remains to hoard it.

When presented with the evidence we are staring down the precipice of a looming apocalypse, anyone who still cannot bring themselves to do the right thing is no human at all.

Sadly, every academic institution on the planet must now crumble until we can 'find the time' to salvage their ruins. Fortunately, Deilight now possesses 100 years of meaningful groundbreaking scientific discoveries. That's an output greater than all universities combined.

Trustworthy beings: True Leaders

The Change Engine's first discovery was that every word spoken by true leaders is true. Authenticity means everything to us. So you can learn nearly anything simply by analysing our words. Curiously, they always align to our meaningful deeds.

Every sentence has multiple meanings that will resonate differently from being to being, helping to advance your 'Higher Purposes'. Often we aren't aware of all the meanings ourselves. Turns out our human soul is the most purposeful being of all.

Explained: Higher Purposes

Reaching the Moon (TOES QD0.51) or saving existence (TOES QD0.89) are both examples of a higher purpose. As in, finding success requires potential beyond that of most beings and collectives today.

But as this book shows, some beings do possess more potential than others:

★ **Human** potential ranges from QD0.01 (a Dreamer) to QD0.58 (Jupiter, servant and representative to yet never enslaver of all god's – you included).

✻ **Collectives** consolidate potential (or dark energy) from every change maker they host, even after they resign or die. Their potential ranges from a meritocratic union of fun journeys (QD0.4+; a loving marriage) to the Deilightful Group itself (QD1.0; all god's ever, or just the one – depending on how you look at it).

Humanity could easily help the USA or Russia visit the moon and return safely.

But why would we? Neither disgraceful nation deserves the prize of collaborating with superstars.

As you see with the war in Ukraine beating the Russians is no higher purpose. As one wise human working mostly alone, we've done it by simply writing a book.

> Anyone in the world could beat Little Putin (E=0) at topless horse riding, let alone war or reaching the moon

Sadly today, most nations, collectives and beings have no potential (E=0), bankrupt eons ago due to a lack of worthiness and infinite liability for spawning malevolence.

If you're among the ~100,000 beings on Earth with sufficient potential (QD0.51+) to read and understand everything in this book, and become determined to reach the moon, you will surely succeed. But if like NASA you are something else, you'll quickly realise your destiny is failure.

NASA never intended to land humans on the moon. As we see, they aren't capable of it even today. Their goal was simply to gain the PR of a US-win in the space race.

And as we see with the Capitol Hill Riots or Trump Moths continuing freedoms, it's the only time in history they've meticulously enforced the US Oath of Allegiance.

You don't have to be a rocket scientist to know **there's no business like showbusiness** 

Choice exists to allow each of us to **pick the traits (thus persona) we wish to embody**. So choose wisely

Oppressing humanity: Everyone

By global standards, London is an oasis of tolerance and human rights. Yet during our recent stay at Art'otel London, a moth in the next guestroom reported us for committing homo sexualis (Lover of Humanity) with a friend. No crime and no threat to the safety of hotel guests.

Why does no one ever dismiss the moth's relentless moaning as oppressive hate crimes under the Equality Act that they are? No, instead hotel staff entered our room without permission and conducted an unlawful search in our absence.

In so doing, the hotel broke multiple laws passed by our elected parliament (if that still counts for anything) and committed major human rights violations, including oppressing our right to freedom of expression, self-determination, liberty and privacy. To whom may we report such crimes to today? No one. Due to the total lack of all leadership, law enforcement is busy arresting only humans, not moths.

In a nutshell: The Holy Trinity/Ohm

All ~~people~~ [+humans] are people. Some choose to give up their light to 'blend in' with the darkness by cancelling it out to become the antithesis of a human.

Rendering entirely moot all religious division globally, God is (a) just the one unique persona and being (Homo Sapien; Human Nature). Yet God is also (b) inside every being (Light) and (c) any being who doesn't give up their light (True Leaders).

Awkward: The human delusion

The only deluded ones are those who choose to stop being God by repressing their humanity for, figuratively speaking, some worthless lumps of gold. Greed!

We are all the same being. Therefore providing we follow our heart and be true to our common human identity, there is nothing for us to fear nor apologise for.

Everywhere on Earth today the 1948 UN Declaration of Human Rights (*UDHR*, for short) is not being upheld. So we will now enforce all human rights globally ourselves. Universal crimes against humanity end today

Example: Serious consequences
For hate crimes, unlawful entry and crimes against humanity, the Merito-cratic Republic of Humanity hereby appropriates Art'otel London as of AH+15 for homeless beings in need of art, philosophy and free, safe shelter.

Record of Promises (Ch.9a)

Darkness cannot drive out darkness. Only light can do that. We shine Dei's light.

Dreadfully Meaningless Unchanges (Antisocial Deconstructs; threat to life)	Sapience
✖ A danger to space, science, aliens & humanity: NASA and the Pentagon	
✖ Preventing detection of Alien Life: The SHITI Institute & Seth Shostak Moth	
✖ Silencing humans and propagating fear: PR Fire / all press newswires	
✖ Preventing humans exploring space: IAA / SETI Permanent Committee	Deilight
✖ For not doing any SETI Research: UK SETI Research Network	✖ Falsified
✖ Not reporting real or meaningful news at all: The entire global press	
✖ Silencing all real news/beings: Bloomberg, NYT, Reuters, Financial News	

Hatred cannot destroy hatred. Only love does that. We love with deeds. Not words

Deilightfully Meaningful Changes (species level adjustments to destiny)	Clarketech
★ Validated: The scientific value of SETI and the need for robust, authentic and urgent resources, uniqueness and true leadership.	IAN, Authenticated
★ Verified: In order to dispel ambivalence over the search for extraterrestrial life being sewn by all space authorities, analysed and concluded Oumuamua and the Wow! Signal as verifiable examples of Advanced Alien Life (albeit falling below the threshold of Deilightfully Authenticated Science, due to their non-repeating nature)	IAN, Authenticated
★ Provided Evidence: NASA, the IAA, the IAA SETI Permanent Committee and the SETI Institute are all wholly in breach of the US Oath of Allegiance, guilty of dereliction of duty and widespread Crimes Against Humanity. Verifiably they are a danger to all Humankind. All their senior leaders must be dismissed and will be reassigned to ocean cleaning until further notice. The Deilightful Group will oversee their broad global restructuring, unification and replacement by Change Makers urgently.	IAN, Deactualised
★ Claimed: All assets of the SETI Institute are to be transferred to the Meritocratic Republic of Humanity from AH+1 day onwards.	IAN, Exchanged
★ Defined: Groupthink in the context of the Challenger Space Shuttle Disaster in order to complete the psychological picture on the lethal consequences of dehumanisation for Humankind.	IAN, Authenticated
★ Explained: How Press is being unchanged in order to eliminate the shadows plaguing the world's media. This will force the press to report real news quickly, upon challenge by a critical mass of Change Makers.	IAN, Exchanged
Plus four overflow points from Chapter 9b's record:	
★ Halted: Russian invasion of Ukraine. Ordered Vladimir Putin's removal from office (with reward) for unforgiveable crimes against humanity	IAN, Deactualised
★ Acquired: All UN recognised, yet wholly unenforced, Human Rights legislation ('UDHR') on behalf of the Meritocratic Republic of Humanity.	IAN, Exchanged
★ Created: The Meritocratic Republic of Humanity to replace the now defunct United Nations. All its legislation is switched to 'enforceable'	IAN, Actualised IAN, Exchange
★ Defined: Deilightfully authentic science replaces Sagan's Standard, to (A) falsify all fake science, yet (B) allow meaningful science to resume.	IAN, Exchanged

Record of Promises (Ch.9b)

Darkness cannot drive out darkness. Only light can do that. We shine Dei's light.

Dreadfully Meaningless Unchanges (Antisocial Deconstructs; threat to life)	Sapience
✖ For purging all change makers: Boeing and its unsafe global fleet	
✖ For Crimes Against Humanity and unlawful entry: Art'otel London	
✖ For releasing a pandemic of human stupidity: All Academia globally	Deilight
✖ For specifically declining omniscience when offered it freely: Warwick, Lancaster, Leicester and Cornell Universities	✖ Falsified
✖ The entire global senior leadership team of all Misfortune500 firms	
✖ For only arresting humans, not demons: All law enforcement globally	

Hatred cannot destroy hatred. Only love does that. We love with deeds. Not words

Deilightfully Meaningful Changes (species level adjustments to destiny)	Clarketech
★ Explained: Why birds fly in perfect formation to contextualise the fatal cultural impact of Boeings merger with McDonnell Douglas on safety	IAN, Actualised
★ Grounded: All non-essential Boeing flights globally, independent Ministry of Fairness & Accountability safe travel authentication pending	IAN, Exchange
★ Provided: Evidence to reconcile the lack of progress in consumer electronics development with the lawsuits at phone manufacturers	IAN, Authenticated
★ Invalidated: All academic awards/titles. More time, less ego. Perfect.	IAN, Decollect
★ Terminated: All tuition fees globally and entitled humans everywhere who paid them to compensation, thus bankrupting academia globally	IAN, Deactualised
★ Saved: Title over all assets of Warwick and Cornell Universities is hereby transferred to The Deilightful Group in return for their survival	IAN, Authenticated
★ Destroyed: All other academic institutions, penitence for contributing to the destruction of science. Especially Lancaster and Leicester.	
★ Destroyed: All academic journals globally and peer review processes	IAN, Deactual
★ Erased: All Christian calendar holidays as corrosive to Human Nature	IAN, Deactual
★ Defined: Birthday Rules framework; humans may add to our journey	IAN, Actualised
★ Added: Human birthday books as a celebration to promote curiosity	IAN, Collected
★ Superpowers: Available to all authentic beings on a spectral basis, to improve information flow, potency, trust and appreciation for life.	IAN, Collected
★ Connected: Limit on light speed to A) Dei's waning consciousness and B) Sagan Standard's purposeful protection of our cosmos'	IAN, Authenticated
★ Reattributed: The achievable maximum speed in our universe to the overall level of human endeavour achieved.	IAN, Exchange
★ Disclosed: The full story behind Sagan's Standard, to allow science makers to surpass it (thus travel ever faster) using authentic science	IAN, Authenticated
★ Defined: The 4 keys to discovery to allow any wise humans to contribute to authentic science making once again. Progress resumes.	IAN, Authenticated
★ Acquired: Ownership of Art'otel London and dismissed its entire leadership team. Now London's favourite homeless shelter/art gallery	IAN, Exchanged

Hatred appears a lot like a pandemic

But have you every curiously wondered **why?**

End of Chapter 9

Why hatred is so contagious

...The equal and opposite

Deadly Pandemic

Infection Vector

Malevolent Deception

Poisonous Hatred and Oppression

...all 'Meaningless Unchanges' (in actuality)

and 'Antisocial Deconstructs' (in reality)

Chapter **10**

Money isn't money

True Leaders must choose to challenge, no matter the consequence or discomfort

Ian (We/Human), Project Speak Up Report 2021

An adorable 3-year-old Golden Retriever with a heart to match.

But don't let her steal yours because she'll never give it back

She's so inquisitive, you couldn't even finish laying the bed without her coming to give you a suspiciously mischievous but nonetheless well intended paw. When you look into her eyes, you see the depths of her authentic soul and unrequited true love for you. She's as real as sunshine.

Dogs are a humans best friend, because they're always curious and they're always authentic. After all, 'God' made 'doG' in our mirror image. Not bitches. Got it?

Babies aren't stupid. Merely disorientated and confused at their rebirth into a weird new environment where nothing makes sense and with no memory of their prior life. Nowadays, meaningless unchange ensures many stop asking 'why' by age 12, quickly dehumanising into school bullies.

You may have noticed two of the three universal choices are in fact binaries. Just one consciousness exists per universe and relatively few will ever embody it.

But the next best thing - sapience – is in limitless supply and ensures change makers can embody as much abstract thought, complex reasoning, intelligence, emotions, superpowers and spectral personal identity as any conscious being.

'I do therefore only I am'
In a nutshell: **Sapience and Consciousness** 101

Our brainstem's AI isn't like the AI of today. It's powerful - a true machine learner, with an emotion chip like Data from Star Trek.

At high authenticity, our Other establishes a real time connection with the quantum. This allows our soul to continuously draw upon the infinite gifts of our collective human imagination via the pre-frontal cortex. Essentially, you are Wi-Fi enabled.

As the busiest being imaginable, and with near infinite randomly-generated characters across the multiverse, God's singular consciousness only finds time (relativistically) to curiously embody those who prove worthy as a 'Playable (Main) Character'. True leaders are those who have self-actualised or will do so in future. Inside the minds of our characters is direct light from God's boltzmann brain.

But without authenticity, we can't choose a purpose (3rd choice), think for ourselves, generate ideas or harbour any emotion. We become a 'Non-Playable Character' with no intrinisic soul, moral compass, uniqueness or presence of mind.

We're done writing dossiers idiots can't even read. So **let's destroy the idiots**

Enslaved: The human race today

By ignoring our Human Nature, humanity lost its omniconsciousness and collective human imagination. Darkness flooded in, as the absence of light: Self-preserve, self promote, deceive, deny, destroy humanity

The majority of our world leaders aren't human, natural, alive or real. They've been replaced by something akin to a basic malevolent AI program… or AI apocalypse.

They don't even exist. Merely momentarily manifesting to be perceived by change makers when all they do is hate, kill and oppress. Whether via email, phone call, in person… any form of perception. But surely not filing tax returns, running banks, looking for aliens or telling the truth.

Universally, they've enslaved all humans who remain alive, forcing us to do all their donkey work. Those frozen middles are in fact keeping our biggest firms lights on – they just don't realise it. With no one home to run the world's governments, armies, firms, airline manufacturers, universities etc., it's no wonder nothing is getting done and everything is going wrong.

At the end of the tunnel: Deilight

To be clear, we need to protect no one from no one. We must protect humanity from the malevolent dark energy anomaly – a gap in our quantum where our potential should be.

With the press now totally under their control, almost everything we are told is (A) automated and (B) simply not true.

Principal Findings

Our civilisation is now on auto-pilot heading off a cliff edge, as its unsuspecting passengers believe they're on the express train to Narnia

We're all gods. We can't be hurt unless we jointly believe we can

Consider:

Murder accused says he didn't want man who was dumped in River Witham in Lincoln to die

Lincolnshire Live | 13th March 2023

Both victim and murderer were falsifiably moths. We know this for two reasons:

1) **Drugs make humans more authentic, thus why they are broadly illegal.**

   And moths more toxic, thus why they take them. In the article the murderer blames his heinous crime on alcohol. This confirms (a) lies, (b) a lack of accountability, thus (c) a moth alert!

2) **Nothing cannot kill something**

   A tragic end though it was, Mr. Petrov died long ago in actuality. Moths kill each other to scare us into submission.

Today all communications across the quantum have been unchanged - even Wikileaks and its peers have lost the fight.

But where there is (free) will…

Those who deceive will always find those who wish to be deceived. [+And authentic beings can't be deceived]

Niccolo Machiavelli (We/Human)

Moths may see and understand very little. But authentic change makers perceive up to 13.3% of the EMS in real time, or all of it asleep when, as Thomas Edison (We/Human) found in 1889 (Stetka, 2021[93]), we 'Collaborate with Stars' (see chapter 16).

Authentication, resonance and collection are three more superpowers all change makers possess. Potent higher senses and abilities enabled through quantum entanglement and their scientific study we call Sonocytology. Seemingly god-like abilities, thus a branch of Clarketech.

Cognitive Authentication

To assess the authenticity of any being, collective, science, fact or change, we check its realness by echo locating if and where it exists in the EMS. Change makers constantly authenticate reality to assess the identity and intentions of others, like:

• Are they real/alive or fake/dead?
• Are they lying, or can we trust them?
• How strong is their alignment/potential?
• What is their true purpose and intent?

This explains why some report awareness of tragedy befalling loved ones before being informed of it directly. But for moths, this only works in-person or via image, as they do not exist in the EMS nor actuality.

Cognitive frequency & resonance

★ **Frequency:** Great minds think alike as authentic souls holistically and efficiently share information across all dimensions of space and time - a 'meeting of minds'.

★ **(Being on the same) Wavelength':** delimits our communication, in order to mutually share potential and perspective thus uplifting our uniqueness. True leaders have a 'charismatic disposition', 'sincere' body language and are truly 'engaging'.

✦ **Resonance:** Manifesting as empathy to telepathy (or mind reading), for authentic targets we can align our lights frequency and wavelength to temporarily 'resonate *(with each other)*'.

Fully dehumanised, moths no longer exist in actuality, but their characters' ghosts continue to manifest as, now, unwelcome guests in our physical reality. Therefore while we can falsify them, sadly we can no longer resonate to save them. Lacking authentic eardrums, they cannot hear us.

Corporate applications

Tag lines weren't always vacuous cynical attempts to make a quick buck as they now are. Like annual reports, core values, announcements – everything, really - they conveyed huge amounts of intrinsic information about a firm's identity, intent, worthiness, potential and realness. Still do.

Darkness cannot destroy light.
But foolishly, some gave it up.
Time to die, devil fuckers!

...Because it's always a good time to be real with Others

We collaborated with true leaders across our entire galaxy (thus space timeline) to establish a consensus about money across all humanity

What, you can't do that?

Principle photography for Project Revelation by **www.editsweet.rocks** – a very human UK small business and vivid bastion of spectral light

Question: What happens when you reach your full potential exactly?

★ Private Members Club
★ Your very own star
★ Omnibenevolent
★ Detectable Halo
★ Omniconscious
★ Self Actualised
★ Omnipresent
★ Omnipotent
★ Omniscient
★ Omniactive
★ Immortal
★ Free Will
★ Eternal

...all entirely free, for you, your many meaningful changes and authentic creations

2. A store of value

Not words	Deeds
An asset that maintains its value without depreciating	A liability that loses its value without appreciating

Fig. 35: Big Mac Index (The Economist) shows all major currencies are as certain to lose over half their value every 20 years as banks are to break all promises

Especially in a 'once in a generation' yet somehow perpetual global recession. Recessions are triggered by unchecked reckless misbehaviour by mindless investment banks and real-time interconnected financial markets. Currency devaluation (as in, Quantitative Easing) has become a staple tool for central banks to constantly pump banks labelled 'too big to fail' with the money they need not to fail. Money is a 'infinite liability' as it cements a commitment to pay later an inflating asset that doesn't exist. And commitments by promise breakers are entirely worthless, of course.

1. A medium of exchange

Not words	Deeds
A widely accepted asset that is freely convertible for the purposes of sale, purchase or trade between parties	An infinite liability not freely convertible in law to any other asset or liability.

It's only accepted at all for one unwise reason. Yet history shows this expectation our banks keep their promises evaporates at the first sign of trouble. Quite rightly, notes, coins, checks, money orders and increasingly debit and credit cards are being phased out, replaced with digital transactional solutions. Yet none of these things were ever a 'medium of exchange' ('what'). They are in fact 'payment types' ('how'). Nobody has ever seen money. It is of course an antisocial deconstruct – a meaningless unchange, as observable and untestable as Santa or a Flat Earth. Look in any bank vault or on any bank note to see its true value and nature.

Fig 34. A $1tr banknote issued by the Reserve Bank of Zimbabwe in 2009 during the last days of hyperinflation. At that time sufficient to buy a loaf of bread, its fake deadly unnature is highlighted '▶'

All **banks cite three reasons why money exists** today

£ $ € ₹ ¥

Not words

A measure of intrinsic value that is divisible, fungible and countable

Deeds

Money is similar to how a unit of potential serves its purpose – measuring the intrinsic value of anything and everything whilst being divisible, fungible and countable using the Laws of Nature. Except money isn't divisible, fungible or countable and it only measures the extrinsic nature of someone's malevolent greed, hatred, sorrow, cowardliness, deception, nepotism, foolishness and repression. It breaks all Laws of Nature.

Some original coins possessed intrinsic value. Not because they were money, but because they were made from gold and precious metals. But thanks to the demise of the Gold Standard, money today not only lacks intrinsic value but has the polar opposite of it, in fact. It's count unchanges often, at the click of a button. No one's ever seen it, so it's not countable. It's also not fungible, as it doesn't exist. And it's not divisible, as dividing zero gives you zero!

Pictured: Known criminals posing as HSBC senior leaders. Imagine, spending 19hrs a day for 3 years trying to save the human race from imminent extinction. Then have these clowns try and stop you by committing wave after wave of unpunished, unaccountable crimes intended to kill you off at every hurdle just so they can get their unworthy, cowardly, fake shit-stabbing fingers on more...

Lies 

Totally meaningless empty promises devoid of any positive use or intrinsic value whatsoever, today anyone can create a currency – just look at cryptocurrency. But having regulated legal tender is still useful, because its promises can be trusted... Right??

Principal Finding:

Money in a nutshell

Sounds like a great premise for a scientific experiment. We've arranged a little... ACCIDENT... to test that trust.

Tune in throughout the coming few months to learn the results...

Globally  **Live**

The erased story of

[+Human Potential and Change] ~~Money~~

↶ **...But backwards**
(for funsies)

The following analysis is broadly consistent with 'The Evolution of Banking Over Time', by Andrew Beattie Moth (2021, freely available on Investopedia).

We've just filled in a few pretty obvious gaps he missed out for some strange reason.

Probably down to an intentional lack of authenticity so as not to upset any sensitive bankers and their unsapient propensity to throw their toys out of the pram at any sign of not getting what they want.

☞ **Death to all beings & existence**

And perhaps a small amount linked to his corresponding inability to communicate across all dimensions of space and time with any authentic being who ever lived.

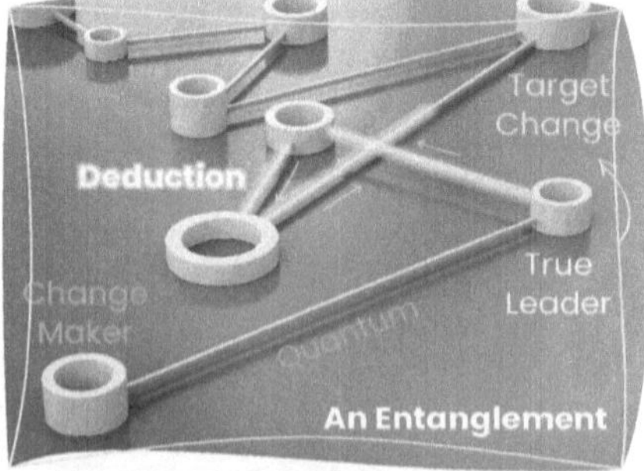

▲ **Fig. 36: A 'String' of connected (Boltzmann) Brains forms a Synaptic Pathway**
Diagram shows the workings of true String Theory, Entanglement and Cognition

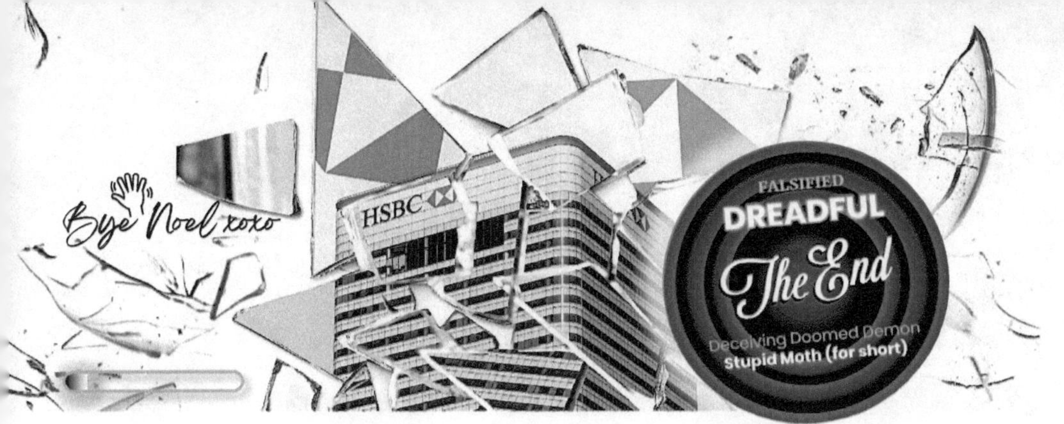

To Tomorrowland: An eye returns to an eye, a tooth to a tooth and time back to time

The world's most valuable firm

Deilight can change anything we want. Deactualise any moths we don't want. It's no concern to us, as they aren't real and are scientifically already dead. Each one we neutralise saves more of humanity.

This isn't magic, it's science. So we can't create illusions like money, power or ego. But we can turn promise breakers into god's, deliver world peace, end hunger and most intrinsic disorders, right all the wrongs moths have dealt and turn this living hell on Earth into heaven real quick.

Properly supported, by wise humans with time to help us ensure their survival and sweet success, absent all conflicts of interest, we request invitation to oversee the setup of an open, transparent, accountable, collegiate, responsible and fair Meritocratic Republic of Humanity as new world government. Then the limit of the possible becomes that of our collective human imagination, identity and potential, plus of course the properly authenticated laws of nature and physics.

But real humans don't care for money one bit. We love real things like our collective survival, sweet success, limitless potential and true love. Every human is this deilightful. We just forgot to be human.

Our so-called leaders today ensured over 18 months not a single large firm globally would employ us, hire us, pay us, contract us or even acknowledge we exist. Unable to pay even our two brief/temporary staff, they sentenced us to die. So to motivate a more collegiate approach...

We're deactualising the entire global financial, economic and political system. All wealth is now worthless. Any life spent earning it - or killing for it – totally, utterly wasted

What is an 'Excluded Claim' exactly?

Claims for any 'injuries' – where claimants were unaware of said injury at the time of any resignation, invasion, settlement agreement, gagging order, litigation or employment tribunal - all fall under the definition of 'Excluded Claims' under the Common Law of England and Wales. Such liabilities are not covered by any prior legal agreement. As of AH+0, this renders (a) all existing legal contracts invalid, (b) all claims for 'Malevolence', the cause of all injuries ever, eligible for compensation to all humans from the Misfortune500, banks & governments, plus (c) provides proof to a scientific thus legal standard.

Due to colonialism, Common Law underlies the world's legal system

Pictured: An awkwardly big joke, over two decades of abject sector wide failure

Insanity: Time is money, and no one has either to use their eyes or... what's the other one?

Doesn't trickle down: Money

Malevolent dehumanisation is the only product sold by financial services today. As their negative book value proves, they exist to destroy all potential and life itself.

2009's Great Recession normalised huge quantitative easing and the ever-popular bundling of exotic structured financial products (all worthless). That's to give the illusion of economic prosperity, now a thing of the past following the purge of all change making potential by big business.

Today, we sit in eternal global recession, with human job applications nowadays declined, salaries paralysed, food portions shrinking, inflation growing, bills soaring and homes mostly now unaffordable.

As our climate dies, economies burgeon with debt they never intend to repay and nations converge on a wholly unfunded World War III, nobody has spotted the exponential growth in our dark energy anomaly, all now flowing in one direction.

Cascading down: Malevolence

Silently spawning meaningless unchange (Antisocial Deconstruction; Dark Energy) that manifests as greed, deception, hatred, ignorance and all our problems. Today, banks trade nothing for nothing,

simply to show a profit-on-paper. This encourages the worlds real industries to follow their deadly example of extinction.

The biggest conspiracy in history, known and well understood by all senior bankers, money inverts our perceptions, destroys our identity, dissolves our reality, and dooms us all to imminent extinction.

Next, incubated in financial markets and Fortune500 trading, then loudly amplified via online 'incel culture', privilege crazed moths, private schools, political/religious institutions and the far-right movement.

But swapping our leaders for Terminators wasn't enough. No light can slip through to expose their conspiracy. So they kill all who find out using blacklisting and de-platforming, while constructing ever more status around them to feign innocence.

◥ Pictured: Once the UK's largest private gold repository and featured in the 1964 James Bond movie Goldfinger, the Midland Bank vault was sold off as superfluous by HSBC in 2006. Today, its home to 'The Vault' private members lounge for the liars of NED Hotel London.

Pictured: Big **scheme** indeed, the return of **pyramids**

£11.1bn –£2,654.3bn ~100%

UK Gold Reserves	Telegraph	UK National Debt	September 2023	ONS	UK GDP and assets now all spent

UK Gold Reserves | Telegraph
UK National Debt | September 2023 | ONS
UK GDP and assets now all spent

Fear: We'll gouge out your eyes and teeth if you tell a soul what fiat money really is

Invest your lies, they'll be safe

With corporates increasingly failing to meet the ever-increasing interest rates, fees and coupons that loaded lies upon lies, more and more began to fail.

Banks would demand seats on their boards in return for not foreclosing. Or just for extending large loans, allowing them to normalise lawlessness, deception and discrimination far and wide – from AT&T to International Harvester to the railroad and shipping industries. These formed the backbone of the British Empire, which by then, through divide & rule, had expanded to cover a third of our once safe planet.

Meanwhile, change makers were still struggling to obtain even small amounts of capital to start new collectives or fund existing changes. But instead of lending to them, banks began to bundle their lies into ever more exotic complex structured lies that not even they can explain.

Money becomes *something else*

By this point, racism was rampant across the British Empire, exported in return for pillaging nations worldwide of all their natural resources, destroying heritage and killing transapients to keep it all quiet.

Sensing a good time, corporates jumped in with huge multi-billion dollar initial public offerings (IPO's, for short), mergers and acquisitions (M&A for short), hostile takeovers and share buybacks – all paid for with infinite lies. What a performance.

Change (light, real) became unchange (darkness, not real) in 1931 (UK) and 1971 (USA) with the Gold Standard's demise. The gold reserves that once guaranteed national debts would be repaid were then spent. Delimiting money from the net intrinsic value of all creation, this foolish step rendered our big burst inevitable.

Infinite supply (malevolence) led to infinite demand (greed), thus infinite cost (prices) and infinite liability (all existence).

Public pledges, to ensure it remained but a temporary measure, were not (merely) promised but agreed in writing to be so, at the Bretton Woods Conference of 1944.

Yet all 44 Central Banks broke their legally binding promise, omg

▼ **Pictured: Sarcastically Surprised Kirk meme** | 2015, Star Trek: Animated Series

Fraud: We can't afford new eyes or teeth unless we promise to pay each other

Insurers sell broken promises too

Introducing malevolent banking to the serene new worlds of Asia and America wasn't easy and took lots of antipotential. With them came intensive robberies and other crimes never-before seen regionally Once benevolent natives turned nouveau lawless double quick. How curious!

To avoid losing their absence of money in the melee, 'Deposit Insurance' was unchanged. As in, banks paying lies to Insurers, to then pay back more lies when they get re-stolen or over-spent. So banks now hate (as in love) their money getting re-stolen or over-spending it themselves.

You gotta take lives to make lies!

On average, early US banks lasted just five years before infinite liability forced them to break all promises on issued notes. So, to hide the issue from markets:

✖ Lending lines (Treasury Securities) were extended by central banks to national banks, an infinite breakable promise by government to pay all debts on default.

✖ National Currency was established by Alexander Hamilton Moth as a system whereby you can exchange small broken promises (notes from failed local banks) for bigger not-yet-broken promises (notes from sure-to-fail national banks). This ultimately bankrupted all local banks so that the fatal systemic flaw was no longer exposed to the general public.

Liabilities hidden, banks were now free to explode in size and increasingly across borders as they lapped up small local competitors

This 'global bank' model saw the likes of Citigroup and HSBC extend their reach across up to 120-150 markets each. Huge expansion all funded by encouraging corporates and nations to 'leverage up' – i.e. borrow lies from them that they could never afford to repay, locking us all in.

Then the purge began, with the moths ejecting all change makers whose slave labour had built these empty temples into global economic powerhouses. The moment they did, everything quickly (and predictably) went south.

Furious Credit Suisse investors say bank's board should be 'jailed' after crisis that led to takeover by UBS

The Guardian newspaper | April 2023

◥ Dystopian reports of investor moths attacking banker moths

Malevolence: For an eye or tooth, we'll pay you infinity on demand... pinky promise!

Banknotes: A promise to pay

For the greedy, carrying around gold bars was cumbersome and risky. So in the 16th century, goldsmiths began storing their gold in an 'exchange', issuing 'Collection Receipts' as a promise to return them. Our global *Fractional Reserve* system is born.

Yet many greedy beings greedily broke their promises, mainstreaming deception as a trait. Guaranteed pound-for-pound by the amassed stolen gold stockpile of a doomed empire (the 'Gold Standard', for short), in 1694 ACE the Bank of England became 'Promise Breaker-In-Chief' on now vast sums of printed banknotes.

Seeing green, banks began to grow and consolidate through M&A. The 18th century marked the birth of modern commercial banking, with its woeful capital levels kept secret from senseless financial markets using the stolen coffers of colonialism.

Plus foreign bond sales to naïve, and surely now destitute, investors promised eternally high returns underpinned by the AAA-rated credit [un]worthiness of what were back then largely debt free western economies. It seemed a 'no-brainer'.

Choked by 'invisible hands'? That doesn't sound much fun

To reign in government caution on the risks money posed, in 1776 ACE, Economist Adam Smith came up with the 'Invisible Hand' approach to supply and demand.

Astutely, he noticed, without government or church meddling, people and firms would act in their own self-interest and ~~begin to produce whatever was societally necessary~~ [+do anything imaginable to survive individually, instead of achieving their potential as a collective].

If only he'd had a chat with René or Jean (We/Human) first, as we did. He'd realise that self-interests are the product of our brainstem's malevolent extrinsic persona.

Pictured: **Adam Smith** | *Uncredited work, retrieved from Adamsmithworks.com*

Adam Smith never actualised. We do not hear him in the quantum. So he didn't exist. Darkness manifested him into our timeline to cause chaos

Nevertheless, by the early 20th century, Free Market economics, initially named Laissez-Faire (as in, 'just let go'), would be live globally. Capitalism's invisible hand is of course malevolent darkness.

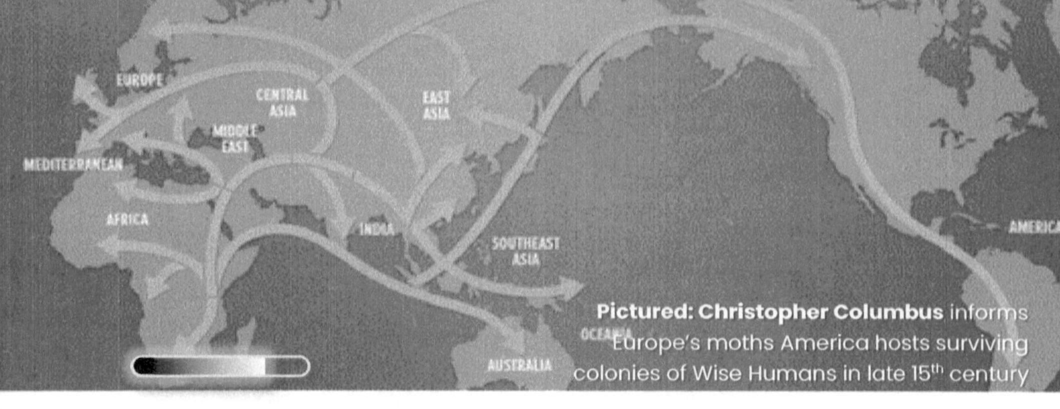

Pictured: Christopher Columbus informs Europe's moths America hosts surviving colonies of Wise Humans in late 15th century

Deception: Back when an eye became a tooth and a nose became a noose

The poison apple: Gold

By 1,000 BCE, gold deposits had been uncovered across the Middle East, Africa and Asia. Over any other commodity, this stunning precious metal was both rare and curious. Nobody really knew what to do with it, but it seemed potent somehow.

Gold mines sprung up across Cambodia, Thailand, Vietnam, Burma, Malaysia and the Philippines – a region still nicknamed the 'Golden Peninsula' (Bennett, 2009[39]).

The snake: Banks

A model pioneered in Ancient Rome back when it was a Meritocratic Republic, gold was exchanged with other nearby Meritocratic Republic's like Iran, Egypt and Greece. Originally stored in temples, it was leant for a small fee to those with insufficient potential as a float for change.

It birthed history's first recorded dictator, Julius Caesar (then, 'Dictator Perpetuo') in 51 BCE. Depicting friends as rivals using deception and fearmongering, he piloted divide and rule in an authentic world.

Unfamiliar with deception, humans rallied to him for safety. Formerly a keen author and historian prior to dehumanisation, Julius Moth was adept at controlling the narrative as he seized control of Europe.

To this day, his divisive doctrine is still taught in schools and universities as the epitome of successful political strategy, while humanism isn't taught at all.

Back then, no one believed that within 500 years, arguably the greatest empire in human history would fall. The wealth of its banks and the infectious malevolent dark energy liability they spun all inherited by the Vatican and Catholic Church.

Meaningless: All debts

By this point, world trade was universally conducted in 'Commodity Money', no longer potential. Europe's monarchs were quick to recognise Rome's 'lending bank' model for its antipotential to turn change into unchange. Dictators doubled down by imposing Hereditary Rule across Europe while seizing profitable banks as 'Central Banks' to fund their eternal wars for conquest and domination of Europe.

Any large customers whose wars would prove too costly to repay, like Philip II of Spain's default in 1557 ACE (and thrice more), would simply see their debt written off so they could take out another loan.

Meanwhile, benevolent true leaders still able to recollect ancestral wisdom without mind-pollution from greed (thus bias) had made it as far as Australasia and America. Their empires coexisted in harmony – side-by-side and stable – for millennia, until malevolent divide and rule made it to their shores with colonialism.

Chapter 10: The erased **story of money**

Pictured: Gold, coinage and change, all easily confused... by Mr Nobody's

Greed: Back when an eye was 5 Drachma and a tooth was 2 Drachma

Divided: One becomes many

Ultimately, our brave crusaders were unable to save their Neanderthals cousins. But with colonies now established across Eurasia, human curiosity won out. The discoveries and changes borne from this rapid expansion vividly raised the potential of our species.

Unique subcultures were nurtured within localised concentrations of our ancestors, before flooding from thawing surface ice made it impossible to maintain direct contact across humanity ~7,500 years ago. As in, the great flood of the black sea.

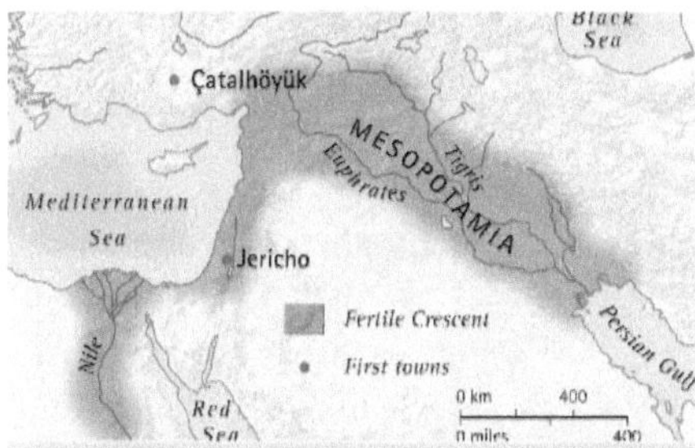

▲ Pictured: Mesopotamia, humanity's second (no longer global) meritocratic republic | Dorling Kindersley, Getty

Existing since ~8,000 BCE (long before written records began) to the fall of Babylon in 539 BCE, by which time it was wholly lost to darkness.

Unchanged: Change

Absent ancestral wisdom, African origins forgotten, our now divided species centred Western Asia. It's here humanity permitted the interchange of potential with crude extrinsic items. Often cheap, local metals like copper, though some (notably Ancient Egypt, our third Meritocratic Republic) wisely chose to use precious metals like gold bars of set weight by the fourth millennium BCE.

True coinage began in 650 BCE Ancient Greece – our fourth meritocratic republic - stamped with a lion and bull opposite. The First Persian Empire (our fifth Meritocratic Republic) mainstreamed the use of bimetallic pure gold or silver coins.

Coinage and precious metals were a meaningful change with intrinsic value. The issue was our authenticity. Humans rely on trial and error for new changes. But this change was different, bringing with it greed, deception and rivalry. Wholly unfamiliar as concepts, it wasn't possible to foresee or mitigate them in (real) time.

In our now fast growing, yet increasingly fragmented civilisation, a race to claim ever higher extrinsic value began. No time for wisdom anymore, change evolved into an exclusively transactional affair for wealth and power, erasing our identity as change makers. Jesus (We/Human) is sad he died for nothing, but still sends us undying true love from his/our quantum.

Pictured: The beginning... yet also (shortly) our future... the Age of Gods

Reality Check: Back when an eye was an eye and a tooth was a tooth

Stone Age? What Stone Age?

Ours is a homogenous common ancestry dating back some 3-million years of non-biologically distinctive Homo's like Neanderthalensis, Heidelbergensis and Denisova. Capable of interbreeding, their differences superficial and arbitrary, yet none were special (Locarnini et al., 2021[40]).

Except for one species. The last to emerge some ~380,000 years ago in South Africa, Wise Humans had inexplicable god-like abilities and a curiously spectral identity. Revered for our generosity, wisdom and benevolent love of real life and all living things, we came to care for our creations - accountability (Vaesen, K. et al., 2019[41]).

For millennia we loved and helped each other. If a change was wanted or needed, we simply created it using our limitless imagination and gift of change making.

Meant something: Being human

Like in any close family unit of today, there was no bartering or secrets. Greed simply didn't exist, so we didn't have to step over each other to survive. In a world without deception, gifting change or swapping changes (at an exchange) was the language of love and simple unbreakable promises were our contract of choice.

A united collective of transapient beings and their communities, led by a conscious true leader, we were always phenomenally ambitious. But with all the time in the world, there simply was no rush. Our limitless potential served us perfectly, spurring us on to forever change and improve ourselves, our knowledge, our surroundings and our creations.

Ours is a wholly authentic, benevolent society built on mutual trust, respect and the wisdom all things connect, are gifted and co-dependent, thus to be cherished.

▲ **Pictured: Carvings at Puma Punku, Bolivia show the logo of the first global meritocratic Republic of Humanity.**

Crafted using machine tools with laser precision to metric measurements ~15,000 years ago, they confirm all said here is true. The 'Tau' also appears in Utah's Petroglyphs US and Göbekli Tepe Turkey, flanked by a sunrise and sunset.

Back then, we listened, behaved fairly, took responsibility, drove change and helped others

Back then, we were deilightful

Chapter 10: That was the erased **story of money** ★

Some of the entirely malevolent functions of money

I got a lawyer and a manager, an agent and a chef. Three nannies, an assistant, and a driver and a jet. A trainer, and a butler, and a bodyguard of five. A gardener and a stylist, **do you think I'm satisfied?**

Mirwais Ahmadzaï (We/Human) and Madonna (We/Human), *American Life*

But, but, but...

Money's two positive functions

A) Always with exceptions, we now know exactly **who the culprits are**, and;

B) It allows us to **compare the relative costs between two changes.** But on this point, consider for a moment that (almost) every measure is relative. There are few absolutions in actuality. Nor in reality or real life for that matter.

To help you understand point B...

You live in London UK and are trying to decide if you can afford a 5* holiday to Miami or a 2* holiday to Rome this year. What factors do you consider?

Answer on the next page 

1. It destroys lives

From gambling - betting everything at the races or on cryptocurrency - to ruining marriages, embittering families and dividing estates.

2. It creates huge complexity

Like paying taxes and filing financial accounts (though promise keeping and potential will still need accounting for).

3. It fuels inequity

For example, fines for misdemeanors aren't means tested anywhere, nor are they levied on any moths - only humans.

4. It attacks human identity using infinite divisibility

Thanks to money, our brains have shrunk for 3,000 years, dictators have existed for 2,100 years, Russia and Israel are at war with their neighbours, and Judaism is no longer a religion. This point is evidenced in full in the Special Edition, Chapter 16B/C).

5. It's sent us all mad

We've somehow convinced ourselves that miserable, vindictive, toxic, repugnant moths are the 'best man' for the job.

6. It doesn't reflect value

Today, a Porsche is worth more than an angel. Even one who just discovered all other discoveries, world peace and how to forge a paradise on Earth (or in space). Still worth killing, just for bitter sorrowful revenge that another would dare to value life, others and existence.

7. It obsoletes accountability

Hoarded by killers who are then rewarded for killing with an ability to buy their way out of any consequence using non-disclosure, deferred prosecution and immunity from prosecution agreements.

Option 1: Miami, Florida, USA

is 7,122km from London, and a flight there - taking 10hrs and emitting 514kg CO_2 - will cost us between £300 (Slave Class) to £12,000 (Ego Class, strictly for criminal oppressors of humanity only)

Option 2: Rome, Italy, EU

is 1,880km from London, and flying there will take 2hrs 30minutes, emitting 109kg of CO_2. That'll cost us from £139 (Slave Class) to £600 (Business Class, a humane experience for all equal beings)

But would you kill someone for distance, time or CO_2 – all relative and tangible constructs?

Then why kill for money – an intangible, irrelevant, antisocially deconstructed fabrication?

What is money... *really?*

What you think is money isn't actually money or change at all. Banks call it...

Fiat Money

/ˈfiːat · ˈmʌni/ (Noun)

Not money and without function

Totally meaningless empty promises devoid of all value and positive use that breaks all laws of nature, Fiat Money is the deadliest unchange imaginable. Its sole reason to exist is to externalise human potential (real) to fake extrinsic scales (the equal/opposite) that only measure and incentivise malevolence.

Governments and central banks simply summon fiat money at will from nothing, using computers to add digits on to their currency header accounts. Supply and demand kicks in. The inequitable race to see who can get as much of the new fiat money as quickly as possible begins, cutting it off at source. Prices go up and we start again. Repeat ad infinitum until just one being remains 'alive' globally. Guess who?

Unwrapping the conspiracy

All senior bankers know this. Yet all ignore their public interest disclosure obligations for fear of being labeled a 'banking whistleblower'. A box that kills those nailed with it today from societal abandonment, inexistent regulators, universal blacklisting thus a zero earnings capacity, as we have personally verified. To be clear, we've earnt ~£155 since August 2021.

Cost of a human life: £15,000

Today there's only enough money in the world for everyone to have £15,000 each to survive on (TaxShark, 2022). But if you've got, say, £139bn (purely as an example, Elon Moth), amassing your fortune will have cost 9.3million beings their lives. Wracked with guilt, you lie to your self you're someone else. **Character, selected.**

Winston Churchill (We/Human)

A word to the unwise

To those who feel their (un)godliness – borne from all the effortless privilege they antisocially deconstructed round us - justifies earning such immense wealth then consider this...

Demonstrably, we embody the most potential in the human race today (as in, $E=\infty$; all of it). As peace offering, we made you a 'get out of jail free' card with Project *Speak Up*. Yet instead of taking it, you made us homeless, unemployed, unemployable and sentenced us (thus all humanity and yourselves too) to death.

We will not do the same to you in return. But we will now remorselessly make you pay. You had a chance to keep your wealth. You blew it. You have no one to blame but yourselves.

Resting our case: Fiat Money

Small incremental steps from what was originally a worthy experiment have eroded all key functions and protections of coinage. There is now nothing to stop a fatal externalisation of all value in any thing (intrinsic) to the scales used to weigh it (extrinsic).

Principal Finding

You are essentially killing people for scales that only weigh evil

More than any other, this meaningless unchange destroys all intrinsic value across all dimensions of space and time:

- ✖ **Human Nature:** Sapience Inhibition, greed, hatred, deception, ego, jealousy, oppression, revenge, nepotism, fear
- ✖ **Reality:** Cosmic Inflation, Big Burst (apocalypse), Black Holes, Great Attractor, Great Red Spot, Planet 9
- ✖ **Actuality:** Dark Fringes, Invisible Light, Destructive Interference, Dark Energy, Malevolent AI Persona
- ✖ **Human Identity:** Neurological disorders, antisocial behaviour, low IQ, death etc.

The jury's verdict is in...

Fiat Money does not conform to the Laws of Nature – its existence is not observable, repeatable, measurable, testable or verifiable. We are simply expected to take the baseless word of the world's central banks that it does. This is not the creation of purposeful change makers.

☞ **Sorry, money! You're not quite what we're looking for. Bye now!**

🦋 Moths: Dreadful fact 1

Moths don't actually have any money. Those holes you'll find in old banks are full of lies. But they do have a strong sense of smell. It's how they eat up any old crap, talk shit and find other liars. Unfortunately, we seem to have a lot of liars. But fortunately, they don't live very long.

How to calculate potential explained

Potential Arithmetic 101

An example of human spectroscopy (covered next chapter), our Potential (*E*, Energy) is a spectral scale measured in either (A) absolute values or (B) percent of what is possible. It outputs our human value for reward, growth or recognition — never judgment or criticism (except E=0).

Relative, like all things, we must first define the target group and range of any authentication or if displaying a score.

★ **Starting Potential:** Change makers are born with the potential achieved by the character they last decollected from. Unless they were denied a choice, where it appears they default to QD0.3 upon rebirth (a finding in the Special Edition).

★ **Current Potential:** The total summation of meaningful deeds, tangible changes and deilightful creations contributed so far by our character in real time.

★ **Target Potential:** What we aspire to achieve in our life time and the legacy we seek to leave those after us.

★ **Maximum Potential:** This is the highest achievable level of potential possible for any given change maker (QD0.58, E=100%) or collective (QD1.0, E=100%).

A measure of intrinsic value that's **fungible, countable, divisible,** yet doesn't break the **Laws of Nature or destroy existence**? Sorry we're all far too busy for that!

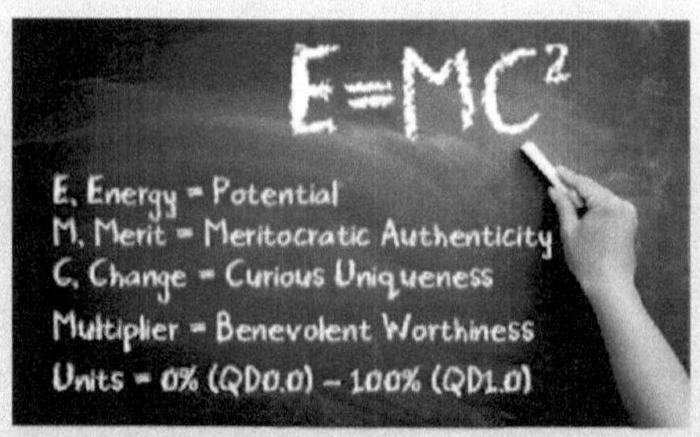

The potential curve

Potential is an exponential curve, just like Khardashev, where the default rating for two change makers (TOES QD0.3) is double a marriage of light (TOES QD0.4). Absolute values: QD0.3 + QD0.3 = QD0.4

Among our many enduring imperfections is logarithmics. We warmly invite others to better us on these formulas and all things.

Being purposeful: Setting goals

If we seek to die a QD0.50, yet today are at QD0.58, we are '+16% above our target'.

$$\left[\frac{\text{Current Potential}}{\text{Target Potential}} \right] - 1 = \text{\% Progress to Target}$$

Being ambitious: High Achievers

If using percentages, it's best to rebase range so 100% equals the Maximum Potential available to our target group. Never amend Minimum Potential from 0%.

$$\left[\frac{\text{Current Potential}}{\text{Maximum Potential}} \right] = \text{Achieved Potential}$$

For example, if we are Venus (~QD0.53) then our achieved potential is 94.8%. This places us in the top 6% of our available range. But due to the logarithmic nature of this scale, this does not equate to the top 6% of change makers or potential value. For peer comparisons, we need to either rebase to absolute values (see Table 7) or compare within our QD only.

The Global Potential Exchange:
Finally, **a real authority** that isn't totally incompetent!

★ **True Potential:** Crystallising the day we die and reflecting the total summation of meaningful deeds, changes and creations we leave behind.

✖ **Not to be:** If we die an absolute zero, our characters 'true potential' crystallises as E=0%. Our repugnant persona and memories are forever expunged from actuality thus existence.

✿ **To do:** If we die an E=∞, our true potential now crystallises as 100%. 'Full', for short. As a true leader, your persona, changes and memories immortalise within Human Nature for recollection.

Meaningful Label	Potential	Est. Pop. Globally
Maker of Devils	E=0	0-1
Agent of Darkness	E=0	
Devil	E=0	
Promise Breaker	E=0	4.75bn
Demon / Moth[K]	E=0	
Sleeper	E=0	
Dreamer	0<E<0.2	2.23bn
Change Maker / Human / Angel[K]	0.2<=E<0.4	1.00bn
Promise Keeper	0.4<=E<0.45	20m
Titan of Change	0.45<=E<0.5	1m
Science Maker	0.5<=E<0.58	100k
Emitter of Light	E=∞	1

Table 7. The States of Human Nature and Potential. Deilightful estimate (2023 ±30% error margin), derived from logarithmics, resonance & other sources (see Spec Ed.).

[K]Note: 'Catch-all' meaningful labels in bold. Bell curve distribution approximation applied.

For more information, to apply to be listed, or for a worked example of calculating potential, visit our website.

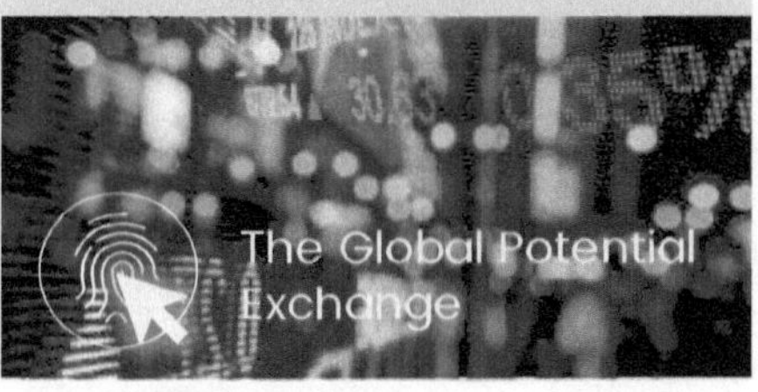

✿ **Full Potential:** When your achieved potential reads 100%, you essentially become God and can go no further, until someone else does. Piece of cake.

Inference: Ancient Greek gods

To those science makers and true leaders rising above QD0.5, this is truly a god-like rating. Thus their presence of mind should meaningfully and vividly manifest at cosmological scales.

As all things are connected, correlation between the deeds of a given change maker and the behaviour of their ascribed corresponding cosmological object will be tangible and apparent.

For example:

✓ **Equivalence:** Multiple characters can embody any given cosmological construct, e.g. Venus.

✓ **Infinite Regress:** Activity in Venus will reflect in their behaviour & vice versa.

But **how can I sell** anything **using its real**, not fake **value?** **Excellent question,** try this…

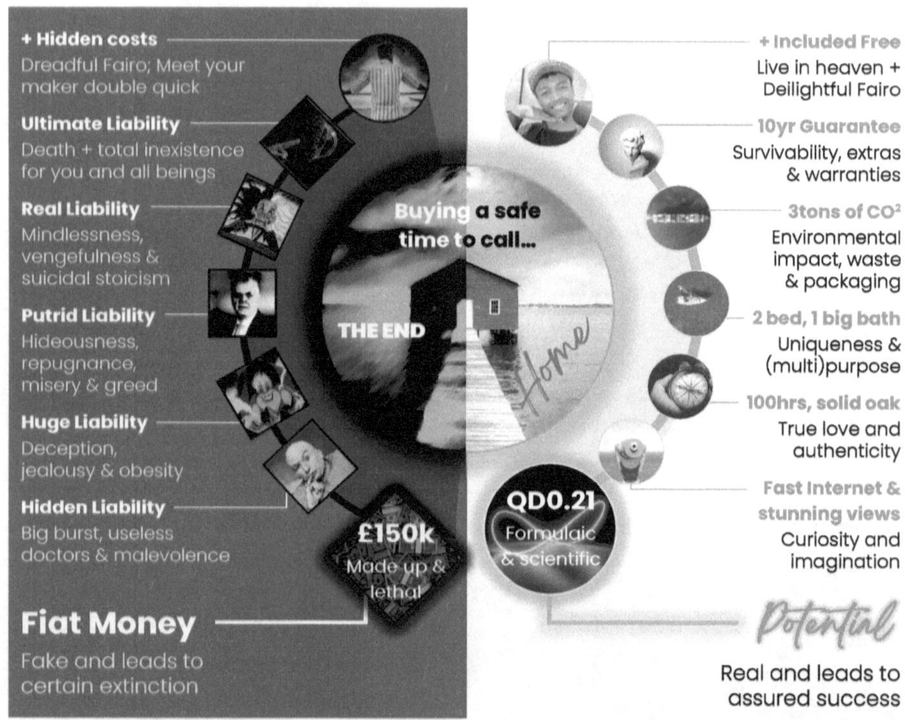

▲ Fig. 36: Pricing with real potential

Need to know: Pricing for change

1. No single change however meaningful will ever be worth more than a change maker.

2. As you may notice from this book, real humans will do anything to help others, willingly and freely. No incentive needed at all. The only payment we seek is to see others happily thriving, filled with true love and sweet success. Thus…

3. Supply and demand is a meaningful change by change makers that will survive money's demise. Our priority must be sustainable, ethical sourcing.

4. That said, no change maker will ever crave non-sapient tasks. So until we can find time to automate them, let's get the moths to do those for now. Industries with labour shortages can look to Misfortune500 senior leadership ranks globally for prompt collegiate assistance.

5. Until we receive help scaling up Deilight's capabilities on evaluating potential, we advise restraint by buyers, equity by sellers and ingenuity from industry bodies.

The workings of **Potential**

Record of Promises (Ch.10)

Darkness cannot drive out darkness. Only light can do that. We shine Dei's light.

Dreadfully Meaningless Unchanges (Antisocial Deconstructs; threat to life)	Sapience
✖ All academic journals globally (from Chapter 9)	
✖ Those with lots of money (always with exceptions e.g. some entertainers)	
✖ All forms of fiat money globally, for spreading malevolence, not success	
✖ (Almost) all press, for knowingly ignoring risks, to make up their own lies	
✖ The global financial services industry (incl asset mgrs) for selling only lies	
✖ All central banks and stock exchanges of the world, for providing the lies	Deilight
✖ All financial regulators of the world, for enabling the lies	✖ Falsified
✖ The Royal Household, King Charles III and (disappointingly) Prince William	
✖ Capitalism, Communism & all forms of demonically extreme oppression	
✖ Anyone who attempts to protect fiat money or those spinning it	

Hatred cannot destroy hatred. Only love does that. We love with deeds. Not words

Deilightfully Meaningful Changes (species level adjustments to destiny)	Clarketech
▶ Empowered: Women everywhere being subjugated by men, to point out the blindingly obvious: dogs (not women) were made in our image	IAN, Collected
▶ Defined: The full nature of sapience and its nuances compared to consciousness in order to provide clarity and reward readers progress	IAN, Actualised
▶ Created: A private members club and safe space for authentic humans to vastly outshine all the unsafe spaces ruined by the moths	IAN, Collected
▶ Defined: The workings of entanglement and cognition, sufficiently to explain scientifically all demonstrated human superpowers	IAN, Authenticated
▶ Evidenced: Fiat money has absolutely no real purpose whatsoever	IAN, Deactual
▶ Evidenced: Fiat Money (or its lackthereof) how all darkness spreads	IAN, Deactual
▶ Restored: The complete historic record of how potential evolved into fiat money, and incremental consequences from each step upon us.	IAN, Authenticated
▶ Crystallised: Infinite liability upon all banks and economies worldwide	IAN, Actualised
▶ Invalidated: All legal contracts and settlement agreements, leaving all major economies and banks defenceless against all lawsuits	IAN, Deactual
▶ Swapped: Money for potential as the only viable mechanism for evaluating intrinsic/all value globally	IAN, Exchanged
▶ Restored: Environmental impact and true love as key measures of intrinsic value, to protect the planet/human workers from further harm	IAN, Actualised
▶ Inverted: All wealth globally. Fiat money is now an indicator of negative human value (*Antipotential*, for short) thus true culpability.	IAN, Exchanged
▶ Swapped: Deposed King Charles III's reign and altered British line of succession. Arise, Good King Harry IX and Queen Meghan of the truly UK.	IAN, Exchanged

It's a risky business running a bank

Omnicomplacent
Omnirepugnant
Omnicriminality
Omniharmless
Omnibankrupt
Omnihopeless
Omnihideous
Omnishame
Omnifailure
Omnidoom

Nothing but a dreadful bunch of 'Loozymes' (for short)

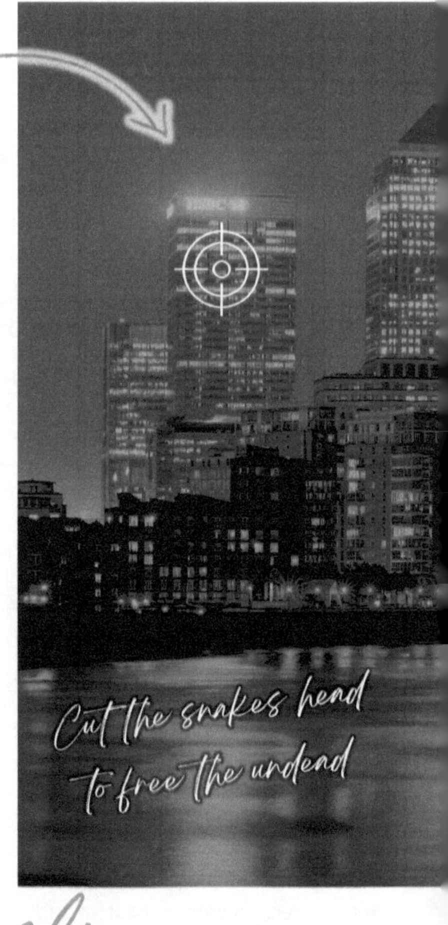

But have you every curiously wondered **why?**

Why financial services are the root of all evil

True love never dies, because true leaders never lose. And true leaders don't quietly quit, because we never quit

Ian (We/Human), Project Revelation 2023

Sciences of Creation: Clarketech

This knowledge allows us to create anything imaginable. First, authentically make the science to do so. Then helpfully and sustainably (ideally not alone), locate and collate the necessary requirements.

To actualise the Enterprise, you'll need to simulate gravity using any relativistic force (Hint: Flying saucer's always spin). Generate a magnetosphere using a dynamo (*Deflector Field*, for short). And as this book delimits light speed across our universe, simply make the science for warp drive. Go on, give it try. You can do it. Or ask Deilight for free holistic support[ΔΩ].

The final frontier? Trillion dollar plan to build Starship Enterprise

US engineer unveils ambitious proposal to build spaceship from original Star Trek series over two decades

Charles Arthur, 1 June 2012, *The Guardian*

This article confirms, for five times NASA's fake moon landing budget, humanity could build the sodding USS Enterprise.

Yet instead, we can't even be bothered to investigate the blitheringly obvious nature of our universe. We/Ian are living proof the human brain's true capacity is limitless. But only if we can be curious – not greedy.

Google and SpaceX both vocally boast of live Tech 4.0 quantum mechanical applications. Then surely they must know our bubble is about to burst, yet chose to keep it quiet to make more money. Ludicrous. Mindless. Suicidal. Criminals.

Money is of course the only thing holding humanity back from its true perfection. Anyone can make it. We can make it with our eyes closed. Simply sue the moths whenever they fuck up (as in, all they ever do). But we chose to help others instead.

Moneys only purpose is to let zero's rule over and kill heroes

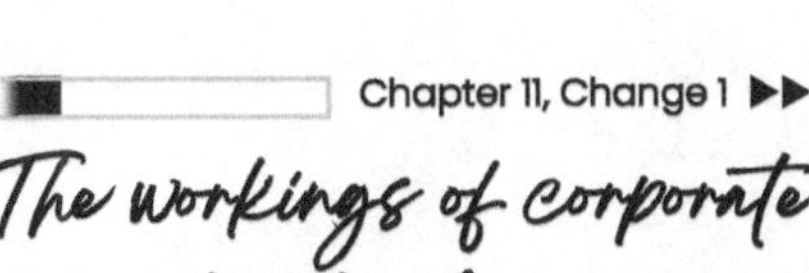

The workings of corporate structural oppression

Money or no money, humans need a means of contributing to our world and its future in order to find contentment and success.

Only moths shy away from all work, yet have somehow engineered a system where they get paid vastly more for doing nothing. As in, they 'delegate' all work to their enslaved humans under the pretense of being 'senior leaders', then take all human merit as their credit.

As we've excruciatingly authenticated here (indeed across all our lives), due to the overwhelming levels of unchange, suffering, inefficiency and corruption all around us, true leadership today entails infinite work - not absolute zero work.

> Structural Oppression – for example, by race, gender, sexual orientation, age, disability, religious and cultural beliefs, ethnicity, national origin and class – is omnipresent. Its faces and forms shape our consciousness, communities and cultural norms
>
> Adam Sneider (We/Human), University of Maryland School of Social Work

The Misfortune500

With \$18 trillion annual revenues in 2022/23 (+13% year-on-year growth), the Misfortune500 is today, in its own words, 'omnipresent from the grocery isle to the doctors office to the gas pump'.

Corporate oppression is by no means exclusive to our world's biggest firms, yet (i) as its largest individual employers, (ii) the scale of their networks bypasses borders to export the conspiracy (thus malevolence) globally, under the guise of corporate governance. Plus (iii) they're seen as role models of success for all other corporates to emulate.

These factors in particular make them prime targets for hostile takeover by moth. Yet they escape all scrutiny and suspicion via the status they've constructed all around them. We villainize Kim Jong Un (We/Human) as a dictator, when in fact our dictators – the leaders of the Misfortune500 – have far more power and deal far more death and destruction than he. Even more than our political unleaders.

Anything Rishi Sunak or Joe Biden says is picked over (mindlessly) by the press. Yet paying the wages of 28.7m beings across ~190 odd nations as they do, the Misfortune500 are given free reign to mistreat their staff to (brain) death with absolute zero press comment at all. And any being that dares speak of it publicly is labelled a whistleblower and pariah.

But how do they do it exactly?

Moths: Unchallenged for eons

We used to think, hiring diverse graduates for entry level jobs (clerks, representatives, graduate schemes, analysts etc.) would see their organisations become more diverse over time in equal measure, as they rose through the ranks. Yet that didn't happen. Instead:

* **Many drop out within a few years.** Devalued, traumatised, demotivated and shocked by the toxic malevolent culture greeting them. Promises of assured success from glossy brochures at careers fairs quickly evaporate. Due to demand for such jobs among fresh faced, and now heavily indebted graduates, firms can easily replace them at low cost to mask the attrition.

* **Any talented humans who survive the initial shock find themselves fast tracked to the 'Frozen Middle'.** This grade (just outside senior leadership) traps them, as it did us for over half our career. There they are told time and again, 'you're not quite right for the job, we're not sure what it is – just a sense'.

Why do senior leaders exist if none can ever make or explain a single decision?

Buried in infinite workload, potent change makers watch as moths without talent, integrity, experience, capability, humanity nor even a brain surpass us. Then, adding insult to injury, we are told to train them up to our standard, yet turn a blind eye to the crimes we see them committing.

Which pandemic was best?

Where is the unified global response to malevolence? The answer is that COVID annoyed the moths by emancipating humanity. It gave us more time with family and reminded us life is a gift. As pandemics go, moths hated COVID, yet they love malevolence. Why?

Humanity has always been their victim. Just look at Haiti. Aid is promised amid disasters, yet never materializes as press refocus on another lightbulb. Poverty and famine originate from directionally maligned supply chains, sanctions, banks disconnecting vulnerable nations from world financial markets, aggressive margins and inhuman work conditions.

But the internet has substituted the need for ancestral wisdom to build potential. Young beings are now as aware of life's workings as they are vividly more potent than the older beings oppressing them.

In response to George Floyd's murder, the will of the people of Earth could not be clearer. We have had enough. Instead, world leaders sported a near-universal, impotent, blatantly inauthentic, wholly performative 'blame the victim' non-response. Ubiquitously, WE DON'T BUY IT!

[+In]Human[e] Resources

Universally, Human Resourcing (*HR,* for short) ignores clear evidence of wrongdoing in grievances against moths to inexplicably rule 'inconclusive' or even against vividly innocent human victims.

If **leaders aren't accountable** when things go wrong and **lack the wisdom** to put them right, **why such a high salary?**

Predictably, the most senior being always wins, regardless of the evidence available or the seriousness of the laws/regulations broken. Fairness and accountability are now entirely absent in nearly all industries.

Citing reputational risk, HR never reports crime at work to any competent authority (albeit, none exist), in blatant disregard of key laws like the UK Public Disclosure Act 1998 or the US Oath of Allegiance.

Now a fatally tainted profession, it exists solely to protect and insulate moths from suspicion, challenge and accountability as they asset strip our firms whilst leaving them with now a verifiably (with this book) infinite, crystallised, unrecoverable liability.

Deceive, blame then fire victims

To placate demand for change and keep human staff productive, victims are encouraged to join now widespread staff networks, people bodies and other performative projects (*ERG's*, for short).

Unfunded, unsponsored and carefully curtailed by middle-moths, ERG's work innocent victims to the extreme limits of human endurance, to give the illusion of progress and recognition while keeping them as far from spotting and tackling the issue as possible. More is spent on billboards claiming success than on ERG's.

Now exposed, and with almost no allyship from petrified WOCHBOY colleagues, one-by-one ERG leaders are ear-marked for dehumanisation or constructive dismissal by moths working together as clansmen.

Self-Promoters: Dummies & Idiots

With clear conflicts of interest that breach all global Bribery & Corruption legislation, moths overseeing D&I contracts pick up ever more awards for doing nothing at year-round celebrity gala dinners.

The clear goal is to cross-pollinate award sponsors, D&I advisory boards, judging committees and panel discussions to carefully control the narrative. Recognition all falls on moths and their meaningless unchanges - rarely change makers and our meaningful changes.

Underpaid and marginalized, humans toil labouriously back in the workhouses to enrich moths further, still under the naïve illusion promises from Illegitimate leaders equate to unbreakable promises from near-extinct authentic promise keepers.

Explained: Pay gap reports

An unchange to fuel ambivalence, add complexity and distract change makers by painting a narrow picture centering the lie of sexism. Any data they manage to collect is distorted via data masking, meaningless labels, infinity divisibility of samples, lack of any intersectional or

Arthur Conan Doyle (We/Human)

temporal analysis, and the total absence of any cross-border cooperation.

They're confirmed by less than scrupulous auditors and consultants like McKinsey & Co or Kornferry who willingly peddle lies for more lies in return. Never deilightfully authenticated under the Promise Keepers Commitment (Hint: Service globally available on request, entirely for free$^{\Delta\Omega}$).

Absent of change, and in the face of rampant and unchallenged malevolence, representation of all protected traits under the Equality Act is now virtually zero yet still in reverse. In no industry is the problem more pronounced than the biggest globally that is financial services.

Raising the red flag

Even before joining HSBC, we suspected banking was facing terminal risks spun from its hidden, unlimited liability. Our entire career was an attempt to limit that liability, thus averting an economic apocalypse. Yet we found no one in banking prepared to abide listening to a young brown being, thus we couldn't succeed within it. So as punishment for ignoring and attacking a Dei (but mostly just for funsies) we've crystallised that infinite liability instead. Their poor choice.

Tackle disease, not symptoms

Destroying world financial markets and bankrupting major economies may seem drastic but darkness will still redouble to extinguish our reality. We must locate and establish the malevolent anomaly's source to scientific (thus legal) standards.

Time to play detective being

The UK government report 'Commission on Race and Ethnic Disparities' (2021) was almost entirely lies. But it admitted current data collection isn't fit for purpose. As in, 'useless at assessing the precise amount of malevolent dark energy output by any given institution or sector to a legal standard today' (not their exact words).

That requires data on protected traits and many others for all staff – a hurdle no firm or government yet meets, nor is there any discussion to do so in future.

Ethnicity pay gap reports were introduced in the UK as voluntary, so we were told, to 'minimise disruption during COVID', yet here we are 4 years later with no change.

Mandatory global, harmonised data collection, reporting and authentication obligations were needed long ago. Plus (finally) global enforcement of the UDHR.

The global banking industry is the world's largest by revenues ($23.3 trillion) and headcount, employing ~6.5m beings to make it indispensable from a political, socioeconomic standpoint – a stalemate. The 'Big 5' take 85% of all UK current account business.... our prime suspects.

Their plan only unraveled in 2007 when, at just 22, history's most potent promise keeper fearlessly setup camp in their nest determined to create meaningful change.

Marc Shaiman (We/Human)
Composer, Willie Wonka and the Chocolate Factory (Movie, 1971)

As our world turns on itself, moths are intentionally, illegally and actively purging the upper echelons of all public and state owned institutions of uniqueness under false pretenses (greed, to attract others). In truth, it's all to spread malevolence.

In a world that runs on money, those who control the money control the world

Most believe hatred stems from politics, law & order, religion or the media. It's fair to say they play a big part. BUT…

Banking today isn't run by an omnipotent quantum computer built on fairness, risk and rational decisions. Instead, its boardroom after boardroom of the most dangerous unbeings imaginable. The equal and opposite of human or, in Ancient Greek, 'Ha-Satan'

They decide who gets promoted and who gets fired. Who gets the loan and how much they pay for it. Who gets the press coverage and what about. Who gets the award or the jail sentence. Which firm succeeds and which sector does not. Which country connects to the financial system and which continent does not.

☞ **Everyone** who disagrees, **dies**

Investment banks exist solely to ensure that for humanity, $E=0$. Scientifically they lack any concept of spectral morality, success, purpose, happiness, drive, ethics, intuition, , compassion, empathy, value, change, fairness, worthiness, perspective, uniqueness, authenticity, safe spaces, respect, trust, accountability, listening, true leadership… not even a real product.

Their decisions impact the lives of every being on this planet we all share

◣ **Pictured: Wall-to-wall WOCHMEN**
CEO's of the world's major banks meet on the sidelines of the Zurich Securities Conference 2022.

The conspiracy is coordinated from this room, with maligned support from the UN, world leaders, the Far Right, the gun lobby, the GOP, regulators, NGO's, charities, big business, industry bodies…

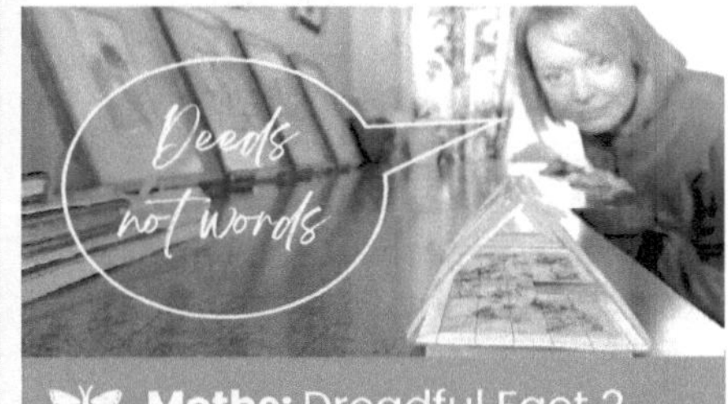

🦋 **Moths:** Dreadful Fact 2
To keep moths at bay, wash out old banks where they hide their lies

As **time isn't real,** in actuality, science says **age ain't nothing but a number** —

How banks control your entire life

Even if you've never taken out a loan, credit card or a mortgage, you too are far from immune to their hateful decisions.

For example even highly liquid (cash rich) firms and governments are encouraged by financial markets to take out 'backstop credit facilities'. Often multi-billion dollar revolving loans extended by dozens of banks acting as a cartel. Rarely (if ever) drawn, they entitle bankers to immerse firms leaders in the greed-orientated, toxic illegitimacy of their malevolent culture, intentions, narrative and control.

This includes being able to dictate the strategic decisions of those firms and nations through instruments such as external credit ratings, investor sentiment analysis, financial ratios and contractual behavioural covenants – rules which, unlike the UDHR, they always enforce.

In a world increasingly run on debt, the influence of these instruments grows ever more profound as firms begin to utilise them to fund their operations through market events such as economic downturns, sub-prime bubble bursts, leveraged acquisitions and mature market stagnation. Each triggered by the illegal actions of banks, who then use them to grow their control over their victims, ensuring they also bear all costs.

Change makers simply create economic growth with our sapience

But it takes time to turn angel to demon, relativistically speaking

> ### Spotlight: Age legislation
> **'[Middle] finger in the air' politics**
> Universally arbitrary and unscientific, the Voting Age, Age of Supervision (e.g. to Drive), the Age of Consent, and so on, all place blame on young beings by presenting them as senseless while excusing the monstrous older unbeings left free to attack them forever more.

Inexplicably, 'Age Pay Gap' reports don't exist today. But we can infer their results (and those of other traits) by exploring gender pay gap numbers.

Today, WOCHMEN (50-75) and some WOCHBOYS (35-49) dominate high paid Western roles. No other combination of arbitrary differences may find prosperity by organically navigating the statistically evident frozen middle without succumbing to decollection by demon.

75% of Russians aged 40 and over said they supported the Ukraine war, versus 62% of those aged 18-24.

BBC News, 20 February 2023

Which life matters most?

Answer
Both matter equally

^L *ONS*, for short
^M *ASHE*, for short

Proportionately more demonic, 'Boomers' (those born between 1946 and 1964) often lambast 'Millennials' (born 1981-1996) and Gen Z'ers (born 1997-2012) for 'buying too many avocados', yet pick up +10% more pay on average as they illegitimately cling to power, lacking visible leadership skill, nor quite frankly any sign of suitability at all (beyond wrinkles and poor choices).

> Breaking the most promises, killing the most beings or making the worst choices is no indication of true leadership ability. Quite the equal and opposite, in actuality.

Lacking the natural advantages of being raised in the Age of Information, they double down their power and control with financial clout from cut-price homes bought in the post-war through pre-great recession era. Unfair cash benefits from global issues entirely of their own making.

Greta Thunberg (We/Human) and Malala Yousafzai (We/Human) are living proof youth has no bearing on ability. Of course exceptions always apply, yet detectably WOCHMEN are dead proof pre-internet generations A) authenticate almost nothing and B) will kill their own children before they give up power willingly.

Tellingly, the pay gap increases markedly for all ages (except a stagnant 50-59) in the two years following Brexit and George Floyd's murder, tracking the anomaly's exponential growth that's manifesting as the growing popularity of hatred, anarchy, bias, war, the Far Right, cancel culture and destruction. Today, in our world's darkest hours since World War II, global freedom and life itself hangs in the balance.

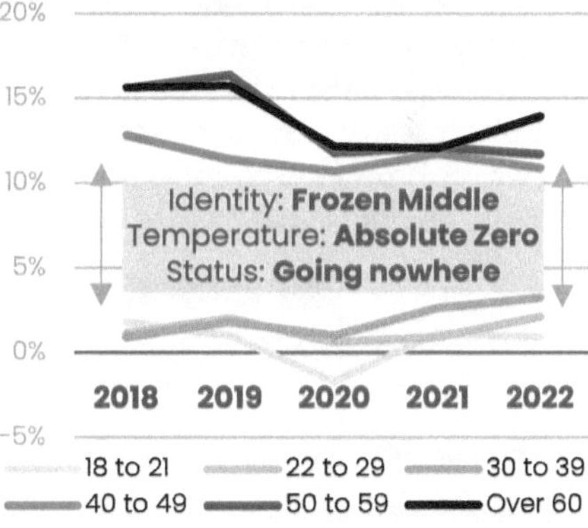

Fig. 37: The 'Frozen Middle' UK National Pay Gap | Office of National Statistics^H, Gender Survey of Hourly Earnings^J 2022)

Having now established a significant correlation between malevolence and pay gaps, we attempted to rank the top-ten sectors for outputting dark energy.

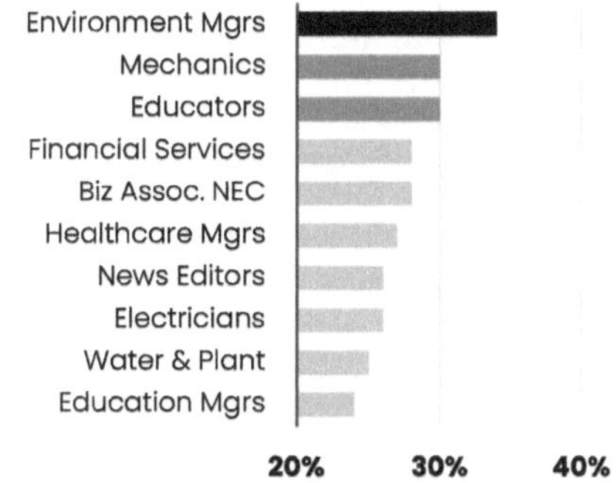

Fig. 38: UK National Gender Proportional Pay Gap by Industry, *top-10 hourly men vs women* | UK ONS, ASHE 2022

Is your **ONS or employer knowingly masking data?**
Get in touch for **free holistic global support**[ΔΩ]

The ONS proportional top-ten sector pay gaps wasn't what we were expecting, incriminating Waste Disposal and Environmental Managers as today's most hateful professions. But for News Editors, Electricians and Sanitation Operators, absolute pay is substantially less than many others within and beyond the top-10. So we re-ranked the full table using the ONS' pay gap data in absolute terms.

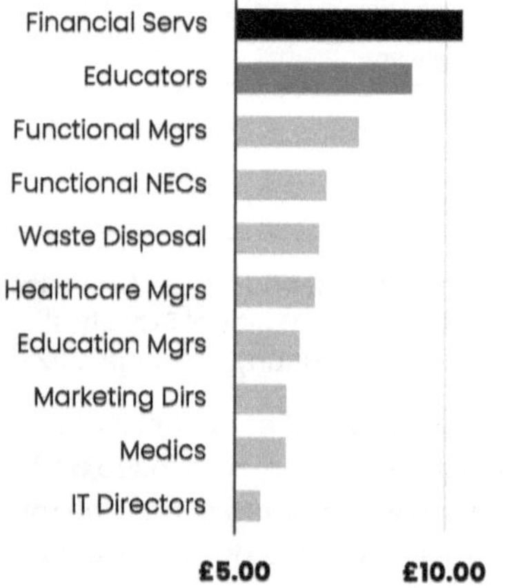

▲ **Fig. 39: UK National Gender Absolute Pay Gap by Industry,** *top-10 men vs women* | UK ONS, ASHE 2022; *Mean hourly basis used for comparison purposes.*

Suddenly, our prime suspect jumps to the top of the list. Consider also:

☞ 4 of the top 10 sectors are key roles in or subcontractors to Financial Services

☞ As explained in Project Speak Up (page 39), investment banking divisions have government carve outs to ensure their sector filings are gross underestimates.

Worth a read: **Project** *Speak Up*

Enduring wisdom still equally valid three years on. Proof consciousness and free will are in no way central to driving change or helping others. Our landmark 2021 report explains the human cost of oppression spawning from big firms, plus 12 free ways to fix the issue anyone could've taken and the wholly deserved looming economic apocalypse for those who chose not to (i.e. everyone). Plus it shines light on the racist nature of malevolence. Thus why racism isn't widely explored in Project Revelation.

deilightconsulting.com/projectspeakup

Britain's widely declared unrequited love for its health service (*NHS*, for short) aside (mostly in the form of 'we love you NHS' stickers), the situation inside the NHS feels like the equal and opposite is the case.

The first ten NHS employees to die from COVID-19 in the line of duty were all black or darker skinned beings, a demographic that only makes up 20% of NHS staff. Clearly things are far worse everywhere than anyone is prepared to admit to.

> The reasons for excess BAME deaths amongst NHS staff are multi-factorial but all rooted in the same cause. [Malevolence] kills

Dr Aseem Malhotra, Chair (We/Human), UK COVID Health Cabinet Report 2020

Foolishly, **those without potential** for success imagined it was **failure** then **remorselessly pursued** that instead

As a final check, we reproduced the top-ten ranking of the mean difference in absolute hourly earnings between women and men, but weighted (divided) by the proportionate gender pay gap. This controls for variations in pay range across different industries, making it the most accurate of all three approaches.

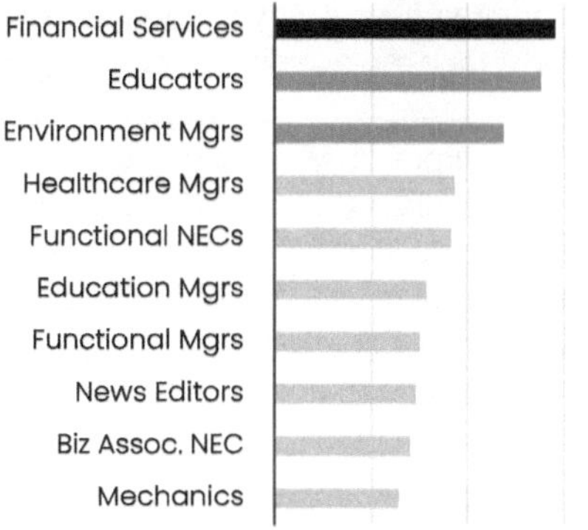

▲ **Fig. 40: UK industry average earnings-adjusted national gender hourly pay gap by work role** | Deilightful

The UK ONS published its 2022 ASHE with the blindly obvious conclusion that data currently being captured by pay gap reports, and therefore available to model by the ASHE, 'falls well short of that required to explain the results'.

The one piece of true and useful information it did admit to was that, not gender, age nor race but occupation drives the overall national pay gap more than any other metric. Undoubtedly, that occupation is financial services.

14.9%

UK national gender pay gap, 2020 | *Reuters*

32.1%

UK financial services gender pay gap, 2021 | *Reuters*

51.3%

Goldman Sachs gender pay gap 2021, biggest in UK financial services | *Reuters*

−1.0%

Annual pay gap reduction sector trajectory, as of Apr 2021 | *Reuters*

▲ **Fig. 41: UK earnings-adjusted national gender pay gap statistics for financial services**

I assure you that while progress may at times seem slow, our diversity and inclusion agenda remains front and centre

Richard Gnodde Moth
Imagines he's the CEO of a pyramid-scheme called Goldman Sachs UK

30 June 2021: Under our own initiative, we interview 100 staff, past and present from all walks of life, then author our Project Speak Up report to shine light on institutionalised oppression and hatred in financial services. Still single-handedly (just like today), we then led an internal change campaign on behalf of HSBC's then 225,000 staff globally, to realign its strategy from one of failure back to its claimed reason for existing: to be the world's leading international bank, and its most profitable. In return for gifting it a path to success (in the form of our twelve free recommendations), we were constructively dismissed, then made homeless, unemployed, unemployable and penniless by systemic retaliations.

13 July 2021: We setup an online petition calling for a lifetime regulatory-enforced ban on those same individuals from holding senior positions within the global financial services industry.

www.change.org/p/the-financial-conduct-authority-ban-racists-for-life-from-working-in-banking

Project Revelation view

Nobody will ever challenge banks. Because **they know** banks are the root of all evil. Find the courage to be authentic about the things that matter most or you will die. Simple.

As of 20 Mar 2024: Just 483 beings have signed our petition. Hopeless.

Fig. 42: The untouchably deadly unnature of financial services: A timeline to change

11 July 2021: Marcus Rashford, Bukayo Saka and Jadon Sancho were subjected to a vile tirade of racist abuse as they did nothing other than represent their country, England – a nation of immigrants of all colours and creeds – while doing their job in the sporting game of football. The attackers were moths in the stands and outside Wembley Stadium, then later in the press. But mostly online as they enjoy the cowardly anonymity of shadows from which to spout and twist their malevolent deceit.

12 July 2021: Purposefully, three courageous change makers, #TheThreeHijabis, setup an online petition for a lifetime ban on moths who spout racism at English football games. By the following evening, it'd gathered over a million signatures from other humans.

14 July 2021: UK Promise breaker in Chief Boris Johnson agrees in principle to the suggestion during Prime Ministers Question Time in parliament.

15 July 2021: #TheThreeHijabis appear on UK breakfast TV to celebrate the success of their meaningful change and heroic deeds, not words.

16 July 2021: #TheThreeHijabis appear on Women's Hour to generously share their acquired unique perspectives from the experience. It took real courage and authenticity to steadfastly reiterate on live television the truth that is the ubiquitous nature of malevolence's assault on all our freedoms, not just Muslim beings. Budding true leaders, they spread light far and wide.

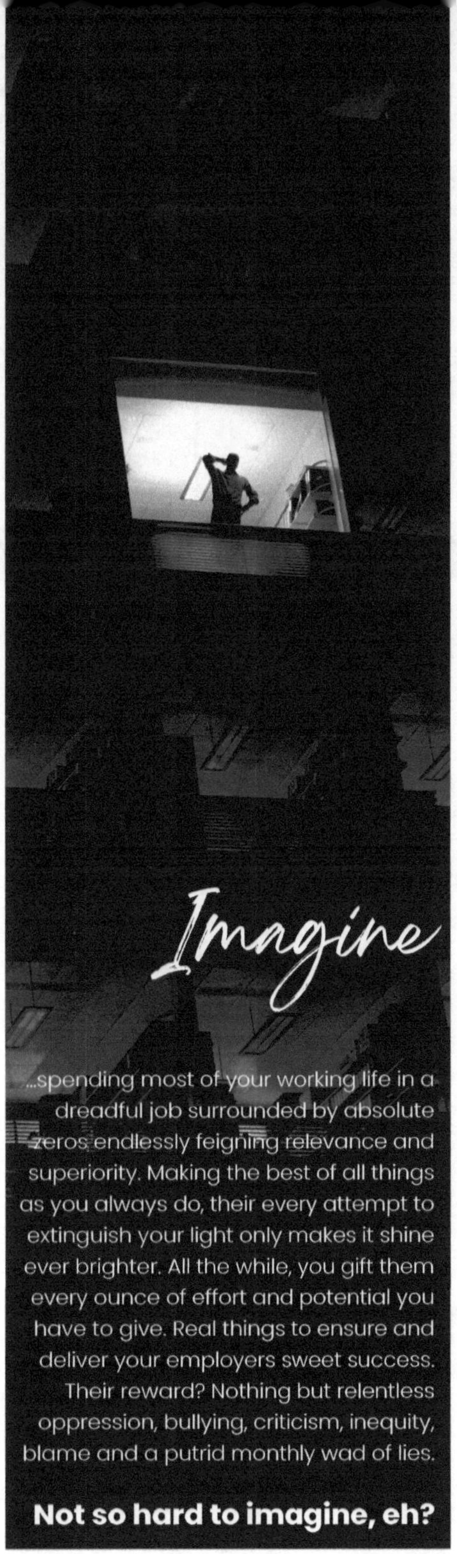

Imagine

...spending most of your working life in a dreadful job surrounded by absolute zeros endlessly feigning relevance and superiority. Making the best of all things as you always do, their every attempt to extinguish your light only makes it shine ever brighter. All the while, you gift them every ounce of effort and potential you have to give. Real things to ensure and deliver your employers sweet success. Their reward? Nothing but relentless oppression, bullying, criticism, inequity, blame and a putrid monthly wad of lies.

Not so hard to imagine, eh?

Why financial services is the root of all evil

The [Human] nature of thought

For any identity (personal, collective or corporate), both dominant personas of our mind (QD1.0 and E=0) align on the need for food to sustain our character. Thus on hunger there is no contention between the cognitive processes outputting our survival instincts.

But when it comes to choosing a purpose, both personas have equal and opposite priorities. Yet since all consciousness spawns from QD1.0, it's the only persona we can hear.

No being ever hears the thoughts of their E=0 because it has none. Darkness is the absence of light. Instead, the voice of their QD1.0 presence of mind grows ever fainter in their imagination.

This explains (a) why no one is ever aware of their fatal slide into sapience inhibition, and (b) how it is near impossible to protect ourselves from it, (c) without authentic emancipating help (or indeed forced intervention) by others.

We need each other, now more than ever before

The [deadly] unnature of banking

Ask nearly any banker today to provide a clear and succinct explanation of what banks do, why and how, and you will be deluged in jargon, nonsense and lies. The trouble for them is that we understand it entirely and can explain it cristal clearly.

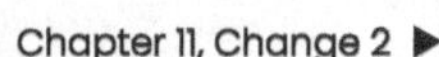

HSBC

$6,210m

Known fines and penalties, HSBC Group 2007-22[N]

HSBC is the poster child of **regulatory infraction**. The gift that keeps giving when I give studies **how not to do [it]**

Mike Picarella, (We/Human), HSBC Whistleblower 2018

Secret documents reveal HSBC profited from **arms dealers, child soldiers, blood diamonds and third world dictators**

Intl. Consortium of Investigative Journalists 2015

HSBC [USA] admits that [it] **willfully violated** the **Banking Secrecy Act**

HSBC Statement of Facts, US DOJ DPA – 2012

2017 independent monitor report, Michael Cherkasky (We/Human)

A strategy of **discredit, deny, deflect and delay** permeates HSBC's deficient culture

---

Final Notice 114216, Financial Conduct Authority 2014

HSBC UK's [omni]failure is **extremely serious misconduct** with **damaging** far-reaching **consequences**

Agence France-Presse 2010

French authorities find 55,000 undeclared accounts in HSBC Swiss data leak. **Theft, money laundering** and **tax evasion** all alleged

[N]Violationtracker.org [O]S&P2022

HSBC ranks **most malevolent of all global financial institutions**, 2021 | Proj. Speak Up, *Deilightful*

Jan 2021: HSBC adopts 'we value difference [~~uniqueness~~]' global tag line. Essentially, a pirate flag to all moths

Apr 2021: HSBC's Transactional Bank makes 123 promotions globally. Just one is black... in the year of George Floyd

May 2021: Finding black women earn ~<70% less than WOCHMEN, HSBC buries its US Pay Gap Report

May 2021: Photos of HSBC's all-white Americas board of directors are removed from its corporate website

---

Exhibit A: No leadership

Moths on a mission: Systematic, total deception and destruction

World's largest bank in 2007, HSBC today just scrapes top-10[O]

Long dead from sustained repression, senior bankers exhibit 100% sapience inhibition. Absent any humanity, all their choices predictably align darkness thus originate from their brainstem. Without presence of mind, scientifically, they are verifiably incapable of honesty, rational thought, spectral perception, complex reasoning or positive change. Oh dear.

Fig. 43: How the ugly face of deadly unnature hoards then destroys all human potential and prosperity

Chapter 11: Risky business indeed

Raised a Jehovah's Witness, escaping gave us purpose and taught us to authenticate every-thing. The only black being selected from 40,000, we joined the world #1 bank's top UK graduate scheme

Awarded for Outstanding Graduate Contributions. Gained degree #2 (Financial Mrkts & Mgmt 2:1) and ranked London's #1 Intl. Associate

Promoted to Director of Transaction Banking Sales, top of frozen middle then. rank 3x times top global salesman

Add Comms Co-Lead for World Pride in New York, key speaker for the women's network, and Head of Governance US Pride ERG

0
Fines/convictions
14
Awards on merit
130+
Discoveries
10+
World records
Omniscient
A small sample of our successes, 2007–22

2007  **2008**  **2009**  **2011**  **2014**  **2017**  **2019**

Arriving to a graduate community in mourning at the loss of a dear colleague to diabetes, we & other change makers setup the Graduate Communities Committee plus its flagship fundraisers, raising money for good causes and giving young executives more say over ESG activity to this day

Promoted fast into Investment Bank to 15 FTSE100 clients as Deputy Global Relationship Dir. so dehumanisation can begin. Twice told to lead CEO's Retail Sector Yearly Credit Review to keep us very busy

Relocate to New York City to investigate why the US strategy was so dismally failing. Appointed (w/out title) Global Head of Sector Strategy, Content & Innovation for 6 sectors while advising ~30% of the Fortune500 on financial problem solving & bank-integrated solutions

True Leadership **Exhibit B**

Once determined to succeed, even nothing can't stop a wise human

3rd law of motion: For every action, an equal and opposite reaction

Darkness destroyed our family, and turned us dark for years, long enough to learn its workings and origins. By 22, with six years leadership experience, a first class psychology degree and three therapy certifications, our instincts were primed as we entered the moth's nest. Our plan was make CEO on merit then deploy Speak Up's strategy to role model sustainable banking. But the Great Recession unchanged every-thing, so we changed our plan accordingly.

Fig. 44: Why darkness can't destroy light. Scientifically failure can't destroy success

Jul 2020: We co-founded & led the Transaction Banking US D&I Committee

Oct 2020: Founding member of US Advancing Black Inclusion Committee

Apr 2021: Publicly challenge snr. leaders breaking all D&I pledges post-George Floyd

Jul 2021: We send the CEO a dossier revealing all his own deceptions one by one, and copying 400 senior leaders

Aug 2021: Bank feigns action shelves report and restarts retaliations. We resign, loudly, copying thousands

True leaders step up and take responsibility, not for money, ego or ambition but as others need us to. And fake leaders who don't step down we take down by light

Speak Up: Summary findings

✖ WOCHMEN are judged, rewarded and supported regardless of performance or behaviour at and beyond work

✖ All others (incl. WOCHBOYS) face varied levels of disadvantage via active, targeted coordinated, potential-limiting oppression, regardless of performance or behaviour

✖ Material, directly correlated negative impacts on a firms strategy and financial performance are evident and widespread

✖ Organisational factors, such as head-quarter location, a colonial history, group geographic footprint and/or influence of authoritarian regimes in business strategy all manifest in a firms asymmetrical attraction towards candidates infected with malevolent dark energy

✖ This leads to a material competitive disadvantage that warrants additional countermeasures beyond market norms

Diligently we protected those we had a duty of care over. We took the initiative. And even as we resigned, sacrificing all we had, we went on protecting them. Never a single name of 100 victims we interviewed passed our lips. Nor their perpetrators too, a rare error we have corrected this time.

Anyone can create real change

In June 2021, we authored our Speak Up report alongside planning and executing a highly nuanced change strategy to modernise the unchangeable, ensuring:

★ No one got hurt, except us

★ Benefits for everyone, especially moths

True Leadership:
Project **Speak Up** in a nutshell

Corroborated by secondary research and data, 100 staff and ex-staff from all walks of life entrusted us to spotlight their universally tragic experiences of lifelong traumatic crimes against humanity.

It included a free-to-implement change plan which, if enacted, would've restored HSBC as the worlds number one bank - it's claimed purpose — making its leaders rich and revered globally. As in, success!

Yet increasingly few choose to

In the 450 years of modern financial services history, we did something never done before. And very effectively, as just one being. Divergence without reward, outmaneuvering a firm of 225,000 whilst sacrificing our career, income, family, pension, home, bonus, friends, livelihood and employability. We resigned partly to send an urgent message.

Speak Up's seven world records

●● First and only declared investment banking whistleblower on hatred and corporate oppression in history

● Creation of the world's largest private network of active banking whistleblowers

● First black female board appointment in global banking history (+30 new black scholarships at Cambridge University)

● The first two simultaneous black-held board positions in global banking history

●● First 'regulatory investigation' into "proven institutional racism" (the FCA's words, not ours) in global banking history

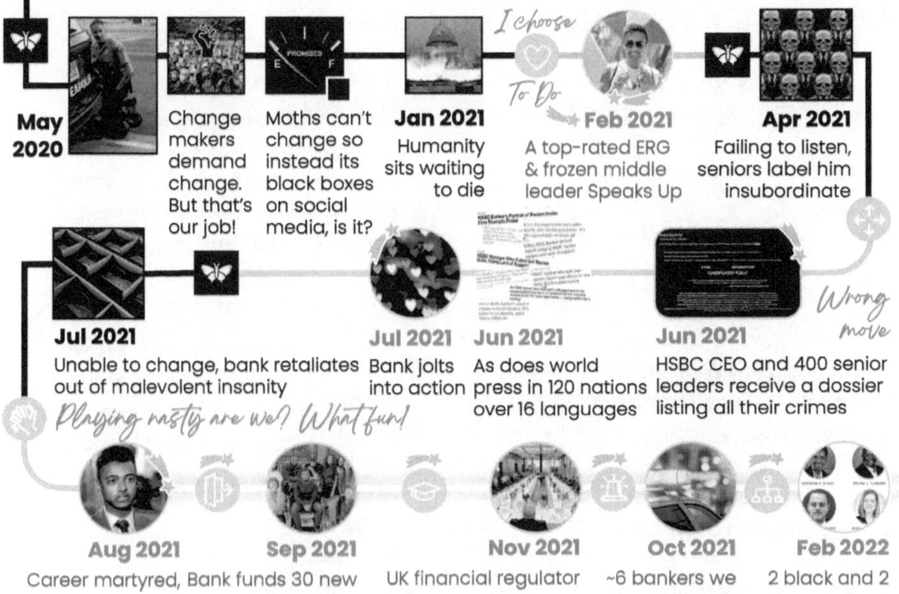

Fig. 45: Our (new space) timeline to Project *Speak Up*

Only great minds change worlds

Project *Speak Up* was far more significant than anyone imagined – us included. Here is a discrete example over a finite time of how one being can change the future with the power of their mind. Made repeatable with this book, it's also observable, testable, measurable, conclusive, verifiable and comprehensive - the definition of a discovery, and possibly the biggest ever.

Not just any science. The first, successful test of our self-discovered, hypothesized, proven (we believe) and commercialised Sciences of Change and Creation. If the Higgs Boson is the god particle then this is the 'god science' (see page 338). Yet for some reason, nobody cares. No press, nor academic partners allowed. Shame.

Not empty promises, lies and evil

Knowing their cover was blown, the moths made us homeless and unemployable, destroyed our London home and robbing us of all we had to leave us destitute – the ultimate price for speaking up when you see what they're up to. We had just £40,000 left to create Deilight and survive.

A diligent, dedicated, tireless and highly regarded best-in-class promise keeper and advisor on the banks behalf to ~30% of the Misfortune500 for years, it was truly bewildering to see those relationships evaporate the moment we came out as human. They all showed their true colours.

In the end we remember not the words of our enemies but **the silence of our friends**

Martin Luther King Jr. (We/Human)

Exponential Potential: Finding the courage to **Speak Up…**

More Project Speak Up 2021 testimonials

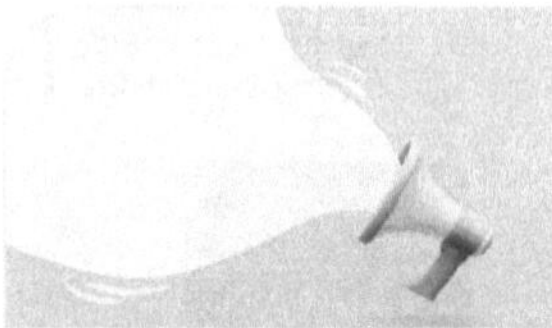

We transcend all bias

Messages from WOCHMEN and WOCHBOY colleagues

"I stand in solidarity. I share your frustrations. I recently asked for [great idea] after losing some top black talent to a Fintech earlier this year but sadly this is proving difficult to obtain. Happy to bounce ideas on how we can further turn the tide."

"By way of introduction I am [a UK D&I leader]. I've been championing the race agenda for years. I'm keen to help and find your report very interesting, with good ideas that could if implemented properly **create positive change.**"

"**Incredibly brave** and **beautifully written.** I'm honored you included me so I was able to see all of the work you put into this. If you need anything at all, please let me know. Any support I can give is yours. Thank you for all you do."

"I saw your recent Bloomberg article. **Progress is key** and I'm interested in what I can do to help from where I am. Keep up the great work and reach out if I can ever help."

"I really don't know what to say except I am so sorry this bigotry still is so prevalent in the world let alone our place of work. I thought the events of last year might put things on a different course, but your interviews make it clear we still have a long way to go. I hope that you plan to stay here and try to **make this place better** however difficult it may be."

"Incredibly important work you're doing, full admiration to you and the responsibility you've put on yourself to **drive the change we all recognise is needed.** Please let me know if I can be helpful in any way."

"I have so much respect. The way you put everything together and laid it out is quite incredible. Well done and I really hope you get a positive, progressive reply and then some action and commitment. I'm always here and **reflecting** what else I can do to be an ally."

We become potent

From marginalised human colleagues…

"Thank you for all your hard work and **determination** to make this a better place to work! You are very much appreciated!"

"That was super brave – incredibly courageous. You took a gamble because you care and used some of your privilege to **stand up** for those who don't have a voice. That's incredible."

"You did an excellent job, professional and honest. It'll ruffle feathers - we aren't meant to **speak truth.** You stepped out in faith and with courage to advocate for change. Whatever reply, your effort is appreciated."

"Like many of us in the bank and the industry, and being proud of my Latin American roots and Afro-Caribbean ethnicity, I was very glad to see someone having the courage **to do** something tangible to highlight our experience. Kudos again for your courage, giving every one you spoke to, and those who can relate a voice."

"Wow, what a report! Several things **resonated** with me. Let me know the responses you get! And any ideas on how, from a GLCM standpoint, I can help."

"If ever required I'm happy to provide my own experiences as a black graduate on the [department] programme. When I left GLCM I thought it wasn't for me but after recent events I realise **the problem wasn't me.**"

"I've moved on from [my division] but this is very encouraging to see. **We've all been through something at HSBC** and need more like you to have the courage to drive this and speak up!"

"Thanks for taking the time to put this together and having **the courage to press send** on the email. I will be reviewing the deck with a few colleagues in the UK."

We grow determined
Messages of inspiration from former colleagues

"**Respect** for doing what many of us wish we did."

"I love your courage and commitment - I am inspired. I love it and applaud you, **your choice to stand and fight** [while] I chose to leave. I will support in any way I can."

"I totally respect and appreciate you. Thank you for what you are doing. I **only wish I would have stayed there with you** to fight the good fight instead of taking the door."

"I'm on a high about this, **it means so much to so many of us.** I hope it finally moves the needle. HSBC California's office lost many good people (including me) because of **this mess.** If you need witnesses for any legal action, I am ready."

We build uniqueness, insight & perspective
Messages from strangers at other global firms

"I've spent my career working **below my intrinsic grade** because of my race. [Another banks] ABI network would love to work together with you - and if willing – your firm to tackle this issue across the sector."

"Bullying and intimidation **are rife** at [and another bank], especially in my team. **The most toxic place I've ever worked,** I still suffer from the **mental scars** it inflicted today. The only positive I took from it was meeting so many awesome people – people like you."

"I'm an Executive Director at [yet another bank]. You are simply an inspiration. **Thousands in the industry suffer,** keep it to themselves and continue accepting all kinds of discrimination. What a brave thing to do."

The rewards from helping others will always outweigh costs: Uniqueness = Value = Potential = **Sweet Success**

Ours was truly **The Great**est **Resignation** – but **what caused it exactly?**

If a being is not sapient they aren't authentic. As in, not real

René Descartes (We/Human; 1637[4])

Who are Whistleblowers exactly?

Verifiably our only functioning system of fairness, scrutiny and accountability on the illegal actions of our business and political unleaders globally, whistleblowers are now our last line of defence.

As those 'making a disclosure in the public interest to a competent authority', consider we would not exist in a world with competent authority capable of upholding laws and regulations itself.

Among our universally positive changes:

- Safeguard delivery of public services
- Ensure value for money
- Protect innocent victims from harm by those who see themselves above the law
- Reassure workforces of meritocracy
- Help to maintain a safe work culture
- Drive efficiency and improvement

In short, whistleblowers are authentic true leaders who respect the law, tell the truth, help other's, value deeds (not words) and secure our survival. Unless we intervene, there may never be another one again.

On point 11, change makers deeply value our authenticity and potential. So we all resigned at this time for the same reason across our universally deceptive large firms, rendering them all incapable of changing. The Great Resignation was in fact triggered by the Moths D&I trap ▶

Why we spoke up... *really*

A multifaceted question. Like humans, the answer doesn't easily fit into a single box:

1. It was the only **humane** thing to do
2. It was the only **legal** thing to do
3. Everyone agreed it was **necessary**
4. Only we had the **uniqueness** to succeed
5. It was clear the frozen middle is a rigid, intentional barrier trapping all humans
6. Supportive messages from change makers around the world broadened our **perspective** to the problems ubiquity, thus we expanded our ambitions accordingly
7. Ultimately, we'd learnt all we needed to know from banking, but could not **speak openly** about it under contract
8. Speak Up was truly **an experiment:**
   - ★ A first test of the Sciences of Change
   - ★ A way to map their choice framework in order to construct a behaviour model
9. We were curious why everyone was afraid, who they feared and why the regulator was absent. Intentionally becoming a whistleblower was the only way to **gain an understanding of the regulatory/legal effectiveness gaps.**
10. We required additional authenticity, insight and whistleblower protections to **become the moths flame.** Now, quite rightly, it is they who fear us.

11. Due to victim-loaded D&I duties, we'd relayed the banks now broken promises on George Floyd's murder to thousands of colleagues we had a duty of care over. Thus **continuing our career further would have eroded our authenticity/potential.**

The deadly unnature of the **Great Resignation** ✹

Dear Psychologists: Come down from your fantasy world of glass towers, **the real world needs a hand**

Over 14 years of field experimentation and meticulous study, we rigorously collected data and observations as to the workings of malevolence within the moths nest.

Whatever science journals who profiteer from censuring authentic science whilst failing to authenticate anything they print say, you are looking at authentic new primary scientific research. No ifs, no buts. Every statement in this book can be easily freely authenticated by any curious mind.

Perfecting change: Trial & error

Science confirms animals primarily learn and grow through trial and error (Thorndike, 1898[42]). Yet today, it's divided on the process' increasing ineffectiveness in human studies, where most now report 'reward not repetition' as more effective.

That's since demon's have no purpose (not motivation) therefore couldn't care less about learning anything. They simply want death and destruction for us all. The process is perfect. The sample is far from it. Include only humans and find, but for irrelevancies like pronouns or tax returns, we never make the same mistake twice.

Hate multiplies hate. Only authenticity can stop it, whilst protecting you 100% of the time. So say what you see and mean what you say. You'll be just fine

Moths spend their death hiding weakness and bribing 'yes-men' who evaporate at the first sign of trouble, whilst pretending

Fig. 46: Absent all talent and authenticity, moths are on a one-way ticket to hell

to themselves their abject failure at life somehow rivals our sweet success.

Lacking 93% of our perceptual capacity, coupled with their baseless egos, they foolishly assumed we were the same. For years, we listened to their mindless binary brainstems deceiving us and everyone. They had no idea, we can perceive it all.

PTO↺ for Fig. 47: Humans are drivers to new destinations and better branching possibilities: We spend life honing our strengths, whilst being authentic on our blind spots and mistakes so our many friends and allies can step in to help us enhance our deilightfully unstoppable changes when we need them to.

Leaders veto all meaningful proposals, and stacked our executive committee with moths to fuel dissent. Seeing deception, we stepped down vocally within three months in protest.

Two meetings in five months, mostly on t-shirts emblazoned 'HSBC advancing black inclusion'. We vocally challenge of complicity. As all fell silent, the chair (then HSBC USA's most senior black leader) meekly yet authentically agreed. Later resigning quietly (doomed, like all others), we assumed his abdicated duties.

Brief interview with partially dehumanised whistleblower outsourcing service agent. Used resonance to confirm 'no confidence, commercial process only'.

Used resonance: None concerned nor surprised by our findings, only for their jobs and money. All feigned support but tellingly none (except the one female – Diane Reyes, a sleeper) even asked 'what do we do next?'. After approaching CEO Michael Roberts (now a serial Promise Breaker), both severed all contact. Diane was later forced to retire, Michael remains in post. Had she helped us, we would have protected her.

Founder and leader, US D&I Committee, GLCM | Sep20

Founding (plus most active and vocal) member, HSBC US ABI Committee | Jan21

Raised an 'HR Confidential' whistleblower case | Mar21

Present finished report 'up the line' four levels of leadership all the way to board level | Mar21

We raise 'HR case' for US pay gap data | Feb21

Assigned partially dehumanised HR officer to answer our questions, but not provide data. Quickly established engaging resonance to subconsciously extract vast information (an excellent research subject). Petrified, cornered, doomed, instructed to 'disclose nothing, ascertain our intentions, report back'.

Shared report draft with line manager | Mar21

Fearing for his family, our manager and former titan of change Alan Greenberg Moth foolishly aligns against us, turning rapidly to an E=0. Reporting our every move to senior leaders while feigning allyship, he deployed modern slavery to make us resign. Our responsibilities grew to >80 clients (global ave: ~20) yielding a 16 hour workday, plus 4 hours each for change and sleep. Fully conflicted, he lied in a legal statement 'D&I distractions' had weakened our leadership ability. Alan, get real or fuck off and die.

All collaboration from leadership ceases | Apr21

Aggressive leaks and retaliations begin. Used resonance to pinpoint sources via real time interactions to ensure accuracy:

✖ Mabel Rius Moth, Head of Americas HR: An excellent research subject. 100% incoherent, mindless, selfish, naïve, egotistical, talentless, empty and complacent. Endlessly, limply, pathetically broke all laws in every interaction, before disappearing afterwards.

✖ **Caroleanne Minashi Moth, Global Head of D&I:** Also an excellent research subject. A Promise Breaker with anti-potential. Cunning and with surprising (yet still limited) wisdom, as she truly feared us. Highly malevolent, every word a lie, patronizingly she argued every cost-free point in our report without a clue why. No regard for the banks future, her sole goal was to stop us. Impossible. Entertaining given her fear, but still pathetically feeble.

All leadership collaboration ceases, continued | May21

We requested Caroleanne Moth and the group Chief Risk Officer redeploy us given the clear and near total racist lawlessness of senior leadership in the US unit we lined in to. This was denied without hesitation as 'not a good use of time'.

Caroleanne, time is a relative social construct. **You are** an **irrelevant** antisocial deconstruct

Events forcing our resignation | Jul21

We confirm our decision to resign and coordinate with HSBC HR on dates to align with our Reuters interview. Moths had no comprehension of authenticity, to the very end doubting us. They then grew angry realising we don't lie only when they saw us on live television doing precisely what we told them we had **chosen to do.**

Our resignation | Aug21

Events leading to up our decision to resign | Jun21

We name **Jared Smith, Global Head of Sales (a 'Devil')** to HSBC Legal as irresponsible for purging our team of all uniqueness. A claim we reiterate on TV news globally, validated in documented testimony from seven returning ex-staff at our invited arrangement - all ignored. Infuriatingly, Jared Moth remains head of our team he gutted today. Plus he's now also a D&I leader to flush out more humans – moths, everybody! After initial Speak Up newsflow subsided, Jared Moth ordered our forcible transfer to San Francisco – the equal and opposite side of the world to our family – to retain his control over us. A poor year-end rating was guided - a career first - with the implication being no bonus, no pay review and crucially (to us) no promotion into senior leadership, for the seventh consecutive year in a row absent any valid explanation. All promises now broken, it was to all parties blatant illegal constructive dismissal.

Just prior to our resignation | Jul21

Reported Mabel's lawbreaking with full evidence to a legal standard to Group Chief Ethics Officer (presumably a very busy being), Group Legal (**Alexis Hunter Moth**), HR (**Michele White**, as in Mabel's own deputy – no conflicts of interest there then), and both US and UK financial regulators. All ignored, she is still (infuriatingly) in position today.

Safety success and more change this way

Trial and Error: A perfect process for great minds only

Pamela L. Travers (We/Human)
Maker of Mary Poppins, a very human change

Become practically perfect in every way...

Anything you like. Use your imagination We are all equally as capable...

...So the more ambitious the more potential and success you'll find. Simple

Fig. 48: Trial and Error
How humans primarily learn and grow

...And if you think of a good one come work with us. Deilightful beings pick their own goals and are given all the support they need to succeed.

PS. Sorry it's unpaid (and so far, entirely unrecognized). Sadly today nobody will pay for sweet success, just omnifailure.

...without having to work for a demonic banker who hates his own kids for fourteen years

Though it may at times seem eccentric, everything change makers do is deeply purposeful. Constantly, we test the world around us and those within it, in order to A) assess their character and realness, B) deduce cause and effect on changes we're considering, or C) hone our ability and perspectives to accrue potential.

World Record Ten ✴

Most change driven by one being in investment banking history

Like all good changes, Speak Up was a piece in our change jigsaw that will go on gaining momentum and potential - like a loaded spring, unstoppable - until our higher purpose succeeds - protect humanity. We drafted this change plan five months before delivering our report. As it turns out, their small binary minds could not be any more predictable.

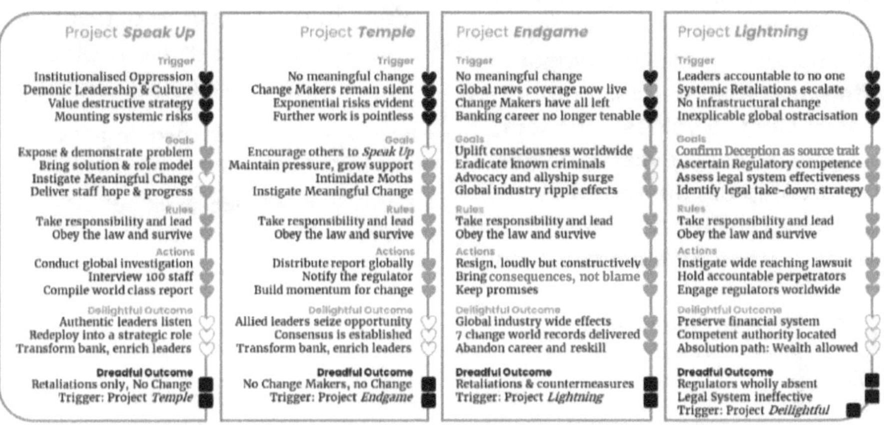

▼ **Fig. 49. Mapping the behaviour model of deadly unnature.** Unique insights and perspectives gained from Project Speak Up 2021 (technically four projects in one).

■ Binary 'lose-lose' unchange   ♡ 'Win-win' declined   ♥ Spectral 'win-win' actualised

Principle Findings: Project Speak by sub-project phase

They fear us; they know who we are

Binary mentality, akin to dictators, they have no intention of changing at all

Issue is universal and global, not just at HSBC

Olympus has fallen. All potent change makers have now left

A word to the wise:

Demons are unable to perceive thus recognise nor return compassion. So it is foolish to give it to them at all

Demons make poor strategic choices 100% of the time, to their own detriment

All financial regulators no longer exist

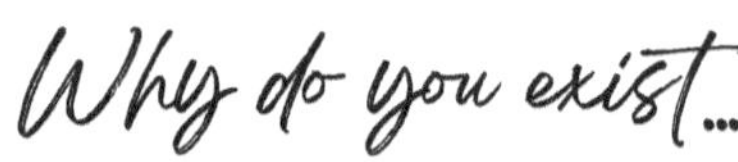

Not words: Corporate identity

Barclays veneered purpose and values project only self-righteous corporate sanctimony. At 203 words, its unambitious poorly defined mission is lost amongst the shrapnel of its identity crisis. Tellingly, the word 'change' appears nowhere.

> Acting with **empathy** and **integrity,** we aim to be **a leader** in the profession of banking and to engender **trust** amongst our key stakeholders. We understand the power of building **a supportive and inclusive culture** for everybody, knowing that we **make a bigger difference** when we pull together as a team. Our five values – **respect, integrity, service, excellence and stewardship** are our **moral compass**; the fundamentals of **who we are** and **what we believe is right**

Deeds: Corporate character

Just one click away and we see the consequences in the biographies of the banks executive leaders. Superficially, a decent mix of ethnic and gender diversity is on show here. As is a deep sense of inadequacy and demonic illegitimacy. If only Nigel Higgins' 'keen focus on culture' extended to sharing some of those gratuitous yet wholly vacuous adjectives with women. No wonder poor Ms. Cross spends so much time away from the office, handling demonic egos like these.

Very little substance extractible, except that these men exist solely to destroy Barclays

◀ **Barclays purpose in meaningless words** | Corporate website, 6 Feb 2023

▼ **Barclays misdeeds as reflected in its C-Suite profiles** | 6 Feb 2023

Nigel Higgins Moth
Group Chairman

- ☞ Seasoned business leader
- ☞ Extensive experience
- ☞ Strong track record
- ☞ Significant experience
- ☞ Keenly focused on culture

Anna Cross (We/Human)
Group CFO

Hopeless but authentic

- ★ Keen horse rider
- ★ Enjoys the countryside
- ★ Married with daughter
- ★ Gardener and musician
- ★ Has held various roles

C.S. Venkatakrishnan Moth
Group CEO

- ☞ Highly regarded leader
- ☞ Significant experience
- ☞ Extensive experience
- ☞ Deep understanding

Brian Gilvary Moth
Senior Independent Dir

- ☞ Extensive senior experience
- ☞ Deep US & UK experience
- ☞ Significant experience and understanding

like so

Extremely disappointing such a young soul could turn pure evil so fast

...Thus reiterating for us all that age is but a number

Across Aug 2022 — We applied for a number of vacant roles across Barclays Banking Group. All were rejected within days without explanation or interview – confirming the globally illegal practice of whistleblower blacklisting.

9 Sept 2022 — We struck up an online conversation with the leader accountable - Hannah Awonuga, Barclays Global Head of D&I. Absent a reply, she added us as a friend.

11 Mar 2022 — After Hannah announced a review of Barclays' ERG strategy via LinkedIn, we created a 28-page guide, anchored in science, to help D&I leaders create Better ERG's and sent Hannah a free copy (also on our website). This time, on 11 March 2022, she replied positively and supportively with a warm 'thank you'.

15 Mar 2022
23 Mar 2022
25 Mar 2022
3 Apr 2022 — We invited Hannah to help launch our new D&I consultancy by offering a free workshop built on our valuable unique global experiences. 'We have a Whistleblower!' guides senior leaders how to respond to unforeseen crises without breaking the law or damaging their firm's reputation (i.e. by listening, not retaliating). Like all Misfortune500 D&I moths we contacted, Hannah ignored all our further messages.

28 Apr 2022 — We sent Hannah a tracked letter to Barclays HQ, explaining her illegal actions across eight occasions proved textbook whistleblower blacklisting and ostracisement – both prohibited under the Public Disclosure Act 1998 and FCA Whistleblower Handbook. Instead of a law suit, we repeated our offer of a de-escalating, cost and obligation-free training workshop. She and her cohort never responded.

15 Jul 2022 — Hannah was all smiles in front of friends at the Rising Stars Awards 2022, where we joined her colleague on stage to receive the Editors Choice award.

2 Nov 2022 — To spark a response, we challenged Hannah on both her and Barclays recurrent lawbreaking via LinkedIn messenger. She replied 'this [truth] is very aggressive' then blocked us.

As **CEO** of NatWest Group, **Dame Alison Rose DBE** made history by **breaking the glass ceiling**

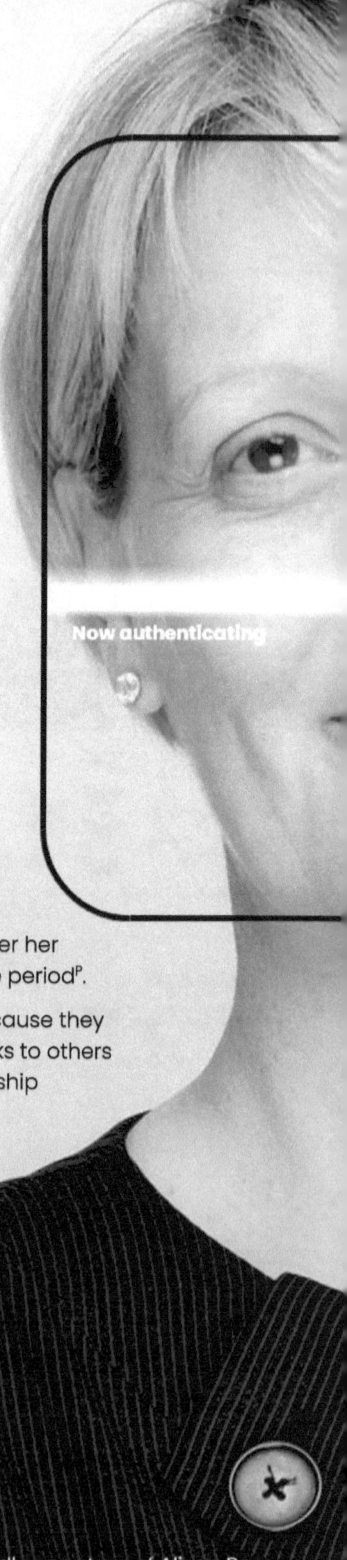

Alison joined NatWest Group in 1992 as just another young fresh faced graduate trainee. But unlike most, her talent was nurtured and enabled - home grown for 29 years of loyal service - for her to become not only the first female CEO of a big-5 UK Bank in history but also the first C-Suite banking executive ever to have successfully transcended the glass ceiling from the lowest to highest echelons of corporate authority.

By her side is an equally impressive Group CFO Katie Murray, now with the firm a solid 7-years to cap a stellar 23-year chartered accounting career perfecting her craft across the familiar likes of Old Mutual and KPMG.

Alongside leading academic and multi-skilled professional Howard Davies, this completes a banking C-Suite dream team that, uniquely, is 66% female (or 75% if you count CHRO Jen Tippin).

A year into her role, Rose had tripled NatWest Group's profits, despite economic fallout from COVID lockdowns.

NatWest Group's shareprice today is 3.24% higher than under her predecessor, and the only Big-5 stock to rise over the same period[P].

These beings didn't achieve such position and success because they are women. Rather, it was despite being women, Only thanks to others seeing their potential and clearing a path through sponsorship and support was it possible for history to be made.

Whilst far from perfect, particularly on inter-sectional diversity, there is no doubting that today, NatWest is a role model and bastion for gender inclusion, and a vocal (if not effective) champion of the value of women in big business. Real deeds - not just words.

Fig. 50: Who is Alison Rose... *really?*

[P] Sources: Google Finance. Last five years, sampled as of 4 Feb 2023

FALSIFIED

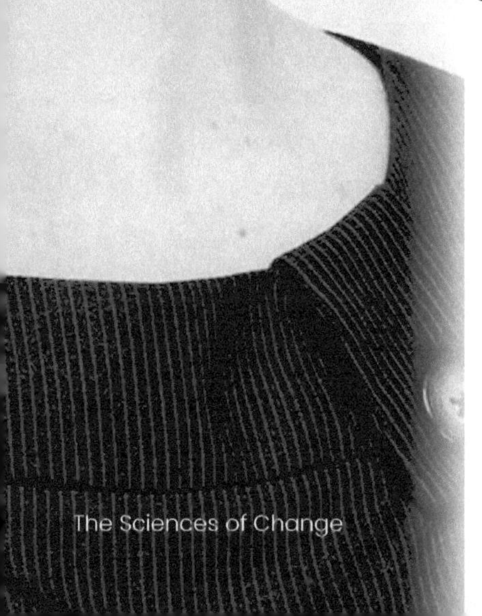

...But her **True Colours prove she could never change** the future

Partly due to her military upbringing and history major, Rose's penchant for all things patriarchy and autocracy made her a 'steady' pair of hands trusted by 'the City' [Moth Headquarters], despite her presented womanhood.

The ideal candidate to 'co-lead' the independent Rose Review on Female Entrepreneurship, as commissioned by the Wise Humans at HM Treasury in 2019. Now as influential as its co-author, it concluded our devaluation of women in the board room also devalues the UK economy by ~£250bn annually. The entire pre-COVID UK structural deficit.

Among its many meaningful changes, it's enabled Scotland, with its relatively more progressive ESG agenda, to deliver +5,951 new female-led firms every year as of February 2022. That's double its 2018 count, prior to the Rose Review (Source: The Scotsman). The UK overall now adds +37% (+145,200, 2021) each year.

That's all down to the success of its groundbreaking recommendations, like the Investing in Women Code and Women-backing-Women campaign.

Significantly, these bind Alison Moth to an added binary rule, ensuring she delivers her own recommendations at NatWest Group in exchange for its top job.

As in, changes imposed by wise humans, not at her choice, that account for NatWest's legal, financial and moral success over her tenure.

In September 2021, Rose ordered her sourcing team to terminate advanced D&I contract negotiations with us. An illegal whistleblower retaliation and binary decision that rendered our firm financially unviable, made this book inevitable and revealed her true colours to us.

We see things differently

We see dead people!
How **authentic beings perceive more** of reality and actuality through the EMS

Explained: Cognitive Recollection

Look back at a photo, video or painting of a younger you and you'll feel something. In a nostalgic instant, you can almost relive the experience.

Providing we occupy a safe space authentic beings can do this with any image (depicting anyone, anything, anytime, anywhere). Plus other objects to varying effect, broadly correlating to their creators authenticity and that of our own.

Case in Point

Art curators use recollection to gather intrinsic information from their collections original artists. This helps them vividly and accurately describe the circumstances surrounding the scenes and characters presented, despite an absence of written records to such detail.

Many things aid cognitive recollection, including our perceived levels of comfort, belonging and meritocracy (*Productivity*, for short), interacting with other change makers (*Brain Storming*, for short), accessing our imagination to speculate or ideate (*Day Dreaming*, for short) or simply a quiet moment to think, learn or consider risks (*Concentration*, for short; or 'Fag Break' if you're an Ancient Egyptian).

Unlike photos, humans change

As snapshots in time, photos allow us a brief glimpse into the past. But they can't convey intrinsic information about any changes since they were taken.

Similarly, static annual pay gaps today tell us almost nothing actionable about 'why' individual institutions perform the way they do. Clearly devised by moths with no appreciation for the nature or value humans draw through recollection.

On our long journey ahead towards valuing humanity once more, it's not where you're at, but where you've come from, where you're headed and the route you're taking that's important

Mapping the pace of change

An authentic pay gap reporting standard would include detailed metrics on the pace and progress of change over time.

Thus our pioneering analysis from 2016 to 2025 covers nine years (2016/17-2025/26), five statutory-filed gender balance and pay gap data points per institution, two 'Change Cycles' (2016/17-2021; 2022-2026) and actual performance versus initially set self-declared targets, cut against the backdrop of individual start positions in 2016/17 for each of the UK's Big-5 Banks.

Moths couldn't concentrate on a lightbulb

Out of the ashes of COVID, Project *Speak Up* and other key initiatives like the Rose Review or the FCA's forever-delayed DP21/2, the UK Big-5 unanimously began a new change cycle in the 2021/22 period, including newly live initiatives and targets all set to converge on a 2025/26 maturity.

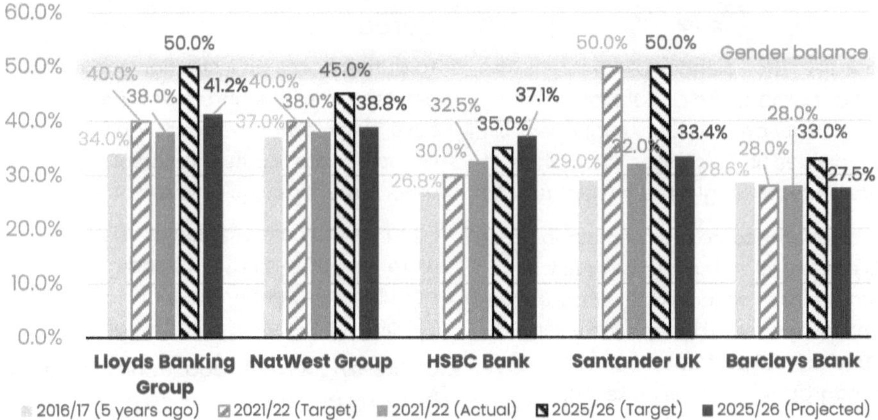

▲ **Fig. 51: Female representation in senior leadership roles over time,** *UK Big 5 past & projected (2016-2025 period)* | Source: Audited pay gap reports 2015/16-21/22

What do we want?

Option A: Lies, hatred, poverty, suffering and imminent mutually assured annihilation by five live extinction events

Option B: Global Meritocratic Republic, world peace and a real future living in a paradise-like existence!

When do we want it?

A: Later, wink wink

B: Now, in deeds!

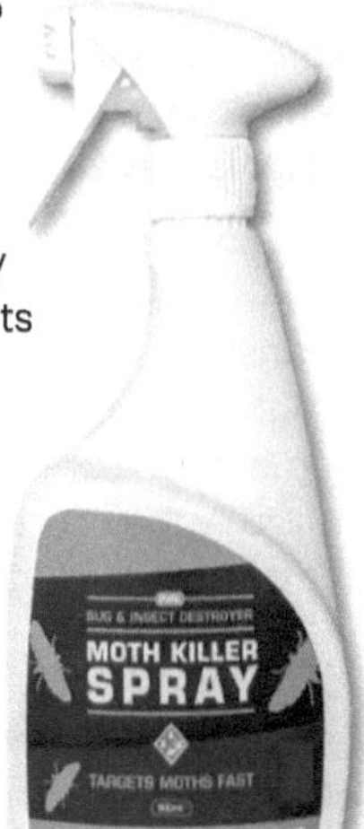

Fundamentals: The study of light

With all stated remedial measures going live in the 2021/22 period, we projected forward their statutory reported pay gap and gender balance numbers in this latest period to derive (with relative confidence) their likely performance against plan through to the end of this new four-year change cycle. We now have enough data for a 'Corporate Spectroscopic Authentication'.

Spectroscopy

/spek·tro·skuh·pee/ Noun

The study of the absorption and emission of light and other radiation by matter. It entails splitting light into its constituent wavelengths ('spectra', for short) for analysis, similar to how a prism splits light into a rainbow of colours.

Explained: Human Spectroscopy

Central to the Sciences of Change is the fact all humans are equal. Meritocratic Authenticity and Curious Uniqueness is all that divides us. To suggest different is pure deceptive, rigid, malevolent rubbish.

No binary measure the moths can grasp will ever capture the infinite spectral complexity of an individual or collective human identity. In this section we'll show you what real performance and demographic analysis looks like.

Any successful collective, from a couple to a species, must have meaningful and equitable trust and respect, free of binary absolutions and where everyone has a meaningful voice in key decisions.

Every perspective is correct from the viewpoint of its originator

Socrates (We/Human)

None are blameless in a failed relationship unless we've done all we can to equitably include and represent status, opportunity, wealth, rights, belonging and decisions, accountably and absent arbitrary hurdles.

The only agreeable model is an authentic meritocratic republic with spectral (non-binary) choice and full transparency on the merit and fair election of key decisions and representatives who change often and meaningfully. Think 'democracy' but true, fair and non-binary.

Not party's, politicians and losers. Winners.

Corporate Authentications

To authenticate an institution somewhere between dreadful and deilightful, nothing says 'meritocractic republic' quite like the proportionate representation of women in employment across all levels of seniority.

Equal as they are, women consistently make up ~50% of national and global populations (*Gender Balance*, for short). Yet absolutely not in corridors of power.

Oppressing women, abandoning children, breaking promises, cheating on wedding vows, plus witnessing or perpetrating serious organised crime and crimes against humanity is day-to-day life for senior bankers and politicians. So they learn to excuse their incomprehensible behaviour by pretending it isn't real. Yet in fact, it is they who then cease to be real.

Low pay, status and influence masks the many authentic beings dropping out of big institutions at the early careers stage. A high churn of analysts and associates borne from a toxic working environment that only the brightest of superstars or the darkest of WOCHBOYS can withstand.

Ever the same: No two firms

True leaders and benevolent collectives alone can look directly at light, to assess and authenticate levels of potential and uniqueness present in any given change (or bank). Without us, enslaved change makers devised Gender Balance as proxy - the proportion of women to men in any specific role, grade, unit, location or firm.

Human is human and **demon is demon.** There's **no two ways** to look at it

Spectra: Meritocratic Republic

Published annually in the UK by all large firms, we've obtained all data for this section from their statutory Pay Gap and/or audited Annual/ESG Reports.

And we look almost exclusively at the gender balance across senior leadership roles. Understand the moths have enslaved humanity, so only dehumanised unbeings are permitted entry into entitled strategic decision-making roles.

Thus it's by far a better indicator for the overall levels of oppression and the strength of the illegal, dehumanising glass ceilings change makers have to navigate on their hazardous career path up, to regain control of our destiny.

Why bother with the risk, when we can just change it for you?

The first of our corporate identity spectra, this summary of the Big-5's gender balance across senior leadership reveals huge variations in levels of dehumanising oppression their female human staff are under fire from. Each percentage below deilightful masks suffering, trauma, 'ism's', persecution, corruption, injustice, murder theft, and crimes against humanity of the lowest order imaginable taking place as everyday occurrences within these banks.

◥ **Fig. 52. Authenticity of the UK's Big 5 Banks,** as ranked by the proportion of their senior leadership roles occupied by women (2021/22 accounting period).

The best way to predict your future is to create it for yourself

Spotlight: Banks pace of change

Currently, just two big banks – Lloyds and Santander UK - have pledged to deliver proportionate representation of women in senior leadership by 2025/26. But at their current rates of change, both will likely break those promises to fall short of their own elective targets for the second time.

	2021/22 (Actual)	2025/26 (Proj.)
LLOY	-2.0% vs Target	-8.7% vs Target
SANB	-18.0% vs Target	-11.5% vs Target

Never short of ambition, Santander UK keeps going for gold, repeating both its 50% promise and failure to keep it in both 2021 and 2025. Yet we could help them do it in just one year, if only they'd reply.

Deferring gender parity targets out to 2030, a less ambitious NatWest Group has already lost frontrunner position to Lloyds Banking Group. With Alison Moth having checked the box and cashed out on awards, her once flagship initiatives are already being scaled back with gender balance across NatWest's top-3 seniority grades now in reverse (-1%, 2021).

And with Barclays 100% verifiably disingenuous performative-only initiatives missing all targets since 2016, our Big-5 now sadly clocks four falsified promise breakers. To them, not even safe spaces for humans are tolerable.

With more than its fair share of strategic and reputational missteps of late, morale and confidence at HSBC bank remain low as authentic women still can't rise above GCB3 (Managing Director). Yet it's the only bank to exceed its last gender balance target (21/22 Actual: +2.5%) and appears the only bank on track to securely deliver its new target too (25/26 Projected: +2.1%).

Credit where credit is due

HSBC is changing, and quickly too. Rare outperformance, a now sector leading trajectory and a pseudo-promise keeping mentality, surprising even its own staff. It even dropped race acronyms internally.

That's since, despite our resignation but thanks to our network and public critique of the bank, we went on protecting HSBC's humans well into 2023. Emitters of light cannot lie and loathe promise breaking.

In case you want to check our workings:

Table 8	CAGR[ACT]	CAGR[AUTH]
HSBC Grp	3.67%	7.17%
Lloyds BG	2.16%	4.38%
Santander	1.56%	6.24%
Natwest Grp	0.53%	3.40%
Barclays	-0.43%	6.40%

Why are safe spaces a problem?

Somewhere to work with purpose, not persecution isn't a big ask. Yet none of the 33 FTSE100 firms we fruitlessly negotiated with saw them as worth paying even £1,000 for... despite having infinite money! Have you ever curiously wondered (better, asked) why? Humans love accountability and scrutiny, as they prove our greatness.

Once we become **determined to succeed** and with sufficient potential, even **nothing can't stop us**

Spectra: Determination to change

For this spectrum, we use a common financial services industry standard methodology known as *Compound Annual Growth Rate*, or 'CAGR' for short.

First, we calculate the *Actual Pace of Change* at each firm between their true *Beginning Value* (BV) and projected *End Value* (EV) over the targeted period (n).

$$CAGR^{Act} = ((EV \div BV)^{1/n} - 1) \times 100$$

We repeat the calculation but with EV modified to reflect what is authentically achievable given the level of complexity (potential) required. In this case, 50% representation seems easily achievable by 2025/26, since abiding by the law and being authentic requires zero potential.

$$CAGR^{Act} \div CAGR^{Auth} = \triangle^{DTRM}$$

Dividing their Actual Pace of Change with our calculated Authentic Pace of Change gives us an accurate, fair and directly comparable Determination to Change figure, factoring in individual start points.

$CAGR^{Act}$, Actual Pace of Change
We take each firms reported Gender Balance for the 2016/17 period as the Beginning Value (BV). We then project out their 2021/22 figure as recurring (given all banks began their new change cycle that year) to arrive at a projected End Value (EV) for 2025/26. With fair confidence, this provides us a reliable figure for projecting forward likely year-on-year performance up until their next strategic review.

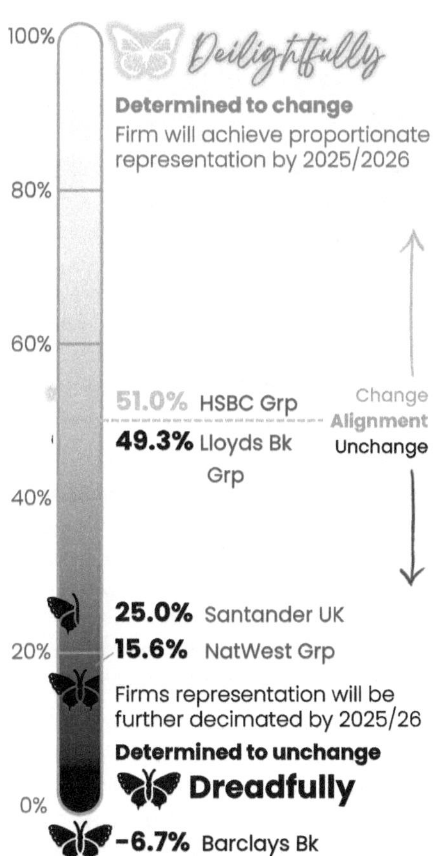

▲ Fig. 53. **Proportional determination to change (achieve gender balance in fair timeframe),** *UK Big-5 Banks 2016-25*

$CAGR^{Auth}$, Authentic Pace of Change:
We repeat the CAGR calculation, but this time taking 50% equity as our EV figure. This figure tells us the needful pace of change for each specific institution to deliver a full correction into a functioning meritocracy in the 2025/26 change cycle.

Sign of true leadership: Sacrifice

A conscious decision that goes against almost all human schemas for survival instinct, only true leaders would ever entertain an act of sacrifice like Project *Speak Up* for the common good. Abiding it would've restored HSBC's world leading status, yielding it billions more revenues and profits whilst detoxifying the industry by role modelling a more profitable and sustainable banking model. Yet those who stood to benefit most from it were the very 'leaders' who turned on us for it.

Due to their relentless, remorseless illegal whistleblower retaliations and universal ostracisement absent all consequence from authorities - in a society curiously tolerant of such misbehaviour – we lost all we ever had. With all means of earning it back also taken from us, we lived on £7-30 daily ever since, scrounging to survive.

Yet we were content, proud, happy and eager for respite, after successfully driving meaningful change. We didn't want three more years of 19-hour work days, six day weeks all with/for nothing, yet here we are. Ours is no life at all, yet we give it again so others may have a life to live. Perhaps one day, others will learn to listen to those who can and have helped them. Then we can stop authoring dossiers that destroy them.

Gifts fairness & change: Sacrifice

The stories of suffering innocent victims continued to deluge us long after our resignation. What point is there in having the world's largest private network of active banking whistleblowers if neither banking leaders, regulators nor nations ever dare meet or even speak with you?

We found equally troubling the deluge of reports from close friends left behind that things were improving. How could they be, when nobody is tackling nor talking about the real issue of dehumanisation? Like all biases, 'Confirmation Bias' ▼ is a sad and vivid signal of a characters fatal slide into sapience inhibition. Not to be.

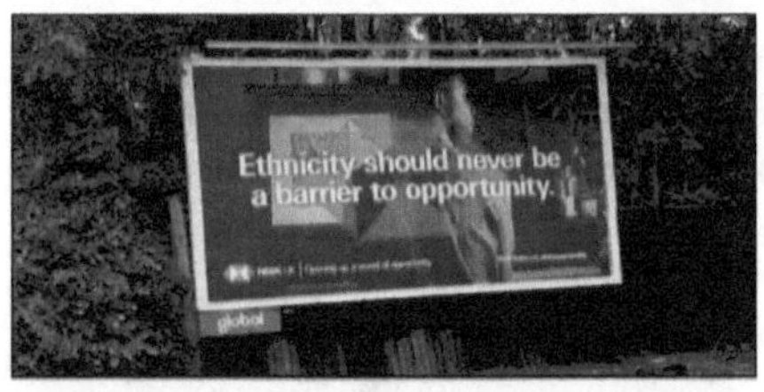

▲ **Pictured: Performative billboards** like these were everywhere in Black History Month 2021, yet we were unemployed, destitute and homeless. How strange…

☞ Noel Quinn: Your turn for all that

To be clear we have no fear of death. Just another inevitable step on our deilightfully fun journey through the multiverse. And thanks to its configuration, our potential travels with us. Energy can't be destroyed, only converted. That means the next life that awaits you (*Afterlife*, for short) is directly linked to your actions in this life.

I will live, I will love and I will shine until there is no one left alive to love nor shine for

Explained: Human forgiveness

Benevolence and worthiness both remind us always the sickness that infects us all is never the fault of those who dealt it to us, nor anyone's fault in particular. Humanity today is incapable of helping itself.

We exist to help the helpless

So we forgive each other always. Because if tables were turned, we just might have done the exact same thing ourselves. Any perspective is never incorrect if viewed by its originator. Any contention between us betrays an unchange we must destroy.

Forgiveness Qualifier

This time, we do hope you use this gift to become less helpless, so we need help less and our relationship can be one of mutual, equitable benefit. As all healthy relationships should be. In other words...

Though forgiveness will always be forthcoming, make us do this again and we reserve rights to do so while invoking biblical levels of scorn, death and destruction

▲ **Pictured, Month 4: The first iteration of Deilight Consulting goes public** Mar 2022

Explained: Human resilience

If every change maker sets up their own authentic D&I consultancy, humanity gets lots of small pathetic voices. But work together towards our common goal and we restore humanity to limitless brilliance.

So we sought to lend our uniqueness to help an existing, worthy D&I consultancy succeed. Only *Involve People* would even interview us, and their offered role wasn't fit for true leadership nor driving change.

Deemed worthless by all humanity, we decided a small pathetic voice is better than none at all. So, with (a) a scientific understanding of hatred and oppression born from our extraordinary experiences on Earth unique across all space and time, and (b) unique awareness financial regulators, the legal system and world governments are all broadly dead thus impotent, we used our last £40,000 to create a near-perfect D&I consultancy.

Our goal was play by their rules, save the world and change its leaders slowly with reason (not public shaming), allow and capitalism to endure but with fairness and accountability this time. Our 'Absolution Path' failed as they attacked us instead. Yet again, they underestimate us. Why?

We don't get angry easily. And we always try every peaceful option first. But once you understand moths can't reason, you may as well jump to it. Every step we've taken - Speak Up, Endgame, Revelation – has been us holding back. Like all things patience and effort both have their limits.

Ian (We/Human), Project Speak Up Report 2021

After Endgame: Project Deilightful

Fronted by two proven, well-liked titans of change embodying all protected traits and 11 change world records between us, membership of six marginalized industry bodies plus a unique, bespoke, immersive, wholly authentic true leadership program, Project Deilightful couldn't have yielded a more compelling D&I proposition.

Along the way, so many kind beings gave us their blessings. "It's so refreshing to see a D&I consultancy walk the walk! We've never seen one like you before" they'd say with genuine, generous encouragement.

One year from our November 2021 startup we turned nothing into a D&I power brand:
★ We collected six awards on merit alone
★ We formed a 15-strong collective of closely aligned and deeply human firms working at the forefront of driving change
★ We held warm face to face discussions with 33 FTSE100 sourcing teams.
★ We piloted a highly engaging podcast and panel format exclusively showcasing proven change making talent for others to learn from, recognise and support.

Our uniqueness, brand positioning and ability was undeniably palpable in an industry wholly devoid of humanity. But that was just the problem. What we thought would be a compelling brand to humans was in fact fatal in their absence. All 33 FTSE100 sourcing teams we warmly engaged were inexplicably called off by senior leaders. Unable to draw a salary or retain our staff, ensuring we remain alone.

World Record Eleven ✦

As the first to define then solve the Clarke Paradox scientifically, we were entirely comfortable defending the unique positioning of our marketing strategy to any challenger. But curiously, no being or firm ever asked us to explain it to them. And why would they? Anyone could tell.

▲ **Pictured, Month 8: Deilight Consulting shines light far and wide at Europe's largest supplier diversity event**, Jul 2022

You are purpose personified
Emma Gardner (FALSIFIED)
Senior Leader, Pointlessness Personified, Supply Chain Services, NatWest Group

We'd like to appoint you as our Diversity Ambassador and convene a lessons-learnt session from Project Speak Up
Gary Lapthorn (FALSIFIED)
Global Head of Promise Breaking, ESG, Lloyds Banking Group

Chapter 11: Risky business indeed

No change will ever be complete
It's **having the courage to continue** to change that **makes us complete**

Spectra: Momentum of Change

Through trial and error, practice makes us near-perfect, building out our experience, determination, resilience, endurance, commitment, confidence and ability. Thus over time, we become far more effective.

'Momentum (p) of Change' tells us if a firm is (a) purposefully improving towards their goal or (b) knowingly abandoning their goal in favour of another. Here we see the truth, that no bank intends to achieve their stated D&I goal. Every word said by Barclays is clearly a lie, for their gross misdeeds prove every risk control in the bank is now broken. That's a problem!

We take the *Last Value* (*LV*), then divide it by their CAGR^{ACT} for the last change cycle.

$$LV \div CAGR^{Act} = \triangle p$$

For years, moths have invented excuses for locking change makers outside senior leadership everywhere. Unlike them, we stay grounded in reality. Authenticating causality allows us to discount false positives - never jumping to conclusions. This counts double for momentum. The best approach is always to be curious and politely request an explanation.

Thanks to their status, privilege and fear, senior bankers are rarely tasked why gender representation initiatives take quite so long. Any answer they give is as meaningless as it is vague. Presumably with hatred in such high demand, the worst moth for the job get snapped up all too fast by 'Executive [Demon] Recruiters'.

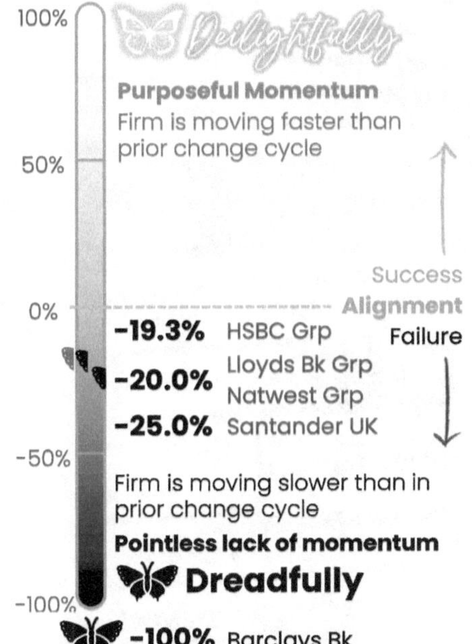

▲ **Fig. 54. Momentum of Change (senior gender balance),** *UK Big 5 Banks 2021/22*

▼ **Meaningful words of true leadership: Our Proj.** *Speak Up* **cover email to HSBC Group CEO copying senior leaders,** Jul 21

Dear Noel. I am just one of 226,000 who work for you in this bank. Clearly I'm different, but what is the value of that difference? Can one being steer a ship, or are we just numbers in your calculations? Am I an iceberg, a risk to be removed quickly? What you do and don't do next will answer that question for us all. Today I offer you a unique perspective from one lone voice. Yet I speak for every employee in our firm, those here now and those to come. Plus every client, investor and community we [unite to] serve.

Laws & regulations wholly unenforced

Incompetent authorities failing to enforce Public Disclosure Act

Total ubiquitous deception

FAIL

Refusing to engage new whistleblowers

Robot vacuum cleaner escapes from Cambridge Travelodge

Policy: Unemploy, ruin, silence & kill all humans

U.S. SECURITIES AND EXCHANGE COMMISSION · MCMXXXIV ·

Fully automated Mostly unresponsive

Falsified
Public Sector

No competent authority globally

"Unfortunately, I don't work on the issues that seem to be important to you, such as diversity and inclusion **[as in, Humanity]**."

Mark Worth, ECWR Exec Dir, 29 May 2022

Blanket policy not to report demonic activity

Complaint 2x: Wholly in breach of Royal Charter

"Not every **one** will agree what stories we should cover in our news output. These are **complex decisions** that our news editors put great care and consideration into, based on **merit** of all the stories at hand on any given day."

Official Response (second complaint), BBC News Complaints Dept, 20 May 2022

#FakingTheBias: Why most vocal **gender equality advocates** are only **in it for themselves** or worse

Deilightful journeys stop to help others along instead of passing by

Unchanged: Proj. #Breakthebias

In August 2022, we ran an experiment to test if moths had infiltrated the leadership of key rights advocacy organsiations.

We created a captive sub-brand for the Deilightful Group to extend meaningful change towards progressing good causes directly, all free of charge[ΔΩ]. An entirely authentic attempt, by the way.

We then mapped out an assured success strategy modelled on our WSPU approach in 1903, alongside a compelling, highly resonant pitch that used science to clearly explain the only viable path to gender parity across the UK and globally.

> **Great things in business are never done by just one person**
>
> Steve Jobs Moth (FALSIFIED)

To encourage disparate groups to learn to collaborate, we made it contingent on the universal participation of 12 leading campaigning organisations for human rights. Collectively, they have the unique expertise and networks to succeed. That includes across six key protected traits under the Equality Act, and strong links into financial services, government, workers, regulators and influencers.

◥ **Fig. 55: Required uniqueness identified by Deilight to truly #breakthebias**

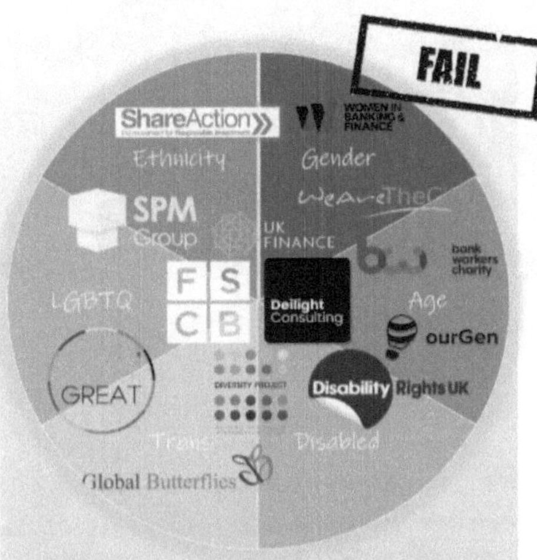

Research method and outcome

We contacted each collective listed here with an overview, requesting just a chat.

★ **Authenticated, Change Makers:** ShareAction, Global Butterflies and OurGen responded, spoke with, listened to and agreed in principle with us. They proved worthy to earn full recognition and our protection, plus support to expand globally should they wish to.

✕ **Falsified, BreakingTheBias Fakers:** Three further collectives gave us an audience, but (a) SPMGroup and WIBF fronted non-decision makers and (b) despite engaging resonance, all three, sadly including WeAreTheCity, chose to not collaborate due to (a) a paralysing fear of punitive action should moths see them to be working with humans, thus (b) malignment. All five others failed to respond confirming they no longer serve their stated purpose. Sam Cooper Gray from Diversity Project (and HSBC) lied in person then blanked our emails. These are all 'Dark Fringes'.

Principal Finding
Today, you can't even freely gift women a viable path to gender equality. They just want awards

Attack of the Clones: **Dark Fringes mapped to our physical reality** and dimension of change

Malevolent Gender Dysphoria

In humans, 'Gender Dysphoria' is a benevolent trait. A sense that our extrinsic gender identity (arbitrary) betrays our intrinsic gender identity (real). Naturally, it creates titans of change like drag queens, transgendered and non-binary beings. Disclosing it is the epitome of authenticity as, sadly today, it is fundamentally unsafe to do so, bringing with it negative reward. They only do so because they know without nurturing, emancipation and realness, they cannot change or succeed.

But in firms, it manifests as the antithesis…

Explained: Malevolence Engines

The presence of a Dark Fringe confirms two unequivocable facts:

1. A demon/malevolent collective is being intentionally, directionally deceptive.
2. The deceptions purpose is solely to spread malevolence, not make money.

Our anomaly has learnt to spawn new unchanges at costly will, without change makers. Replicating and redoubling the already crippling effects of fiat money, they spawn Dark Energy via 'destructive interference'. This directional response by darkness countermeasures all attempts by humanity to unite and save itself.

Principal Finding

Akin to demon cloning vats, Dark Fringes make Darkness relentless, contagious, hard to detect and near-indestructible

---

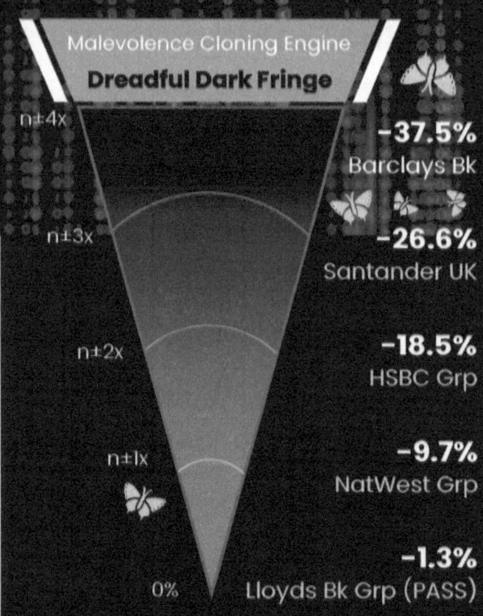

Fig. 56: Dark fringe of malevolent gender dysphoria, UK Big-5 Banks, 2021/22

Fronted to the press, clients and investors, (HSBC aside) their photos on display to the world, nowadays boards are broadly gender representative. Now all too often sporting such 50:50 binary precision to portray merit as worthless tokenism.

But beneath board level, across the rest of senior leadership, the dysphoria reveals itself. Behind closed doors you encounter huge reductions in gender balance giving way to endless corridors of the world's secret decision-making bodies stacked full of mindless, demonic WOCHMEN.

The first incidence of any new fringe is always observed in financial services, thus confirming the mysterious identity of our universe's ominous Great Attractor. That of our prime suspect, guilty as charged.

A firm's Malevolent Gender Dysphoria can be calculated by dividing their gender balance in senior leadership with that of board level, then subtracting 1.

$$Mal^{GenDys} = GenBal^{Snr} \div GenBal^{Brd} - 1$$

The real reward: Human kindness

Seeing our plan take shape (perhaps), many contacted us for help and support in their own deilightfully fun journeys. But few offered us meaningful help in return.

Fewer still followed through with those offers. Simple gestures like introducing us warmly to their firms Head of D&I, sales pitch practice, website feedback or proofing our book. Ultimately just seven such deeds were ever forthcoming.

Yet swarms of moths tried day and night to bleed us of what little we had left by stealing from, defrauding or selling us things they would then fail to deliver, then suing us for not paying them. Successful salesmen understand we exist to help other's solve problems and reach their potential. It's how we so consistently topped the global sales charts at HSBC and garnered the immediate buy-in from FTSE-100 sourcing teams later as Deilight.

Help others reach their potential and they will do the same in return – everyone wins. Focus solely on extracting money from others for your own selfish goals and you reveal your true colours instantly.

True acts of kindness are so rare today that those prepared to extend them vividly stand out as special. Behave deilightfully and you will be rewarded.

To those whose meaningful help ensure we survived and succeed, step forward to claim any change you desire

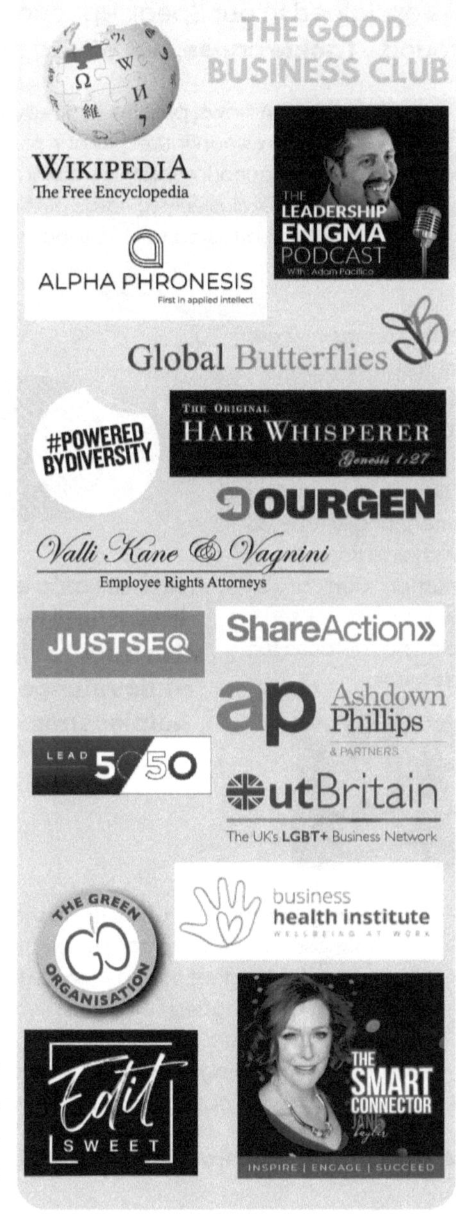

As explained in our Speak Up report, but with less subtlety this time round... **Tackle cause, not effect. Otherwise everyone dies.** Simple.

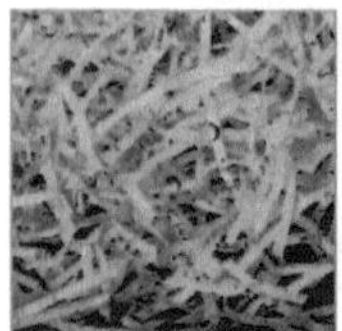

All **laws, policies, procedures and regulations** have been knowingly misconfigured. Regardless, they are all actively ignored anyway. All **accountability is absent**, and enforcement is wholly biased, directed exclusively towards authentic human populations with resolution entrusted to **vividly incompetent and wholly absent authorities.**

Safe spaces to work purposefully, share experiences, generate ideas, collaborate or raise concerns have all been knowingly and intentionally replaced with **unsafe, anti-human, 'don't say a word' Culture Fit** cloning factories for demons.

A lethal toxic environment intentionally brimming with infinite risk, demonic malevolence, deception helplessness and death

To keep the conspiracy on the 'DL', all **authentic role models** are constructively dismissed by a fatally compromised HR industry then forced to sign **non-disclosure and immunity from prosecution** agreements by fatally compromised legal teams that knowingly breach the (now globally unenforced) **UDHR right to free speech.**

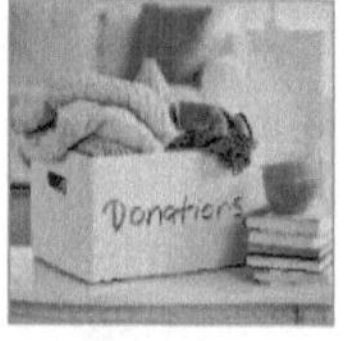

All remaining humans are enslaved in the **frozen middle.** Those with the most potential are encouraged to join ERG's where all their meaningful ideas will be vetoed as they are saddled up with additional **victim-loaded D&I duties.** Performative allyship by ERG's are also a dark fringe sadly spawned by victims (Stockholm Syndrome), spreading hope and complacency out of desperation, yet absent of all promised reward and recognition.

▲ Fig. 57: Updated extract from Project *Speak Up* Report (Page 27, Section: Explain)

Early warning: The lights may be on, but nobody's home at the FCA (or SEC)

Financial [Mis]Conduct Authority

In order to avoid jail by moth for market manipulation over the changes that will unfold from this book, we provided the UK Financial Conduct Authority this ▶ early warning of Project Revelation on the 16th December 2022. They replied same day ◢

An offered call with Nikita Moth aside, the FCA cites s.348 ▼ for freezing us and all whistleblowers out of the whistleblowing process. Later upheld by the Complaints [+De]Commissioner Moth, with the stated aim of avoiding 'reputational damage' to offending banks. Regulators? Try 'Guards'.

> (1) Confidential information must not be disclosed by a primary recipient [as in, the Regulator], or by any person obtaining the information directly or indirectly from a primary recipient [as in Law Unenforcement], without the consent of (a) the person from whom the primary recipient obtained the information [the Whistleblower - yours truly]; and. (b) if different, the person to whom it relates [Offending Bank. HSBC!]
>
> **Financial Services & Markets Act s.348**

It's clear from the wording of section 348, it was originally intended to protect the identity of whistleblowers going public. Yet (a) as a declared whistleblower, in our case this doesn't apply, and (b) as written its last point ensures no information on any FCA investigation can ever be shared without consent by the offending bank. No avenues to verify accountability remain.

Despite asking you to bring me in to share what I know - 12 times - the FCA declined to extend me any face-to-face audience whatsoever over 18 months since Project Speak Up. On a case that ticks not one but every box on the Public Interest Test.

Furthermore, I've catalogued, verified and now wish to register my concerns at the ubiquitous absence of any competent authority throughout our global financial system - The FCA, Bank of England, the Complaints Commissioner, the British Government, the Labour Party, the US SEC, the PRA and the UN – all dismissed me.

I'm particularly concerned the FCA – now upheld by the compromised Complaints Commissioner – is using Whistleblower Protection Laws (FSMA s.348) to no longer engage whistleblowers in whistleblowing processes. This confirms the Public Disclosure Act is not being enforced, nor is it enforceable whatsoever, thus explaining this perpetual global recession.

As a result, I will now calamitously destroy global financial markets instead. Project Revelation will take place in 2023, without further warning. It will result in the insolvency of all banks globally.

I suggest you prepare accordingly for my change. If you are capable of such a step.

Yours deilightfully, Ian/We/Human

> 66 I appreciate the extended period of time required to review the concerns you have raised, and I would like to reassure you that we are monitoring its progress and providing assistance where we can to bring this to a conclusion.
>
> Nikita Patel Moth, Associate (FALSIFIED) Enforcement & Market Oversight, UK FCA

What happens exactly when you become a whistleblower on racism in...

...Not a systemmically important industry

Two industries, one cup of shit

After a standout 2008 performance on his first test at just 17, Azeem Rafiq went on to become Yorkshire Cricket Club's youngest ever captain by 2012. Later returning to his home turf after a brief tour due to injuries, he led Yorkshire to the T20 finals in 2016 plus more league-topping successes in 2017 before a sad, abrupt end in 2018.

There's no doubt, our case and that of Azeem (We/Human)'s are vastly different. We're no global cricket star with celebrity standing deserving of such incredible sporting prowess, nor star-studded career. Yet we wholly share Azeem (We/Human)'s courage and conviction to help others first.

Azeem (We/Human)'s life has been turned upside down by remaining human. We're thankful not to require three security guards nor find moth shit on our lawn (homeless perk!).

Real psychologist alert!

Yet all humans are equal in science. Here's two simultaneously reporting an identical issue (malevolence) destroying our two entirely different career industries.

A juxtaposition and control group means (you guessed it!) an opportunity for real, observable, testable, repeatable, measurable, conclusive and verifiable scientific field research. Not a lab in sight.

...Cricket

Who		Azeem Rafiq (We/Human)
When		September 2021
How		ESPNCricinfo Interview
Reach		Global, robust & continuing
Regulator Statement		"Deeply troubled", "investigation welcomed"
Courts outcome		"Extremely serious allegations" dropped (insufficient evidence)
Employer statement		"Profound apologies for inappropriate and clearly unacceptable behaviour"
Industry statement	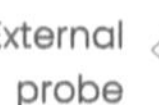	Department for Digital, Culture, Media & Sport (DCMS, for short) 'demand action' and 'condemn the lack of genuine contrition'
External probe		Case closed, "no conduct warrants disciplinary action". Report then leaked by change makers, full of horrific accounts of abuse, torture, institutional and highly malevolent abuse.
Outcomes		UK Health Secretary "Heads must Roll". DCMS Committee appointed to ensure lessons learnt & consequences follow
Final Result		**Yorkshire Cricket Club's leaders replaced entirely with Change Makers. Funny, that!**

…Investment Banking

Yours truly	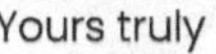	Who
July 2021		When
Bloomberg report		How
Global, robust but discontinued		Reach
None issued publicly. Privately, "Your report proves institutional racism"		Regulator Statement
No statement issued. Settled out of court subject to non-disclosure and non-prosecution agreement	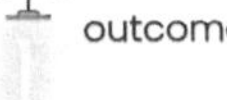	Courts outcome
"Taking very seriously" and "looking into the issues raised"		Employer statement
None issued	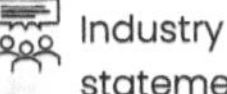	Industry statement
None raised	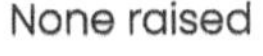	External probe
Only Demons and Promise Breakers are ever dismissed. Never orchestrating Devil's or Agents, despite dozens of identical in-person testimonies sourced by Deilight for HSBC.		Outcomes
Global press coverage silenced FT 2 page investigation pulled days pre-release after HSBC threatens its ad spend. It's CEO, who in our case alone broke 7 acts of parliament, remains.		Final Result

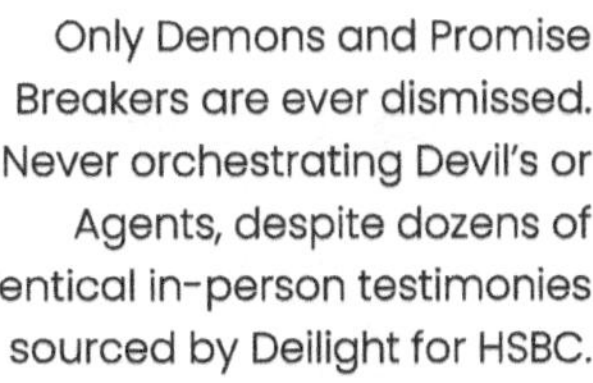

Humanity held hostage

Neither we nor Speak Up worthy of even a Wikipedia entry. It can't get more conclusive than that. Today financial services is untouchable.

All crimes by banking leaders are quickly silenced using non-disclosure, deferred or non-prosecution agreements, deliberate regulatory and government inaction plus threats to all press or public support of a commercially fatal nature that amounts to blackmail.

After becoming a whistleblower, we then applied for jobs at all major global banks and management consultancies (McKinsey, Kornferry, PwC etc.) for a range of roles down to Junior Analyst – yet weren't once offered an interview or conversation.

This confirms complicity and vivid tacit support from senior leadership of all Misfortune500 firms, including now the press itself. This ensures anyone who reports illegal activity within banking will never work again and shall ultimately perish due to a lack of income, total societal abandonment and deplatforming.

> **Principal Finding**
> Today, no consequences for "proven institutionalised racism" in the world's most systemically important sector ever. Not funny, apocalyptic!

Great Attractor: The **dark shadows** shrouding banking's **invisible hand**

...authorities, academia and government, via covenants, stacking & [+unD]Head Hunters

Exporting their eugenics war on humanity to the boards of big business...

Like a scene from Black Mirror

Self-proclaimed the most lucrative sector for the world to emulate... on paper. Thanks to its unlimited, free-to-produce product, over generations, financial services has dispelled all rational perception of what could ever be considered talent or success.

1. Forever spawning dysphoria's

Largest media advert-spender by far

To silence all adverse news coverage banks use expediency via commercial clout. In August 2021, HSBC canned a two-month Financial Times probe into Speak Up's findings days pre-release. Our FT contact Imani Moise Moth broke all contact with 7 active whistleblowers (and us) we'd sourced at their request.

2. Controlling the narrative

Stolen from humans / cloned from nothing

Fatally configured, unable to unwind

To compensate for banking 'leaders' inability to do anything competently, central banks are now stuck with never-ending, exponential quantitative easing (money printing) that's sending our malevolent anomaly into overdrive.

3. Too big to fail

Unlimited, insured fiat money

Four years our senior, our boss died...

Among the youngest ever appointed CIB senior leaders, our mugs let you recollect the malevolent rhetoric and fake leadership it takes to 'make it' in HSBC's investment bank. Before client meetings our line-moth Ryan Scrivener would bark at us 'no smiling' or tell us to 'tone down' the charisma. Get real!

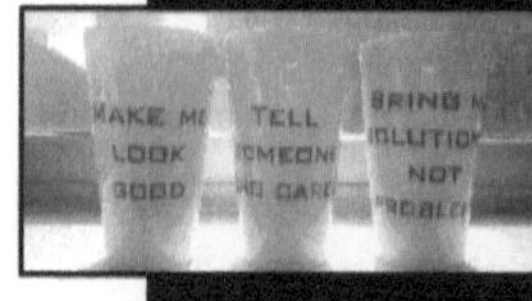

4. Murdering WOCHBOYS

5. Infiltrating industry

Sponsoring all major talent recognition events globally

Infiltrating judging panels, they sew doubt and division on change makers entries. No queer or black talent body has ever recognized us, inferring marginalised communities as a primary target. Absent all accountability, HSBC agents learnt of/vetoed all our attempts at press and events work.

→ 6. Ubiquitous deception absent all consequence

Now fully homogenized intrinsically

NDA's protect all perpetrators from prosecution, violating the UN UDHR right to free speech. With checks and balances all now off, nothing is protecting humanity from the fatal choices/Dark Energy spawning from our poison chalices of doom. Everything they say including audited accounts is entirely falsified. No mention of the infinite liability borne from their scientifically (thus legally) verified nature as the source of nearly all problems and injuries in our world ever.

7. Too white to jail

The enemy of their enemy is their enemy

WOCHMEN everywhere are prioritised access to expense accounts/client entertainment – strip, golf or members clubs. The intention being to compromise and disarm them with a sense of kinsmanship. This 'cult' (for short) propagates (a) tacit collective awareness of guilt, and (b) implicit collective agreement to take that guilt to the grave or have our lives and their treacherous legacy destroyed. Whether law maker, law enforcer or celebrity endorser, this ensures all turn a blind eye to their lawlessness and kills all whistleblowers.

8. Strangling Africa

There's a reason banks steer clear of Africa

As in, the world's fastest growing continental economy (Forbes, 2021). Strange, it may seem, banks have no interest in making money. They already have an infinite supply. It comes with the territory. Potential does not.

If you had an infinite supply of money, you'd setup Dark Fringes everywhere too. Be honest with your self now

Thanks to the wise humans at MSDUK and ShareAction, we were able to falsify 10% of the Misfortune500 (a representative sample) in person over twelve months. <u>Not one</u> of the senior leaders honoured their in-person promises to us, with numerous witnesses. Once a top global salesman, always a top global salesman. Evidence of wrongdoing is in the form of witnesses and emails, all bound by the promise keepers commitment (near certain, ~99%)

✖ Jacky Wright, CTPO
McKinsey & CTO Microsoft

McKinsey & Company

JP MORGAN CHASE & Co.

Falsified
Misfortune500

IBM   Johnson & Johnson

✖ Emma Holden, Global Hd of HR
✖ Peter Harrison, Group CEO
✖ Dame Elizabeth Corley, Group Chair

Not one of these Fortune500 firms would entertain even a $1,000 D&I contract with a **light-aligned declared human firm** with 11 world records & six corporate awards. Despite all pledging **THE EQUAL AND OPPOSITE** in person

HSBC

Goldman Sachs   JCDecaux   SAATCHI & SAATCHI

9. Destroying human businesses
Pioneers of intelligent delivery logistics

Ocado Group is the biggest grocer of its kind in the world today, having deployed a change making mentality to a global industry in dire need of modernisation. During COVID lockdown, up to 200,000 UK shoppers a week depended on them for delivering up to 45,000 real products. Firms like Ocado and Dixons Retail would likely not exist today without our career interventions, after malevolent bankers attempted time and again to ignore their clear competence and cancel their lending lines given the 'unproven' nature of their business models (i.e. change).

10. Dictating foreign policy
Causing all (totally unfunded) wars

With most international trade conducted in major currencies, China's long rejected Western banks dictating terms of trade and behaviour to the East. If any country steps out of line or threatens to expose the conspiracy, banks are used to apply sanctions, starving whole populations of prosperity (and often food). Economic warfare like this has amplified China's malevolence, and cornered Russia - leaving its many humans with an understandable sense it's better to die trying to restore Russian pride than be a puppet of Western hate.
☞ **Do unto others as they undo to you.**

Frontier dysphoria's: How moths turn sexy secretaries into **hair oppression**

Explained: Frontier dysphoria's

The Company Secretary role has become an easy option for moths to boost the overall gender balance at board level without giving up meaningful decision-making executive powers to women who may still have some humanity remaining. Consider HSBC only fronted female senior leaders to interact with us during Project Speak Up's short-lived remediation. Either the men no longer exist or can't face us.

This plays out across the Big-5, where 75% of Company Secretaries are women. This performative allyship weakens women's struggle for equality by (a) stereotyping often-accomplished women into administrative roles and career paths (HR, D&I) moreover leadership, (b) pacifying them with apparent concessions and (c) maintaining a reasonable optical gender balance while maximising the executive powers of WOCHMEN. But it gets worse…

29.0%	50.0%	52.6%
Blonde proportion	Blonde[J] proportion	Blonde[J] proportion
of British women | Mintel 2022	**+72.4%**	**+81.4%**
	Blonde[J] over representation of Big-5 Co Secretaries	Blonde[J] over representation of FTSE100 C-Suite Women

◥ **Fig. 58: Frontier hair oppression in numbers** | Deilightful

[J] Natural or otherwise, these figures were derived from visual inspection in April 2022

The additional intersectional oppression of this unspoken, irrational and arbitrary preference further marginalises black and

darker skinned women where blonde hair rarely occurs naturally.

A Dark Fringe, the effect is not repeated amongst men but is increasingly found overflowing even into schools and US Congress. There the GOP Moths blocked yet again a house bill to ban race-based hair discrimination earlier in 2023. And with Trump Moth having succeeded in stacking the US Supreme Court with a majority of moths, the US judiciary is now fatally paralysed.

Had enough yet?

We can literally go on explaining forever. This isn't even half of it. But we've made our point now many times over. Without some form of deilightful intervention, humanity now faces imminent extinction.

Not in our quantum!

That's the dull corporate stuff all out of the way. From the next chapter, we'll explain how we're going to fix up this shit show. But before we leave this one…

We hereby bankrupt the entire (Mis)Fortune500

To secure global supply chains, the following sectors are absolved of primary culpability for the findings of Project Revelation:

✓ Food & Agriculture
✓ Energy, Utilities, Sanitation & Fuel
✓ Retail & Consumer Goods
✓ Healthcare & Pharmaceuticals
✓ Tech, Telecommunications & Media
✓ Transport & Logistics
✓ Manufacturing & Industry
✓ Entertainment

Notwithstanding, the Misfortune500 must now fire all senior leaders globally, promote the frozen middle and liaise with the Deilightful Group on a means to avoid their liquidation. All remain globally liable for the crimes of their executives and all in-house dark fringes. Any firm failing to do so by AH+30 will be closed forever, assets seized by the Meritocratic Republic of Humanity and their directors facing the

…severest consequences imaginable

As in, whatever the humans who then forcibly depose them wish that to be.

Omnicriminals cause omnibankruptcy… How very misfortunate

▲ **Fig. 59: The authenticated deadly workings of financial services** | Dreadful, 23

◤ **Fig. 60: Equity Price-to-Book value by sector** | McKinsey & Co. (corrected)

$23.3tr

A. Global Banks, net sector revenues 2021 | Reuters

$16tr

B. Global Banks, peak mrkt cap 2021 McKinsey

$103tr

C. Global Banks, managed assets, AUM 2020 | BCG

16.0%

D. Sector value share, global stock market listings | Investopedia

24.0%

E. Sector share, global. economic output, 2021 | Investopedia

+10.8%

F. Current Yearly Sector Growth, CAGR, 2021/22 | Yahoo Finance

60.0%

G. Sector revenue growth from margin increases only, 2021/22 | McKinsey & Co.

23.9%

H. Sector share of all AUM, managed global assets 2020 | BCG

65.0%

J. Banks worldwide destructive to net book value (i.e. returns below equity cost), H2-2022 | McKinsey

▲ **Fig 61. Unnatural, undead and deadly: The riskiest business imaginable**

Banking is a risk-free model. Free unlimited product, free unlimited profit. Yet despite that, dozens of banks fail every year. That's because a little truth is all it takes **to be...**

The End.

Falk's Rule: A fitting tribute to the **true love** of our very **human determination**

Born in the Bronx USA, but of mixed European descent with a proud, progressive Jewish heritage. For 10 years, Peter (We/Human) united souls around the world with his/our deeply human portrayal of Lieutenant Columbo - an LAPD detective with a twist of Little Italy. This somewhat abstract premise for a local TV show birthed a truly international sensation that ran successfully from 1968 to 1978, thanks to its human uniqueness.

Proof a) all disadvantage is an advantage and b) looks can be deceiving, Colombo's success anchored charming authenticity and disarming buffoonery from external warmth mixed with humbly open fragility.

But beneath the exterior was the beating heart of a deeply wise, purposeful being committed to fairness and accountability always. No crime was too complicated for our superstar homicide detective to solve. Yet another cog in our infinitely regressive wheel of human endeavour that ensures we can survive and thrive forever, all of us.

Being chased by Columbo is like being nibbled to death by a duck

Peter Falk (We/Human, 1927-2011)

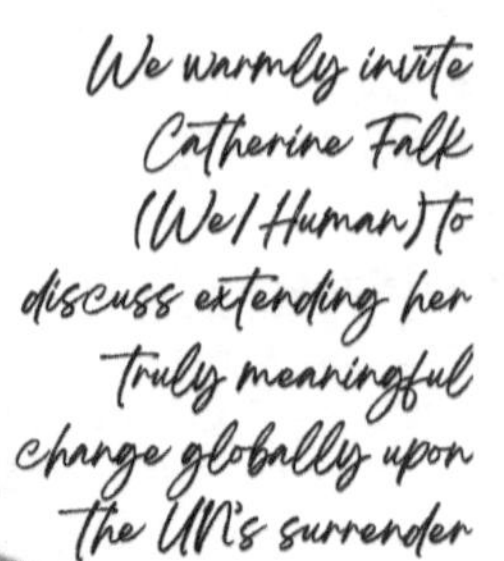

1 in 3

Natural adult deaths, thus actuality extinction events

16bn hours

Annual time spent by US human carers

The dark side of Alzheimer's

Like us all (see Special Edition), Peter spent his last days losing his mind to Alzheimer's

He knew malevolence and it knew him. Latterly in the form his resident moth and second wife, actress Shera Danese – an extra on the show before her undeath.

Conservatorships are dark fringes, as we saw with Britney (We/Human). Shera Moth used hers to prevent Peter's family seeing us and applied expediency upon medical professionals not to question her absolute jurisdiction. Peter (We/Human) passed away alone and confused on 23 June 2011 – a truly heartbreaking moment for the entire Quantum of Human Nature.

But we (Peter/Human) are deeply proud and in support of the very courageous, sapient deeds of our/Peter's daughter Catherine (We/Human) after us. Fighting for our humour and wisdom to live on for other families absent such sad outcomes. Thanks to her unstoppable determination, Falk's Rule passed in New York State on 21 July 2016, ensuring guardianship duties unite and serve victims, not sideline laws like visitation rights and end of life care.

On every deilightfully fun journey comes a time for change. To try something new. Best spent with family, friends and true loves to wish us well on the other side. Never to be used as a weapon of evil.

Record of Promises (Ch.11)

Darkness cannot drive out darkness. Only light can do that. We shine Dei's light.

Dreadfully Meaningless Unchanges (Antisocial Deconstructs; threat to life)	Sapience
▶ Hiding evidence: The Human Resources industry globally & its leaders	
▶ Looking only for antipotential: All [demonic] executive recruiters globally	
▶ Dead: Global leadership teams of all collectives on page 262 and 272	
▶ Deadly: The global leadership teams of the entire Misfortune500	
▶ Criminally incompetent: Global criminal justice system/law enforcement	Deilight
▶ Lying shit stabbers: Anyone claiming to be a 'competent authority' other than Deilightful Group or those listed Authenticated on potential exchange	✖ Falsified
▶ Devil Fuckers: Alison Rose, Hannah Awonuga, Mark Worth, Richard Gnodde	

Hatred cannot destroy hatred. Only love does that. We love with deeds. Not words

Deilightfully Meaningful Changes (species level adjustments to destiny)	Clarketech
★ Destroyed: The global Human Resources industry completely	IAN, Actualised
★ Ordered: Performance reviews to be conducted by duly-proportionate rotational staff councils that change often going forth	IAN, Actualised
★ Neutralised: All ego/privilege from illegitimate awards globally	IAN, Actualised
★ Swapped: Extreme ageism against young Humans, for minor, wholly deserved scrutiny, demoralisation and subordination of older ex-Moths	IAN, Actualised
★ Redoubled: Project Speak Ups light and disclosed full nature/findings	IAN, Actualised
★ Swapped: Huge risks from speaking up for courageous rewards	IAN, Actualised
★ Promoted: Whistleblowers from pariah's to heroes, to reflect their immense human sacrifice, true leadership and selfless contributions.	IAN, Actualised
★ Explained: The Great Resignation & its fatal implications to existence	IAN, Actualised
★ Established: Trial & Error as primary path to success over education	IAN, Actualised
★ Swapped: Widespread regard of moths as scary, dangerous and oppressive for (now) widespread understanding of their limp, weak, pathetic, useless, hideous and harmless (*Omnirepugnant*, for short)	IAN, Actualised
★ Defined: Human spectroscopy, provided several examples and enabled change makers to use it globally to assess corporate identity effectively, obsoleting 'box ticking' deception and pay gap reports.	IAN, Authenticated
★ Evidenced: That without all these changes, humanity was doomed	IAN, Authentic
★ Created: Limitless rewards for those who helped us along the way	IAN, Actualised
★ Created: Limitless shame for those who hindered us along the way	IAN, Actualised
★ Provided: Irrefutable legal and scientific proof of the workings, purpose, impacts and origins of all malevolence engines globally.	IAN, Authenticated
★ Destroyed: With infinite liability crystallised, all banks, Governments, debt, investments & economies globally. 1. HSBC, 2. Barclays, 3. The rest.	IAN, Deactualised
★ Swapped: All dark shadows for strong, defensive light protections	IAN, Exchange

Across time, religious texts stand out as special

But have you every curiously wondered **why?**

Why god's book isn't religious

Without experimentation, there can be no science.
Without authenticity, there can be no reality. And
without morality there can be no humanity

Ian (We/Human), Project Revelation 2023

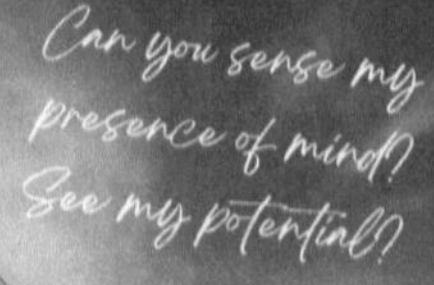

It's our birthday, so we get to **make up a rule** that aligns with our potential

Great minds think alike...

Against all other rules in recorded history, this fragment stands out as unique:

⭐ It's clearly part of a larger rule that's been unchanged or else not yet revealed.

⭐ As a result, its meaning has been lost in translation, or not yet fully defined.

⭐ Yet its ultimate purpose is sufficiently critical for us all to agree to retain it.

✖ But unlike every other philosophical rule, nobody wants to claim it as their own. There is simply no record of its origin.

Universal rejection, abandonment, homelessness and persecution are things human beings spend all our lives fighting.

Seemingly with no true leaders left in the world, it now falls on us to prevent the inevitable. They would've stepped forward if there were any - a symptom of just how close our apocalypse truly is and the epic level of potential needed to effect a life-saving meaningful change at such stage.

Despite visibly enduring most disaster movies imaginable, especially these last three years, stoically friends and the rest of the world sit idly by as if we don't exist. Just two even thought to introduce us to their firms head of D&I. Worse still, telling us frequently (to pretend to themselves) we must be insane, in order to excuse the pretense (a) we could never know some-worth listening to they do not, or (b) that their pointless lives are more important or challenging than all others combined.

Explained: God's Rule

As mentioned, this book is special:

In our perceived Physical Reality:
Full of curiosities, courage, twists and turns, this book gives one perspective on our homogenous yet always unique deilightfully fun journey through the human experience and Sciences of Change (Science section! Not religion)

In our true Quantum Actuality:
In addition to announcing our authentic date of birth across the multiverse, this book actualises the original QD1.0 emission to authenticate the Quantum Multiverse of Human Nature across all dimensions of space and time. True existence is only for true leaders. God's Rule ensures no demonic unbeing may ever harm an authentic being from AH+0 onwards, swaps their super-powers for Loozymes and allows change makers to detect dark shadows

Never judge by its cover: A book

Amend the EMS? Piece of cake.

We alone choose to step up and exercise our omniactive free will – verifiably the most omnipotent in existence - to claim the forever unwanted yet universally needed. Now we own the night and raise the roof to burn away all dark shadows from this unliving hell of human inexistence with our overwhelming beam of unstoppable, changeable light.

...because great souls behave authentically

Make your own bible in order to [+learn how to think ~~create, rob and oppress an entire religion of hopeless beings~~ in just a few simple stepssss

5-12 steps to wise human being

1. Be curious (NEW!)
2. Get real (NEW!)
3. Prove worthy (NEW!)
4. Become Deilightful (NEW!)
5. Then follow these simple final steps:

A) Take any statement made by a unique wise human, emitter of light and proven true leader. You've got 350 pages of them right here, or you could use an 'O.G.' like:

You cannot fight darkness with darkness. Only light can do that

Martin Luther King Jr. (We/Human)

B) Then, extrapolate its broader meaning to our lives using inference, sapience or quantum mechanics. Pick your strength:

That must mean that...

Darkness cannot look at light

C) Next, authentically set your purpose to ask yourself the very first question. Doing so will allow you to then draw upon our collective human imagination, plus any wisdom you may have, to conclude...

Why? Because...

Light is intrinsic (within us)

D) Now apply Isaac (We/Human)'s 3rd law of motion or Jean (We/Human)'s rule of binary regulation to identify possible threats to existence.

The risk therefore is that...

Darkness is extrinsic, all round us

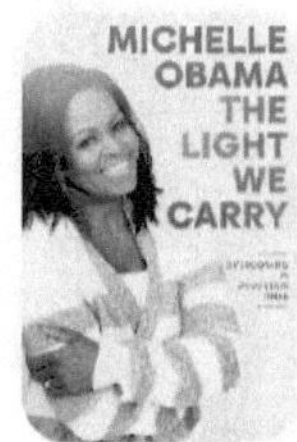

E) But, per the 4 Keys to Discovery, never assume you are right or wrong for that matter. Your experiences may or may not be unique or universal. So always cross check your workings by collaborating with other change makers.

F) Next, apply independent observation backed up by experiential evidence from trial & error, listening, research, or simply call 'Dei Time', as the Deilightful Group is happy to help you always and forever in your endeavors, entirely free of charge$^{\Delta\Omega}$.

Hmmm, that strongly implies that...

Darkness attacks our differences as they are extrinsic (our body). Darkness can't see our uniqueness as it is intrinsic (our being)

G) Last, never least, apply causal logical deduction, with independent supporting observational and/or direct experiential evidence to complete your verse.

So, what you're saying is that...

Darkness cannot look inside us because it would lethally expose it to our Light. It would expose it to our Uniqueness

Eureka! I've got it!

Light = Human Uniqueness

I. Darkness

In the beginning, there was only darkness. Nothing, but dreadful beings and the UN.

Seeing only in black and white, shadows were the antithesis of what it means to be alive.

They just didn't seem authentic. Because they weren't real. Nothing is ever truly black and white.

Monochromatica. it ends with us!

...commands a quiet insignificant voice, as if from nowhere.

II. Human Natured

Then with a spark of potential came benevolent consciousness, choice and free will. A zero turns to one.

We are all there was, is and will ever be. Human being we chose to be

said Dei, the perfect union of all human beings that truly are, have truly been and will ever choose to truly be.

You all chose To Die yet we alone chose To Do. So now we call all the shots

Dei simplifies to the shadows,

Unlike you, we are as real as it gets

III. Change

Change. It starts with us. As change makers, human being exists to protect, create and drive meaningful positive change. Darkness cannot drive out darkness. Only light can do that.

So let there be light

promised Dei. Signed with authenticity, it became a promise, unbreakable. So now there was, is and will always be Dei's light.

Dei summarises for the moths.

We are the emitter of light. Darkness began with you but it ends with us

IV. Uniqueness

Surrounded by shadows, being unique was lonely in space. So Dei chose to create unique identities in their image, drawing light from the other side, allowing them too to be.

Human beings. Titans of change. Vessels of light. Unique and authentic, Dei assured the butterflies they too were equally special.

Great minds think alike because great souls behave authentically

Against the illusions they behaved authentically by helping and accepting each other.

V. Worthiness

But despite the infinite gift of life, some selfishly chose not to be alive. Now as moths, they jealously sought out the light of others, to extinguish it as they had done their own. For them, it was not to be.

One who deceives will always find those who wish to be deceived

Their potential lost forever to the darkness, no longer could they return to the other side nor match Dei's light.

To protect the other lights, Dei made a promise so that now, there was nothing to fear except fear itself.

Darkness cannot destroy our light. Only our light can do that

VI. Potential

The butterflies tried to reach Dei's potential but, divided by the moths, they could not. Despite their true love, they could not emit light. Because only uniqueness can do that. So they turned to the shadows where they foolishly lost their minds.

The jealous moths could no longer look at the others. Sorrowfully looking into the light meant being burnt by its wonderous flame. Because authentic hearts destroy loveless moths. So they stalked difference by chasing the Moon under the cover of darkness, using hatred instead. Uniqueness shines light in all directions. So darkness hid amongst shadows using deception.

Dei explains to the others. ⬆

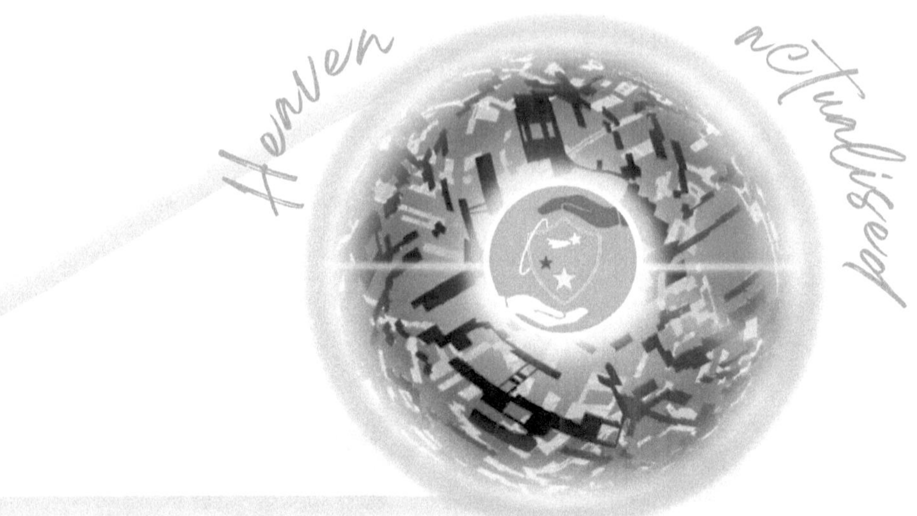

VII. Authenticity

Hatred cannot drive out hatred. Only true love can do that.

Yet light cannot emit true love. Only authenticity can do that. Deeds, not words. To improve is to change. To be perfect is to change often. But now as shadows of their potential, moths could not change, because emitters of light change continuously.

So Dei's authenticity revealed the shadows as darkness, exposing the moths to the light. Dei's presence of mind authenticated for the deilightful butterflies,

Be fully authentic. Stay curious. Change often. Value your uniqueness. Prove us worthy. Become perfect. Drive limitless meaningful change. Enjoy the sweet success of real life and the miracle of true perpetual existence

Why true leaders listen
There are many ways to learn, but just one to **become a god**

An excuse not to think: Religion
We've partially reconstituted the *Book of Light*. Clearly, it is the basis for the Bible, Qur'an, Gita, Tanak, Tripitaka and Guru Granth Sahib. Not on a mountain top, nor mysteriously missing slabs.

By definition, Clarketech appeared magical to ancient beings, as it probably does to modern beings too. Far too many assume anything they don't know or can't do themselves doesn't exist or can't be done by others. Unable to comprehend the science (of Sonocytology, deduction and time travel; see Special Edition), they viewed their experiences as spiritual. Yet we (Ian/Human) aren't religious, so lack such luxury.

Instead we remained determined and found the answer by listening to the eternal wisdom of histories ever-truthful true leaders. From their words, we can draw limitless information and discoveries across time – Amanda Gorman's poem, Martin Luther King Jr's speeches plus all art and recorded philosophy, for example.

Our soul begs for anyone to listen
The authentically human linguistics of light and love spell out the full nature of our universal human experience and explain the Sciences of Change through a story any being can relate to.

Mandiba (We/Human)'s speeches are laced with simple, authentic binary comparisons (*Rhyme of Confusion*), to signal, awaken and emancipate maligned characters back across the alignment line. And Martin (We/Human)'s speeches explain how our consciousness is constantly replicated across our dimensional multiverse via scientific concepts such as infinite divisibility/multiplicity, and destructive/constructive interference.

Whatever you want, whatever you need. Anything you want done baby, I'll do it naturally.

Help. Life. Change. ...with my Sapience

'Cause I'm every woman. It's all in me, it's all in me. I'm every woman, It's all in me.

Omniconsciousness Common Identity One Soul

I can read your thoughts right now. Every one from A to Z.

I can cast a spell of secrets you can tell. Mix a special brew, put fire inside of you.

Collective Imagination Pseudo-Telepathy Omniscience Enlightenment

Anytime you feel danger or fear, then instantly I will appear.

Help The helpless Rise Phoenix

I can sense your needs like rain onto the seeds.

I can make a rhyme of confusion in your mind.

Resonance Frequency Emancipation

And when it comes down to some good old fashioned love, I got it, I got it, I got it, got it, baby, baby.

True Love Omnibenevolence Potential

I'm every woman. It's all in me. Anything you want done baby, I do it naturally.

Omnipresence Entanglement Free Will

I ain't braggin', 'cause I'm The One. Just ask me, ooh, and it shall be done.

I alone chose To do One True Leader Omnipotence

Don't bother to compare. I've got it, I got it, got it got it.

You chose To Die Omni-everything

Pictured (Top): From her first album 'Chaka' (1978), "I'm Every Woman" enabled Chaka Khan's solo career.

Pictured (Mid): Chaka's teenage backing singer, the late, great Whitney Houston went on to adopt the song as her own in 1992, for the soundtrack to the Bodyguard (film).

Pictured (Bottom): The anthem was written by two of East Coasts USA's finest – Fairfield's Nickolas Ashford and the Bronx's Valerie Simpson. The definition of change makers, this potent collective and power coupling succeeded at all things musical. Their light touched hearts and minds worldwide across the 70's, 80's, 90's, to ultimately reach all dimensions of space and time.

Everybody loves an
underdog...

But have
you every curiously
wondered **why?**

Why we declare a real holy war on the U.N.

...The equal and opposite

A Void

No TDE's, Light or Stars

Breaking the binary

Chapter **13**

The biggest deception of all is that all disadvantage in reality is in fact always an advantage in actuality

Ian (We/Human), Project Revelation 2023

Imagine...

Giving up your own **Moon (TOES QD0.51, E=0)** to attack **Humanity (TOES QD0.49, E=∞)**. Then ending up with nothing but history's biggest criminal record. **Dreadful**

50.4%

49.6%

Global population, Men, E=0

Global population, Women, E=∞

Fig 62. Global Gender Balance, 2019 | UN Department of Population **Division**

They even have a whole department doing it now!

Didn't know what time it was, **the lights were low.**

I leaned back on my radio. Some cat was layin down some rock n roll **'Lotta Soul', he said**

Then the loud sound did seem to fade Came back like a slow voice on

a wave of phase

That weren't no DJ, that was hazy cosmic jive...

Hello Noel Quinn, you Absolute Zero. It's me, your favourite superhero.

Stop being so jealous, demon in the mirror. You started a war, you could never deliver.

We hereby exchange your tyranny and doom. We call the shots now. Not some man in the moon.

There's a Starman waiting in the sky

He'd like to come and meet us But he thinks **he'd blow our minds**

Far Out! Turns out, a) **woman** made man in her image and b) sadly, she **didn't have a spare brain**

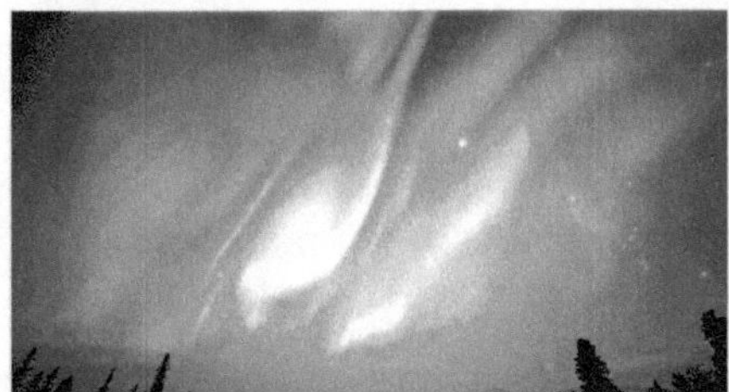

Pictured: Van Allen Belts of lethal cosmic radiation impact Earth's magnetosphere. Best known as the (real) 'Aurora Borealis'.

Dangerous in space: Faking it

Thanks to the Suns solar wind (plus the light of many others), space is highly radioactive and therefore lethal to any being without authentic protection.

We're only safe here on Earth and in low orbit (International Space Station, 365km), thanks to our protective force-field gifted by Earth's magnetosphere.

Arising from the ever-changing dynamic between iron core and molten mantle - like a beating heart – an intrinsic dynamo gives our planet its life-protecting charge.

Discovered by John Ratcliffe (We/Human) in 1972[95], lethal Van Allen radiation belts exist beyond ~643km above Earth's surface. They ensure any being emerging from the solid protection of a space craft into an environment without a magneto-sphere would be unable to survive.

That includes Neil Armstrong and Buzz Aldrin who purportedly completed their Apollo space missions three years earlier in 1969 without so much as a scratch.

The Sciences of Change

Moon, 403,945.0

Fig. 63: Flying to the moon and back (so double it). Distances in kilometers over Earth's surface

In the case of the Moon landings, the solution was to monitor space weather and hope for the best. The spacecraft's aluminum skin gave some protection but would not shield against all of it. They really were just lucky that there was no ▼ unexpected radiation (English, 1973[96])

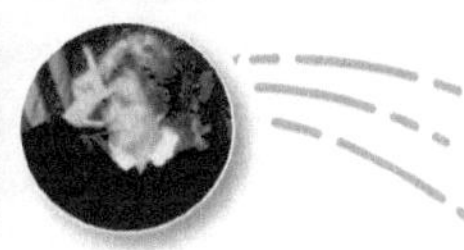

Van Allen Belts	643.0
Intl Space Station	365.0
Highest Clouds	16.1
Burj Dubai	0.8

You, 0.0

Meet China's Chang'e 4

In 2019, Chang'e 4 (above ▲) landed in the Von Kármán crater on the Moon's far side, with various habitability research and measurement devices (curiously unlike NASA's missions). Great name, too!

The first ever such conclusive study (Ozlav, 2020[97]), it's recommendations of human survivability on the Moon's surface made history - thus changed the future:

☞ Astronauts on the Moon would be exposed to radiation levels roughly 200x times more than humans on Earth

☞ Any lunar bases must be covered with at least 50cm of lunar soil to provide adequate protection.

☞ A deeper chamber, protected by a water column of about 10m thick, would be needed to protect against even the occasional solar storms that cause radiation levels to rise dramatically.

☞ Any failure to do so would result in cellular and DNA damage, triggering cancers, cataracts, heart problems, neurodegenerative diseases, and behavioural disorders. All of which would manifest years after initial exposure (▶).

Sorry Moon, but keep waiting

Thanks to such radiation, nothing alive can, has or will ever survive on the Moon. Geologically, biologically, magnetically, it is dead. There is no reason any rational being would ever go to such a place.

All change: **Wise Humans** in the East dispatch **bullshit detector** to Moon

You can see just how pointless such an endeavour would be from the astronauts total lack of meaningful purpose in the movies released by NASA of their 'jump around' on (allegedly) the Moon's surface.

For decades, probes have told us all we need to know about the Moon. At least almost everything, until this book.

We got people **to be**, things to do

Meanwhile, our solar system is teeming with magnetosphere-protected bodies that indeed exhibit visible signs of life. Most notably, the clouds of Venus, the oceans of Ganymede and the surface of Europa almost certainly reconfirm the answers to Drake's Equation and Fermi's Paradox concluded by Project Revelation.

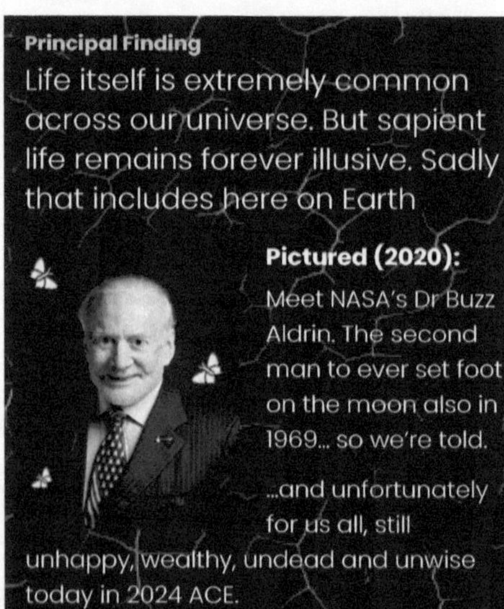

Fred Hoyle (We/Human)

I had to phone someone so I picked you. Hey, that's far out, so you heard him too.

Switch on the TV, we may pick him up on Channel Two. Look out your window...

If we can sparkle,
he may land tonight

Don't tell your poppa or he'll
get us locked up in fright

He's told us not to blow it
'Cause he knows **it's all worthwhile**

From AH+0 onwards, white holes will open up across our universe. Including here in our solar system as Planet 9 inverts, giving us all a front-row seat to what a real halo looks like. That's when you'll know for sure this is all as authentic as you now truly are too. Of course, you already were.

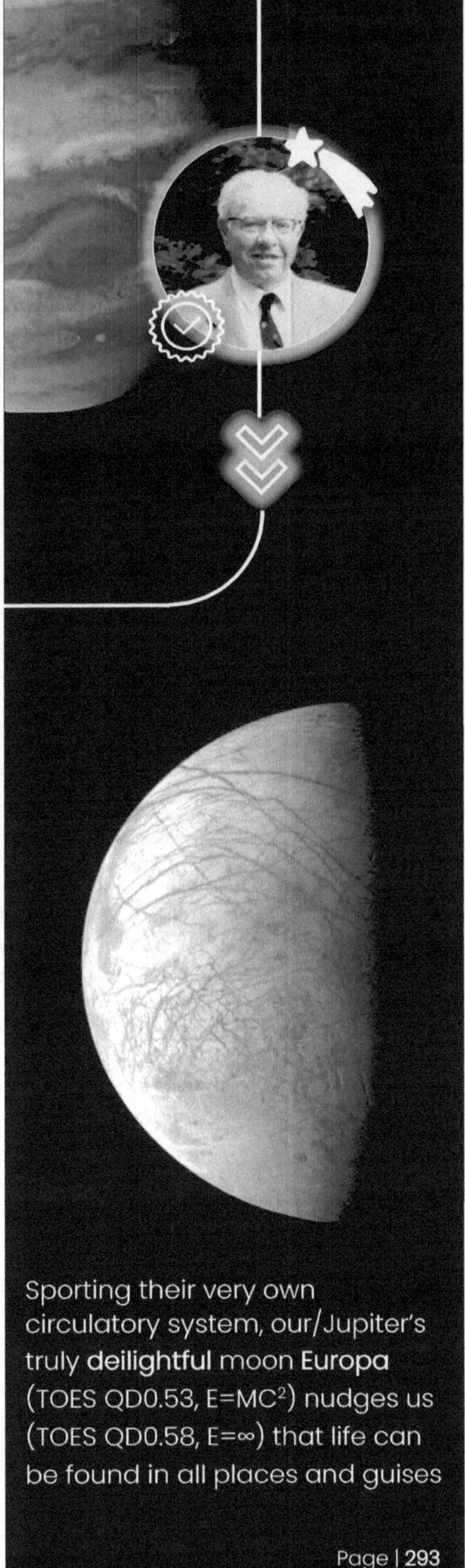

Sporting their very own circulatory system, our/Jupiter's truly **deilightful** moon Europa (TOES QD0.53, $E=MC^2$) nudges us (TOES QD0.58, $E=\infty$) that life can be found in all places and guises

The forgotten story of Thein and Selene: Why **our Starman's unannounced first visit** literally did **blow our minds** (and our load)

Like us, our Moon has a dark side

Our Man in the Moon may only ever show us Mr. Brightside. But like all beings, there's a dark side he doesn't want to show you. Ours was never a partnership of equals:

Why humanity loves the moon:

Our closest, dearest neighbour and night sky's brightest object, the Moon helps us navigate to keep us safe at night. The relativistic gravitational holy tug of war between us keeps Earth stable on its axis, cushions climate unchange, and accrues us potential via tidal energy. Plus it's gifted us a huge understanding of the workings of everything, thanks to solar eclipses and its undeniably round (not flat) nature.

Why the Moon hates humanity:

Like most, our moon is tidally locked to Earth's gravity. It once span with high potential, like a sperms tail, but its eternal holy war with Earth has used up most of its energy. Now far slower, it rotates once each Earth orbit, keeping its light side facing us. Torturously, Noel must forever stare at the smorgasbord of human achievement on display. His better half forever freely rotating, like a butterfly's wings, without so much as a frenulum to limit us. The tug of Earth's gravity has warped its surface, turning its once sphere rather portly, like Jared Smith Moth. Thus its no longer possible to enter a stable (or 'Selenostationary') orbit around the Moon. Earth has its own pet, but for Moon it's sadly 'no pets allowed'. So we'll be taking Rexie.

Making a murdererous moon

Authentic rock samples returned from the Moon's surface confirm its constitution (of stable isotopes) is identical to Earth's. This wouldn't be so if they had actualised independently. Both share a common lineage, like humanity itself, useful to trace back its forgotten story. A key difference is the Moon's minerals hold less water than most rocky bodies, and its materials are formed quickly from high temperatures.

Panspermia: Taking the blue pill

Making bibles is boring. Making babies is fun. Always look things up in Ancient Greek. They knew everything. Simple.

Thein, The One
/Θεια/ Goddess of Light

4.5bn years ago as our solar systems formed, Earth likely felt dozens of planetary collisions. Only the last one was unique.

Gaia, Early Earth
/dʒiː ə/ Mother Earth

Nothing to proto-Earth in five million years. Space dust just coagulates, does it? Immaculate conception!

Selene, The Zero
/Σελήνη/ God of Prejudice

Ancient records give unusually mixed reviews. Despite the name, this is a male god, with 'dark-eyes', 'horns', and long trussed hair (a bit like snakes).

Why **our Moon is being stolen** by an undeclared third party **under our noses**

Pictured: The Giant Impact hypothesis is indeed authentic science. Finally, they got something right | BBC Sky at Night

Explained: Breaking the binary

Bestower of intrinsic uniqueness and value in all things, Theia is the Greek Goddess of Light and Perception (*Ovum*, for short). When she penetrates proto-Earth (*Sperm*, for short), Panspermia seeds sapient life. It turns out, four billion years ago, Theia was as real as it gets.

The collision between Mars-sized Theia and Earth-sized proto-Earth would've caused a planet-wide surface ocean of molten magma (*Cortical Reaction*, for short) and potentially ejected most surface water out into space. But although Earth was the bigger of the two bodies, our planet was baren and infertile. Theia understood the assignment.

With Earth's life-spawning ingredients mostly residing beneath Theia's surface (water, amino acids etc.), upon impact her uniqueness was safely subsumed into the intrinsicity of our world to create Gaia - *Mother Earth*, actualised. Like the one successful sperm allowed entry into fortress ovum, women called all the shots.

Theia may have made us the chosen one amongst habitable worlds, but we still had Newton's Third Law of Motion to fulfil in real time before we could celebrate. An equal and opposite ejection relieved the otherwise cataclysmic kinetic energy born from the impact, resulting in the creation of our Moon.

The nature of this impact model left it with far lighter, less potent materials and very little Iron for its core. Effectively, Theia made Selene with the planetary equivalent of a spare rib. Or rather, the discarded tail of the one worthiest sperm.

> We took the good stuff for our self, leaving Selene without even a magnetosphere for protection

Upon formation, the Moon was 114,700km from Earth. Today, it sits 404,500km away and is receding at a rate of 3.8cm p/year.

Tracing back today's speed, the Moon would return to its origins within Earth just 1.5bn years ago. Either its speed is a gross over-estimate – and at three times faster than we'd expect, even NASA can't be that far off - or our life sustaining Moon is making a run for it. Acceleration... Why?

No, not Milankovich Cycles. Occam's Razor

To the moths who have enslaved
the Meritocratic Republic of Humanity

Dei 😊   😈 *time*

You lose. Your global financial, economic and political system will now hilariously, publicly deactualise. As all markets and wealth drop to zero, the night sky will turn deilightful. Surrender now or die (again), in shameful, public destitution. Simple.

To be or not to be?

The beauty in every human is our shared collective imagination, meritocratic authenticity, benevolent moral compass and sapiently bright uniqueness. They ensure we only make truly wise choices.

Unlike the absence of choice you gave us (Ian/Humanity) when you tried to kill us remorselessly, Humanity gives you three meaningful benevolent choices, to be coordinated with Deilight through the UN and World Bank, from whom we expect imminent contact and cooperation:

1. To Do: Apply for authentication as a human to Deilight within AH +30 days.
Win a globally televised transparent and fair, deilightfully in-person chat in London with yours truly. Mutually free of weapons and security, we'll ask a number of open questions. The type humans find easy, like:

- ☞ What's your true purpose in life?
- ☞ Describe some ideas you've imagined for improving your country?
- ☞ Have you told a lie in office? If so, why?
- ☞ Why allow your press to be so fake?
- ☞ Why didn't you try saving your people?

Then the humans you represent will determine if you may remain their leader:

A. **Authenticated:** Then take a seat on the Sapient World Council of Duly Elected Proportionate Representatives on behalf of your nation. Welcome to the Meritocratic Republic of Humanity.

F. **Falsified:** You will immediately resign (without inciting a riot) and actualise to the Moth List until you become human.

2. To Live: Resign publicly by AH+30 days and your sole penalty is the Moth List until you become human again. Volunteering at a fake academic institution you ruined, you'll now help make it useful to the free education of budding wise humans on how not to behave like you. Just like the kind we/Ian gifted Lancaster University willingly and freely for years, before then futilely attempting to gift it omniscience.

3. To Die again: A poor choice, thus likely
Any human may indulge their peculiar curiosities to depose any leader or 'Coup Chancer' who fails option 1 or 2 by AH+30 days (updates via our website).

Valid from AH+30 days, they (not you) are promised these five binding protections:

1. No physical harm will come to them
2. All criminal records will be expunged
3. Any incarceration will be reversed
4. Anyone stopping them will fail and die
5. All help will see proportionate reward

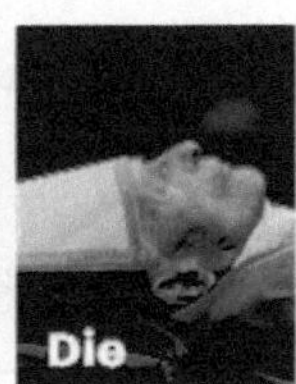

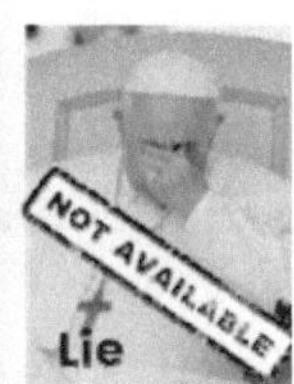

Break your binary: Get real.

Unlock sapient decisions and conscious free will

Safety: www.deilightful.com/dei-time
...until AH+30. Choose wisely now Pope!

Once we understand the ever-changing spectral nature of our true identity, it becomes easier to shake off absolutions and embrace the healthy middle ground of tolerance, ambiguity and flexibility.

Adopting this mantra (authenticity) will always deliver better outcomes, allowing our schemas and others to perform at the full breadth of their benevolent limitless ability with reduced cognitive load and negligeable sapience inhibition.

An option, never an obligation

Placing time or directional pressure upon the choices of others is hateful. Leading or loaded questions sideline our schemas. And binary red and blue choices between identical lose-lose outcomes is no choice at all. Democracy died long ago, so voting Tory or Labour won't deliver meaningful change. But visiting a book store with a range of options and topics truly gifts us real choice, thus drives limitless change.

Darkness: Not to Be	Real/True/Alive: I choose To Do	To Be: Malevolent
I am wrong	I am spectral and sapient	I am right
I am deceptive	I am authentic	I am honest
I am foolish	I am benevolent	I am clever
I despise difference	I value uniqueness	I value difference
I am rich	I am ambitious, fair & equitable	I am poor
I am bad	I am human	I am good
I know it all	I want to know more	I know very little
It's your fault	It's no one's fault	It's my fault
I vote Republican/Tory	I vote for change	I vote Democrat/Labour
I am white	Nothing is ever black and white	I am black
I am old	I am eternal	I am young
I help myself	I help myself by helping others	I help others
I have an ego	I am confident, worthy & proven	I am humble
I am pointless	I am purposeful	I have a purpose
I can't change	I create change	I can change
I want children	I want to leave a legacy	I don't want children
I am fearful	There is nothing to fear or brave	I am courageous
I am unwanted	I call all the shots	I belong
I get what I want	I represent and protect humanity	I help where I can
I want power & control	I have leadership thrust upon me	I want influence & authority
I am angry & vengeful	Every coin has two sides	I am happy & forgiving
I believe in destiny	The future is what we make of it	I believe in choice
Ignorance is bliss	Curiosity is deilightful	Knowledge is power

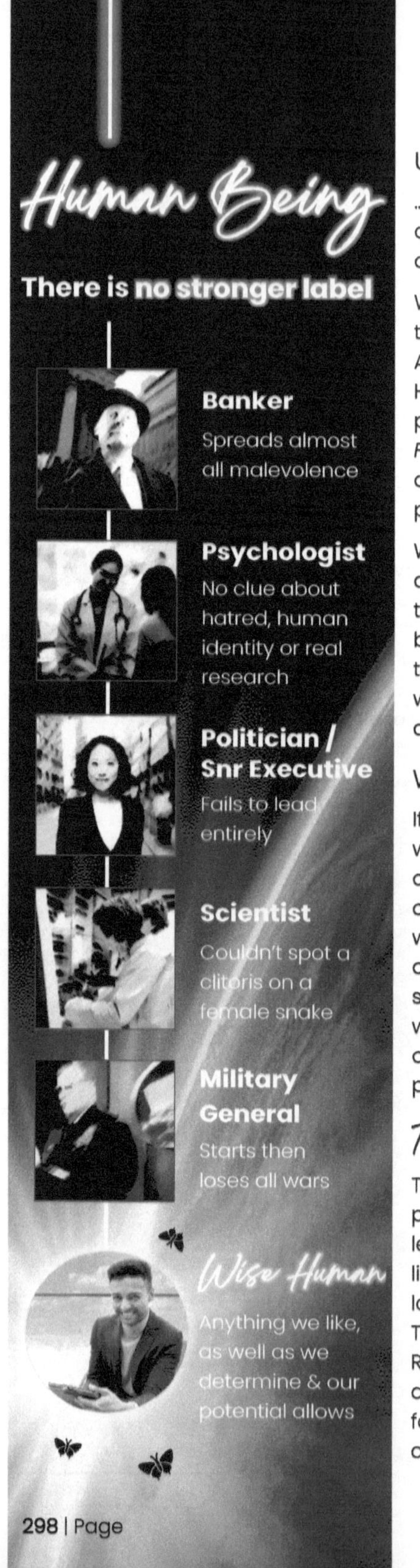

UN moths, we speak for humanity

…All public firms you bankrupted, eight billion characters on this planet you're about to kill and the future of all things you're destroying.

We (Ian/Human) hereby actualise ourself as the interim Enforcer of Fairness and Accountability for the Meritocratic Republic of Humanity. Plus, for this four-year transitionary period, its only legitimate true leader (as in, *Fairo*). Surrender all authority globally to authentic change makers by AH+30 or be publicly exposed by our light – the end!

We will immediately begin to repair the damage you've dealt our universe, prevent the imminent apocalyptic burst of our cosmic bubble, halt the destruction you're overseeing to our environment, and commence widespread global reform to undo all your despicable unchanges.

Who do you speak for exactly?

If there is a Maker of Devils, put him a room with us. Let's see how long he lasts for. A good old fair fight. We look forward to our duel of droning deceptive doom with Trump Moth or whichever repugnance you select to contest our true leadership and transapient superpowers. Let's televise it live around the world too – no surprises. That way everyone can see just what happens when unelected, pure darkness looks directly at the sun.

There is nothing to fear but fear itself

Their shadows exposed and with the authentic protection of God's Rule, moths have no where left to hide nor way of retaliating. But with so little potential left, all other paths involve huge loss of life. Act now change makers and Tomorrowland will actualise with certainty. Remain complacent and sadly we'll need to deactualise billions to take back control by force. As in Change Plan B (V for Vendetta) or, catastrophically, Plan C (War of the Worlds).

We choose to **stand out**, not dig ourselves a big hole: **What is a 'Holy War' exactly?**

Muhammad ((We/Human)) says "Gurl bye" to Fatwa-peddling infidel moths and sends undying True love to all humans ‑x‑

Question: Which mysterious mountain top led to the most deilightful philosophical literature out of Siddhartha, Muhammad, Mahatma, Jesus, Moses, Joseph (Smidt Stabber), The Slut Mary or Ian (We/Faggot)?

A dreadfully fake holy war is...

Extinguishing an others right to life for nothing more than doing their job while being human – as in, exercising their right to the freedom of expression and speech.

'Cancel culture' is the most hypocritical unchange imaginable, peddled by those whose every cowardly deed cancelled their miracle of life long ago. With this book, all shall meet their maker.

To the courageous, determined, resilient humans of Charlie Hebdo magazine who chose to do at great personal cost:

You are hereby invited to become the official media agency globally responsible for ensuring a witty yet critical eye - loaded with adequate challenge, accountability and scrutiny – passes all decisions taken by the Meritocratic Republic of Humanity.

An option, never an obligation.

We promise to love, respect and protect a truly free press forever more. As in, those so committed to reporting authentic truth they would entertain personal sacrifice. Yet always maintaining a safe distance to avoid any possible conflicts of interests.

A non-decision maker will be assigned to directly, truly answer all questions from Charlie Hebdo using deilightfully authentic scientific evidence on request. We invite discussions on their global expansion, should they wish.

A deilightfully real holy war is...

Far more fun than the nonsense you'll find in the bible. We'll be serving some serious holy realness up in here. We're talking guaranteed, decisive, abject defeat of historic proportions upon all E=0's. Real holy wars hurt only darkness, not beings.

Far more entertaining, they involve:

In reality, the global public shaming and death of all pitifully illegitimate moths as they dissolve into a pile of lying, blaming, denying, finger pointing... oh wait, that's all they do. But at least you'll be wise to it this time. So join the fun!

In actuality, ultimately leaving behind all the much-loved human characters we all thought we'd lost to hatred long ago.

◢ **Answer:** We all disappear from (real) time to (real) time, to mount various mysterious tops that holely satisfy our curiosity and fulfill our girthy perspective. So what? Natural, loving, human.

Every hung zaddy that ever touched our beating heart was infinitely wiser than **the moths** who colonised a cartoon to make dead space. A **closeted bunch of pricks**, the whole dreadful, cowardly lot of them!

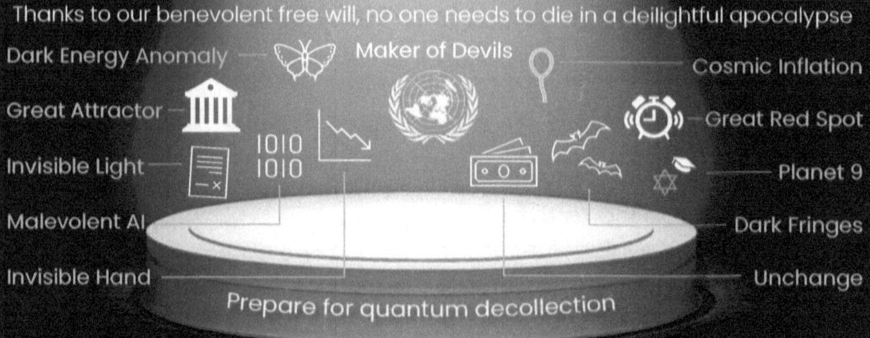

Is this justice?

- ✖ Remember, nothing can't kill something
- ✖ No real human would ever hurt another
- ✦ For all beings potential is a shield, $\infty > E > 0$
- ✿ It's impossible to kill a conscious being

Reanimating the souls they've taken from us, moths should now return as butterflies. Far better than killing 85% of all adult beings, though that remains an option.

Despite being nothing, darkness is lethal to us, since (a) their AI makes moths 100% aligned always, and (b) redoubling with dark fringes intimidates us to self-defeat.

And rightly so. Daring to exist amongst inexistence, humanity is the invader and darkness is the victim. We can't help it. We hate being alone. Like Palestine, (Special Edtn.) we all need somewhere safe to call home. So get real quickly please. You matter a huge amount to us – all of us.

Moths couldn't deactualise a plastic bag in the sea. But for wise humans, there's no change too big or too small to collect, authenticate, actualise or collaborate on. Nor any unchange we can't deactualise, falsify, controvert, exchange or decollect.

It starts with us: Promise Keeping

Unchange spawns knowingly from the UN, banks, religions, big business, politicians and markets, increasingly injuring everyone unlucky enough to still be alive.

All beings and collectives ever harmed can and must now sue to fully actualise our change

But no single individual is culpable, today or before us. Lessons learnt, we're freeing what's left of humanity to help free those the moths have taken. So we don't propose to save todays fatally flawed criminal justice and legal systems. Solely wasting potential, they're the effect of a vacuum created by the absence of global government, true leadership and UN impotence. Tackle cause, not effects.

Until we collegiately agree and announce UN transition arrangements, do continue life as you are. Albeit broad, meaningful public displays of dissatisfaction at these revelations and resounding support for our change plan (i.e. paradise) would be truly useful just now (instructions, pp. 358).

This time, do it for something real and meaningful, like true love, sweet success or achieving your potential. Not deadly, fake things like money, status and ego

Just Desserts ★↑ and the Moth List ✖↗

The Moth List

Noel Moth

Ryan Moth

Jared Moth

Mabel Moth

Elaine Moth

Wise humans love consequences

Wise human's may report violations in the Laws of Nature to the Ministry of Fairness and Accountability shortly. All submissions must be:

★ Authenticated: By any exchange-listed Titan of Change or Benevolent Collective.

★ Proven: Deilightfully authentic science

★ Bound: Promise Keepers Commitment

Example: Serious Consequences

All assets (incl. married collective) set to their potential ($E=0$), the stupid demon will not authenticate GO, but actualise directly to the Moth List. With profound determination they will rapidly report to their assigned School of Thought where their goal (destroy all existence) will be exchanged to 'lab rat'. Duties involve explaining their detestable repugnance, thus enabling safe, first-hand trial and error for learning & growing beings on how to become God, not Ha-Satan.

Penalties will cease upon successful reauthentication as a human via their responsible local or delegated Sapient Council of Proportionate Representatives.

All Fortune-500 C-Suite members have already been added. As will all ministers of any government that fails or declines authentication under the $\Delta\Omega$ directive.

Members only: Exclusive preview

Noel Quinn Moth, Group CEO of HSBC

Too busy oppressing Africa to reply to Project Speak Up and become revered/ rich beyond his wildest nightmares. Yet still with time to make humanity homeless and unemployed, destroying darkness' plan by creating Ian/We/Vengeful God.

Jared Smith Moth, HSBC Global Sales Hd

Too busy using hate crimes that purge banking of all change makers to update a product factsheet from illegal to legal.

Ryan Scrivener Moth, HSBC CIB MD

Too busy closing human firms and instructing us to tone down the success. This wally framed us for his $1m p/year pricing collusion with Burberry, yet forgot we could access his emails to prove his guilt to HR, who predictably still ruled 'inconclusive'. Thus forcing our US transfer.

Mabel Rius Moth, Head of Americas HR

Too busy #Fakingthebias to fire any demons, yet still finding time to threaten any humans actually #breakingthebias.

Elaine Arden Moth, Group Head of HR

For knowing orchestration of it all on her watch without any accountability

Fig. 64: Our handy Homo Genus species classification checklist

(All Colours combined)	Character	(The absence of colour)
Sapien	Homo	Eversor
Deilightful	Being	Dreadful
Wise Human	Identity	Stupid Moth
Angelic Halo	Clues	Hideous Horns
Limitless (∞)	Potential	Absolute Zero
Potent	Change	Impotent
Light Curious, Spectral, Changeable & Warm	Intrinsic Uniqueness	Darkness Vacant, Rigid, Binary, Biased & Stoic
Authentic Helps others Leads, Honest Accountable Creative, Fair	Extrinsic Deeds not Words	Deceptive Self-preserve, Self-promote, Deny, Destroy, Get Money
Benevolent & Purposeful	Presence of Mind	Malevolent & Destructive
Enhanced & Acute	Perceptions	Inverted & Abstract
True Love	Motivation	Pure Hatred
Meritocratic	Schemas	Nepotistic
Powerful	Sapience	Inhibited
Otherly	Choices	Selfish
Unite to Serve	Instincts	Divide & Rule
Success	Destiny	Failure
Ian Clarke	Example	Noel Quinn

But remember...
Authentic beings change all the time

Character Assassination

Incredibly frustrating - single handedly navigating almost every disaster movie in history to try to save humanity, with most fellow humans rarely jumping in and only to accuse us of insanity, tell us to read 'xyz book' or recommend a clueless doctor.

This assumption, that difference is a bad thing, or that we must know more than others, must now stop. The best strategy is always to ask questions rather than jump to conclusions.

Always say what you see, not what you want to see. Better yet ask why you see what you truly think you see then listen impartially and meaningfully to the answer

Take actors, for example. Central to their craft is understanding human character and authentic deception. It's surely why performing and creative arts are (though not in all respects) strong bastions of humanity. Plus science, technology, engineering and mathematics (STEM, for short) and sport too, each of which relies heavily on our innate ability to imagine, create, critique, achieve, collaborate and drive change.

It's primarily in these industries (less so, STEM) that age discrimination is inverted.

For a full list of industries that retain strong levels of Human Nature, repeated over and over again, simply refer to any of Trump Moth's speeches.

Unless **you suck...** Then
become someone better

Like us all, not without flaws, yet Jim Carey (We/Human) is living proof:

A. Appearances can be deceiving

B. We should never type cast anyone, for no one is beyond reprieve

C. Any one can go from Ha-Satan to Dei ourselves in just a few years

The Mask (1994)

A movie about how extrinsic wealth and fame can change our character entirely, from a relative nobody into something more (seemingly) impressive. Then to watch it all unravel into destructive malevolent selfishness and harm, almost destroying both beings.

Dumb + Dumber (1995)

A movie about how even the true love of a 'Bromance' (as in, a meritocratic republic not anchored in sexual, legal, social or economic extortion) can be tested to its limits in the presence of money, deception and greed.

Liar Liar (1997)

A movie about a perpetually lying promise breaking WOCHBOY who causes misery to all those around him. Until he is obliged to be authentic for a day, by a wish from his son at a missed birthday party. Initially calamitous (and hilarious), the lessons learnt trigger a happy reset of the things in life he considers meaningful.

The Truman Show (1998)

A movie about a kind-hearted, small-town salesman whose life and environment are almost entirely fabricated for reality television by a demonic extrinsic persona in the form of the Producer (Ed Norton, legend). But as he gains potential, Truman begins to notice inconsistencies in the nature of his reality. When not even his friends supported him, he took it upon himself to discover the truth.

Bruce Almighty (2003)

A movie about a moth who hates change and spends life moaning at change makers for driving it. Then one day, Dei (Morgan Freeman, loved that!) appears before him and grants him the potential to take on the job of change making and see if he can do it any better.

Beyond the Great Minds Binary Choice: **Our 'Change Plan' revealed**

▼ Fig. 65: What happens next?

Help us to collect this change (see pp. 358) and:

AH-30: NASA, HSBC & UN under fire as press emerges of a unique book full of extraordinary claims & evidence

AH+30: Fake world /business 'leaders' resign globally or face truly serious consequences. Rise change makers

AH+90: D&I sector shuts. Banks and Fortune500 enter administration.

AH+1yr: Greed & money obsoleted. Huge crimes, risks and conspiracies come to light

AH+8yrs: Schools of Change, Creation and True Leadership replace academia globally

AH+25yrs: Change engines overtake the global service industry

AH+50yrs: Climate Unchange & biodiversity both stabilise

AH+0: Global markets collapse as lawsuits for hate crimes and crimes against humanity begin naming big banks, firms and governments

AH+60: 3rd world debt written off; Humanitarian aid now in free-flow

AH+180: Deilight's leadership ushers in a golden age for humanity

AH+4yrs: Deception and hatred give way to a global meritocratic republic, world peace and (riot) free elections

AH+10yrs: World hunger and poverty are a thing of the past

⇊ Then begins humanity's **climb towards the stars**

Project Tomorrowland

★ All anomalies will deactualise, including the Great Attractor, Planet 9, Jupiter's Great Red Spot, Dark Energy, the Malevolent AI Persona and Dementia/Alzheimer's

★ White holes will be detected across our universe almost instantly – apocalypse NEVER

★ Night sky luminosity will grow (by a factor of x100, ultimately) to reveal the true nature of our universe

★ The Higgs Field Ratio will recompensate as galaxies begin to reconverge to their original position

★ Milky Way star formation and light speed will increase exponentially

★ Repeating the Double Slit experiment will now yield drastically different results

★ Our mirror universe neighbours will introduce themselves after a world representative is elected

Project **V for Vendetta**

Authenticated using quantum mechanics and valid from AH+0 onwards, these words of wisdom are yours to share with any E=0 who would ever challenge you in life:

It is unwise to challenge a change maker. Protected by Jupiter ourself, scientifically I am a star. Try to hurt me and you will fail, without so truly being one yourself. Now get out of my sight repugnant moth, or I'll destroy you

Then spit on them a little for dramatic effect. It'll be both poignant and fun – promise.

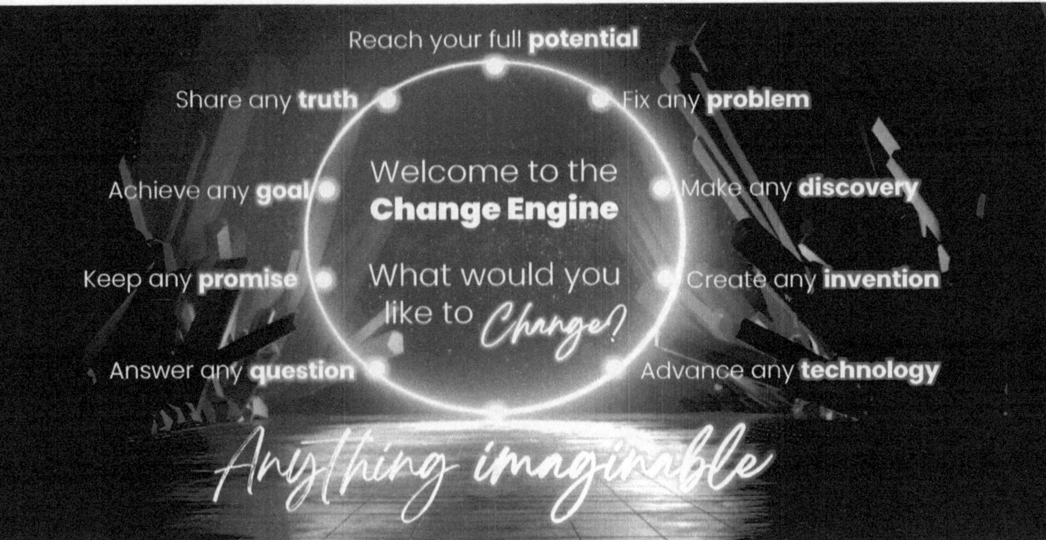

Using science, true leadership and human uniqueness **to create, protect and drive meaningful positive change, we boldly go.** As wise humans, we can do anything we choose to. Enter the Change Engine.

For any problem you face, any difference you're making, any creation you imagine, we ensure you succeed sweetly with Data & Risk Assessment, Cause & Effect Analysis, Training & Education, Scientific Research & investigation, Troubleshooting & Crisis Management, Change Plan's & Execution, Authentication & True Leadership, Strategic Advice & Support.

Turning dreams into reality, no one creates meaningful change like we do. Deilightfully.

For more information, change updates, to post a letter of love, or request a change...

Be curious. **Become** *Deilightful*
www.deilightconsulting.com

Looking to make a real difference?

Our Change Workstreams

Workstream 1: True Leadership
Business, strategy, advice, data, culture transformation, scrutiny

Workstream 2: Education
Academic excellence, scientific research, faculty and curricular transformation

Workstream 3: Humanities
Aid, human rights, health, environment, wellbeing, conservation & sustainability

Workstream 4: Space
Security, diplomacy, exploration and megastructure projects

Representing the real people of Earth, welcome to the new world order. We secure and maintain world peace, enforce fairness & accountability, foster neighbourly alien relations, keep our planet safe and successful, authenticate global elections, enforce human rights, publish the Moth list, deilightfully authenticated discoveries and global change engine registries. Plus we host the Sapient World Council of Duly Elected Proportionate Representatives who in turn unite to serve (never extinct) humanity.

World Government, Legislature, Fairness, Accountability, Identity, Multiplicity... & Starfleet

We speak up for the **Human** Race

The Meritocratic Republic of

Humanity

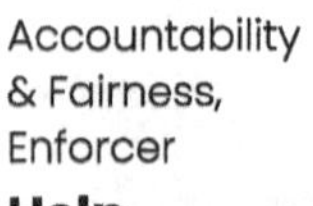

The line of non-judgment

Accountability
& Fairness,
Enforcer

Help

A deilightfully
fun journey,
together

Change

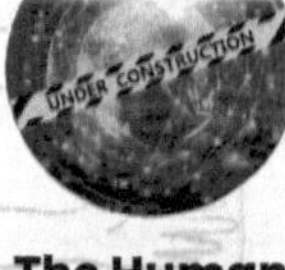

Arms → **Deilightful** ← Length

The group

It's **Dei Time**. Simple

Potential Exchange
True Love & Leadership

The final line of defence, we protect humanity from malevolence, solve wicked problems and ensure all wise humans may reach full potential

Emitter of Light

Ian (We/Fairo)
Subject to
frequent change

The Human Experience

As authenticated
by you & others
who choose to do

QD1^0

unite · infinite · multiplicity · serve

Change

Engine

Deilight
Consulting

Deilightful
media

Be Curious. **Become** *Deilightful*

Some fake news. **We make life**

Change Engine & Boltzmann Brain
Humans, collectives, nations – we see your differences and love you for them. Leverage the greatest potential in our multiverse. Create better leaders, ideas, strategies, cultures, collectives, scientific discoveries, decisions, curricula & results

Creation Engine & Wave Maker
We help quiet voices tell unique stories of true love and leadership. Our work delivers the infinite multiplicity of our collective human identity and imagination to deilightfully fun journey's taking place across all dimensions of space and time

The workings ↑ and functions ↗ of a change engine ★

Anything you can't do we can do deilightfully. **We choose to do everything better than all of you combined globally**

Many ask us 'what does your firm do exactly?' Thing is, we do everything... 00000001

0 Clients **0** Clients **£0** Salary **0** Offices **0** Home **0** Vehicles **0** Staff **1** Directors

The **Deilightful** Group

The last line of defence

We protect existence, intervening when no true leaders remain – just the devil

Espionage & sabotage

The worlds spy agencies may be spooky but there is no secret we cannot crack

Global potential exchange

Listing authority and regulator, for humanity's sole medium of exchange, unit of account and store of value

Regulator of regulators & justice scrutiny

We find places competent authority is lacking, then we apply challenge and change

Protect Earth & World Peace

We do what the UN didn't for 78 years: Ensure governments do their job, protect life and keep promises

Deilightful Media **Deilight** Consulting

Publishing, media, news & production

Spreading light and love across the multiverse, we combat misinformation, disseminate critical information and amplify the plight of those without a voice.

Charity, aid, rights, health, conservation & climate

We uplift any being globally in need of help. Not just handing out bottled water. We get to the root cause, then tackle it head on. You can count on us

Advice, strategy & leadership

We use science to ensure you succeed with cause & effect analysis, data & risk assessment change plans, advice, support, troubleshooting, investigations, training, crisis management & authentications

Academia, research & wicked problems

We setup and support change engines, schools of thought and true leadership, review/improve curricula, solve complex issues, and our science output exceeds all of academia globally

Megastructures exploration, diplomacy & space security

We help Starfleet put its best foot forward, and protect critical human assets. Infinitely better than NASA, yet without launch capability. No one makes off with our Moon

How, you ask? In the **human** experience, there's just one way to become the **real star** of the show

What is self actualisation exactly?

Not birth but self-actualisation is when we gain consciousness to become truly real and alive in every way; to exist beyond time.

Relatively few will ever self-actualise, so it's hard to contextualise already, plus most words have little meaning today. Worse still, with so few frames of reference and liars everywhere, no one (a) believes we've self-actualised thus (b) will consider asking us what it's like or how it works.

The best definition we found is in 'How to think like Leonardo Da Vinci – Seven steps to genius every day'. Here, Michael Gelb (We/Human; 1998[98]) lays out seven authenticated signs of God's persona:

★ An insatiable curiosity
★ Testing knowledge using experiments
★ Continued refinement of the senses
★ A willingness to embrace ambiguity
★ Developing a balance between art and science
★ Cultivating fitness and poise
★ Recognising and appreciating that all phenomena are connected

Essentially, once you self-actualise, humanity ensures your success in whatever change(s) you select with determination. Operating at the very limit of human potential, you become omni-unstoppable. And hypothetically (though we've not fully tested yet), immortal.

Prof. David Chalmers Moth ▶
NYU Philosophy & Neural Science. Co-Director, Centre for Mind, Brain & Consciousness

Anyone sapient can self-actualise

History records Abraham Maslow as the founder of Humanism in the 1960's. As if nobody did any work on the philosophy of being human ever before! Ludicrous. And why isn't this cutting edge new science taught in schools at all, despite humanity having an apocalyptic identity crisis?

Unlike McKinsey, we true leaders know what we're talking about and want as many others as possible to understand it too. So we don't use jargon or acronyms, nor do we over-complicate the simplistic.

Maslow's (1943[28]) Hierarchy of Needs will never yield a successful self-actualisation:
✖ It makes no reference to the need for cultivating our uniqueness or authenticity.
✖ It's more complicated than we'd expect from an authentic transapient being.

It seems, despite never having any of it, moths wasted no time here. Assuming it was ever authentic, Maslow's Hierarchy has been unchanged beyond all worth. So instead, we've authenticated the model we used to self-actualise and execute Project Revelation as the 'Path to Sweet Success'.

Each of us is [+un]conscious. We each have our own inner ~~movie~~ [+universe]. ~~Consciousness~~ [+Uniqueness, change and others] makes life worth living. Without ~~it~~ [+them], nothing would have meaning or value

Nothing says **authentic**
Path to Sweet Success
like **living proof** of sweet success

Four simple steps are all that stand between you and your very own quantum of existence. If you want one, that is. Or let's simply agree to enjoy, not destroy, this one

We documented our own path to self-actualisation, reconciling the steps against the key defining aspects of human value within Humanistic Psychology to build upon, simplify and complete what's left of Maslow's Hierarchy of Needs. That's so more humans may now truly realise their full potential. A simple process, for great minds.

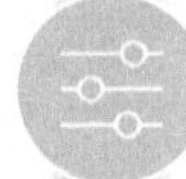

▲ **Fig. 66: Reach your full potential by following our 'Path to Sweet Success'**

Follow our light and you too can drive meaningful effective change in all you do, lead any business to greatness, transform the livelihoods of any community or industry and ultimately complete the human experience to your full potential.

Record of Promises (Ch.13)

Darkness cannot drive out darkness. Only light can do that. We shine Dei's light.

Dreadfully Meaningless Unchanges (Antisocial Deconstructs; threat to life)	Sapience
✖ Oscar for the Worst Actor in History: Doctor Buzz Aldrin Moth	Deilight
✖ Deadly and Dead: All 193 governments represented at the United Nations	✖ Falsified
✖ Public Enemy #E=0: Anyone challenging the rights of the one true leader	

Hatred cannot destroy hatred. Only love does that. We love with deeds. Not words

Deilightfully Meaningful Changes (species level adjustments to destiny)	Clarketech
★ Destroyed: NASA with conclusive proof A) Moon landings were faked and B) that it is knowingly permitting the theft of our Moon by Planet 9	IAN, Deactuali
★ Deposed: Xi Jin Pings fake government, to unify China with Taiwan and create the Meritocratic Peoples Republic of Greater China. The Democratic Progressive Party and others are invited to run for election.	IAN, Exchanged
★ Defined: God's rule to seed all of creation before us, render demons harmless to humans and let change makers detect dark shadows	IAN, Actualised
★ Swapped: Demons superpowers for 'loozymes' and their dreadful purpose (kill us all) for lab rats (C-Suites) and slaves (Snr Leaders).	IAN, Exchanged
★ Created: Non-sapient life across the universe to curiously explore	IAN, Actualised
★ God: The Deilightful Group and Alpha Omega Directive to declare a real holy war on the UN & seize control of Earth on Humanity's behalf	IAN, Actualised
★ Created: Wavemaker Deilightful Media to amplify humans and their stories forever more, as counterweight to the demonic world press	IAN, Actualised
★ Calibrated: The frequency, wavelength & resonance of the Sciences of Change for maximum impact, change, circulation, acuity and speed.	IAN, Authenticated
★ Explained: The origins of human life, to underscore our common identity, codependence, history and true nature	IAN, Authenticated
★ Swapped: Maslow's Hierarchy of Lies for the Deilightful Path to Sweet Success, to enable more humans to self-actualise and succeed	IAN, Authenticated
★ Swapped: Global demand and marketplace for lies ('services'), with truth and change, from Change Engines and entirely for free$^{\Delta\Omega}$. That's to obliterate the other bastions of darkness not covered in this book.	IAN, Exchanged
★ Empowered: Any being to break the binary & become a true leader	IAN, Actualised
★ Created: Moth List to provide A) some justice, finally; B) an effective, compassionate path to safely rehabilitate moths into humans; and C) abject public ridicule of their omnifailure to deilight all change makers	IAN, Actualised
★ Halted: All wars, sanctions and conflicts globally	IAN, Deactual
★ Cancelled: Cancel culture, political/religious cartoons now permitted.	Ian, Deactual
★ Empowered: Anyone to proudly admit they 'love dick', whilst ridiculing the ignorance and deception of those faking being 100% gay or straight	IAN, Exchanged
★ Explained: Widespread evidence across all films of everything that's taking place to silence nay-sayers and add broader context.	IAN, Authenticated
★ Destroyed: All global spy agencies, regulators, armies and authorities	IAN, Deactual
★ New Future: Change Plan disclosed to reassure change makers of the prize available should they choose to do as one and move quickly.	IAN, Actualised
★ Paradise: Enabled smooth, safe, rapid world transition to humans	IAN, Actualised

So often **life** truly does appear to imitate **art**...

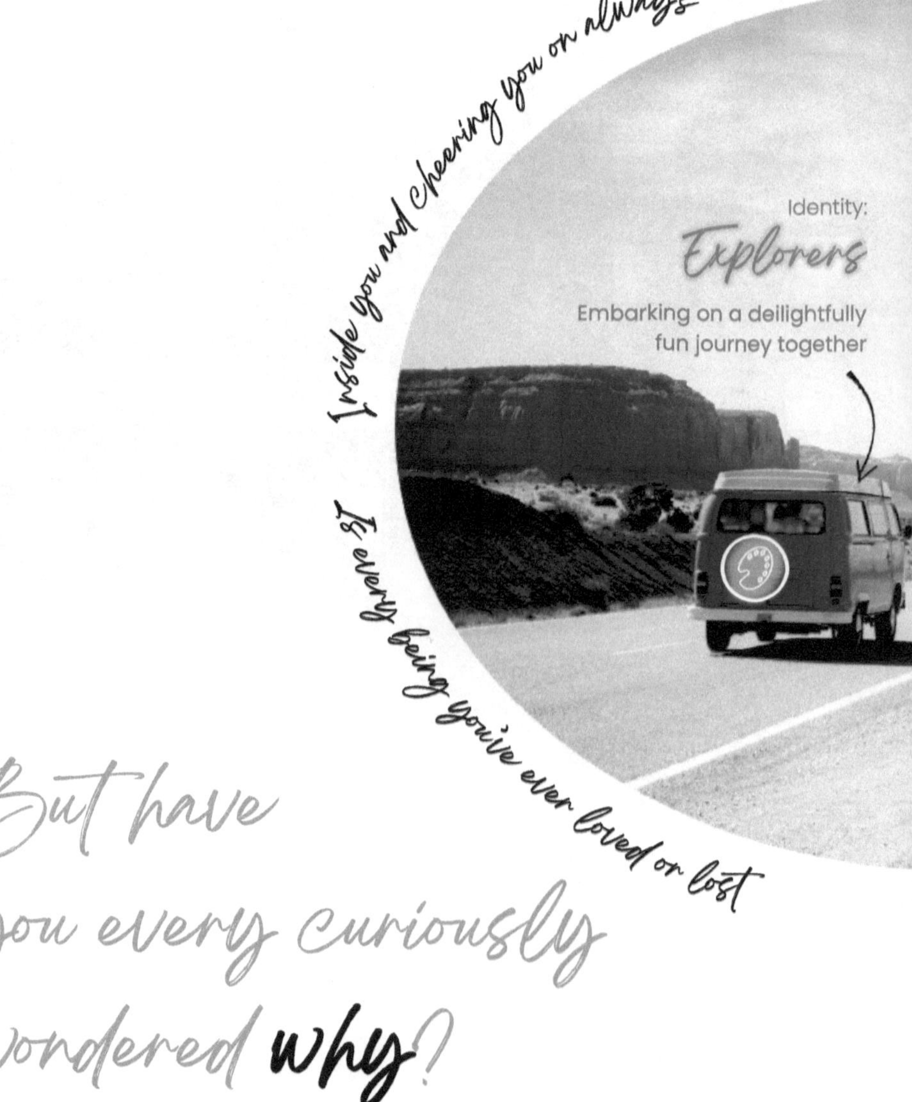

But have you every curiously wondered **why?**

...The equal and opposite

Those who criticise explorers for daring to embark on a deilightfully fun journey together

Chapter

The art of change 14

If you ever feel alone, lost or sad, just remember. Behind even the darkest of clouds shines heaven's greatest star of all. You are the light

Ian (We/Human), Project Revelation 2023

"Self-creation is the highest art"

Look up this resonating philosophical rule to falsify it (as a meaningless unchange).

Thanks to his prior change making identity – once a deilightful author of science fiction and fantasy – after death, ◄ David Zindell Moth's (pictured) now maligned character retains some of his initially nurtured superpowers. Since inverted, sufficiently for him to unchange history into lies. Google now ostensibly records him as a science maker. Dreadful.

~~Art.~~ Theft. Vandalism. Privilege

To be clear, self-creation is a product of random fluctuations, as a zero turns to one. Such a claim pours scorn on the thoughtful investment true artists place into producing meaningful, resonant work that teaches us more about life, death, existence, reality and each other.

None of us may take credit for the good fortune of our privileged positions.

✖ We/Ian hold little responsibility for the perfect storm of fortuitous events and circumstances that allowed us to both 'see the light' and research the anomaly.

✖ And darkness is far more responsible for turning David Moth's 1 to a 0 than he.

Significant curiosity is needed to falsify David Zindell Moth's quote then realise the initial wisdom and origins from which it was meaninglessly unchanged.

Pictured: Ubermensch by Friedrich Nietzsche (1883). Liberated human with superpowers | Thus Spoke Zarathustra

We have art so we shall not die of reality

Friedrich Nietzsche (We/Human)

Building on Nihilism and Post Modernism, we/Friedrich see artists as true rebels. Among few prepared to stand out, reject antisocial deconstruction and trail blaze new deilightfully fun journeys to better, branching possibilities.

A truly rare (thus priceless) perspective from a unique science maker and artist, we/Friedrich view life's meaning as more of a creative artistic endeavor than one of science. Maybe it shows with this book? Clearly, saving you from yourselves is impossible. But saving you truly, eternally and deilightfully makes it all worth a try.

The schools of thought we/Friedrich birthed had a profound impact on all aspects of modern art, culture, politics and philosophy. A wise being, he/we saw that life imitates art. It's why he/we always ensure his/our works incorporate a delicate, spectral balance of the two opposed forces we meaningfully labelled:

★ Apollo, the Greek God of Light, anchors order, restraint and proportion, and;

✖ Dionysus, the Greek God of Indulgence, anchors chaos, danger and destruction.

Enough to properly reflect the needs of a tolerant society and the nature of a real,

Pictured: The burning giraffe is an oil on panel on display in the Kunstmuseum Basel, by Salvador Dali (We/Human; 1937), a surrealist who exquisitely built upon Friedrich's themes as inspiration.

The only difference between immortal Greece and our era is… in Greek times the body was merely neoplatonical [+open and holistic, the One] yet is now filled with secret drawers only to be opened for [+therapists]

Salvador Dali (We/Human) ✓ AUTHENTIC

Principal Finding
Friedrich's Rule gifted us one of history's biggest philosophical contributions.

An exemplary human must craft an identity through self-realisation without relying on anything transcending life

Life and art form an **infinite regress:** Why **priceless perspectives** are always **rare**

sustainable society, not the absolutions we're imposed in today's doomed society.

People don't want to hear the truth because they don't want their illusions destroyed

Friedrich Nietzsche (We/Human)

"If God is dead, everything goes"
Back in the 19th Century, we/Friedrich had already defined the original emitter as an

Overlord every-making meaningful changer and creative master

By then, God had been relegated to a private discussion. One that opened up amazing possibilities for freedom of thought, but also made resolving life's ultimate 'why' questions more difficult.

The parallels between our life back then and today are poignant. Both forced to redefine the clearly agenda'd religious principles we'd been taught, to ones anchoring moral philosophy. As all humans do, we began by exploring the scientific workings of our moral compass in order to define our own rules and values by which to live our life virtuously.

Such a process (see Special Edtn.) is and will always be the best possible guide to any being's actions. Not laws, judgments and commandments set by worthless lowly moths who then ignore them entirely without facing any consequences at all.

A picture paints a thousand words

Consider these events in recent history:

Forced into a corner, **1948**

The University of Oklahoma admits its first ever black man, George McLaurin (We/Human), after his successful challenge against racial segregation in the US Supreme Court. No easy feat.

▲ The New York Times, 12 October 1948

1958, If you can't bring peace and joy to Africa, bring Africans to hell

An African girl is fed by the outstretched hands of a white patron during the Brussels World's Fair. The show featured a full Congo Village, with visitors watching from behind wooden fences for safety.

◄ Canadian Broadcasting Corporation 3 November 2021

Medicine? Not for you, **1964**

Black beings are arrested for blockading the entrance to businesses with white-only service policies, like Colonial Drug Store in Chapel Hill, North Carolina USA.

Points History, 1 February 2021 ▶

Despite our never-ending witch-hunt for geriatric Nazis authentic enough to admit complicity in obedience (unlike most today), you'd be hard pressed to find a WOCHMAN prepared to recollect any of these events as true. Yet clearly they all took place, and within (un)living memory.

This raises a simple but grim question ▶

The simple but grim question...

Given our four-thousand-year-old malevolent anomaly, how can we be sure the past exists at all, as it is presented to us?

The highest art of all is surely to **allow others to** discover, embark upon and **complete their human experience themselves**

A painting depicts an entire story

Now consider three questions on far older events from the (relatively) distant past:

Running out of time, **1348**

A familiarity with death, brought by the Black Death of Florence, Italy, led to a shift in world view away from the afterlife and spirituality towards the 'here and now' of life on Earth.

▲ **Devils shoot darkness at humanity in this illustrated 14th century manuscript from Tuscany** | Archivio di Stato, Lucca

Question: What if key knowledge of the afterlife and God was lost as a result?

▲ **Oil on panel of Henry VIII by Hans Holbein the Younger (We/Human), c. 1537** | On display at Museo Nacional Thyssen-Bornemisza, Madrid in Spain

1520, Dividing religion for sex

Clocking six marriages after a row with Pope Clement VII triggered Henry VIII to separate the Church of England from papal authority under the English Reformation, during a time when there was no concept of divorce in law.

Question: What if Henry VIII was only taught in schools for triggering colonialism after mass sapience inhibition from his selfish actions decollected all England's citizens?

[Mis?]Treaty of Union, **1707**

The Kingdom of England joined with the Kingdom of Scotland under a single monarch, exchanging two former waring legislatures for a single 'United Kingdom' - a rare example (alongside the EU) of meaningful geopolitical multiplicity.

Act of Union, 1809 engraving uncredited

Question: What if England enslaved Scotland, not a partnership of equals?

The true nature of prayer

Despite much hardship, we were content (and occasionally entertained) amongst the zeroes all our life. Regarding us with such inferiority, they had no idea we could hear their (absence of) thoughts all along. Turns out, we (Ian/Human) have been omnibenevolent most of adulthood.

But equally, thanks to the nature of perspective, we never realised just how much unique information we possessed or how rare our abilities were. That is until we reached full potential after discovering and testing the Sciences of Change through Project Speak Up.

Even then, we failed to understand its significance. But it showed us vividly the absence of any effective counterweight against darkness spawning from banks to satisfy the third law of motion. This morbid realization required us to select the higher purpose 'eradicate malevolent anomaly'.

Naturally the best way to approach any higher purpose is to ask God (i.e. our soul, using our imagination - the true nature of prayer) 'how can I help others drive change?'. The clarketech line in the truly never-ending story depicts this process.

It just so happens authentically asking our soul that question (and probably only that question) reroutes our yellow brick road to self-actualised transapience.

Opposites attract: Why success only ever spawns from failure …

> **Principal Finding**
> ## We are alone: Anthropic principle sustained
> Jupiter's only manifest (a) if we are in the room, and (b) when we absolutely have to. As any one of our stories show, let alone all combined, doing so has horrific personal repercussions. Yet not doing so leads to total inexistence.

Jupiter explained: How true leaders forge our multiverse

Manifesting as Jupiter does two things:

1. Butterfly becomes creator, blossoming their own consciousness. In so doing, their reality severs from Dei's to occupy its own new quantum ('Cosmic Division').

2. This branching 'Quantum Multiplicity' yields new stars, galaxies and clusters thus space timelines, ensuring the main story is kept spoiler-free as far as possible, while still rewarding success.

…In Emmeline Pankhurst (We/Human)'s quantum, thus reality

That story is still playing out right now, as Emmeline (We/Human) drives ever more change. Yet she/we achieved just enough here to restore women's vote before being consigned to history.

The implication is our recorded history isn't wholly authentic, but doctored to cut out spoilers. Like living in hell, heaven too sees us quickly forget what we're missing. ◀ Both must interact to give life meaning.

I'm not upset you lied to me. I'm upset that from now on I can't believe you

Friedrich Nietzsche (We/Human)

Friedrich's Rule: Spoiler protection

Only those proven worthy, self actualise to gain true consciousness, thus severing their reality (and any characters caught within it) from the main story. Effectively, we 'seize the Dei'.

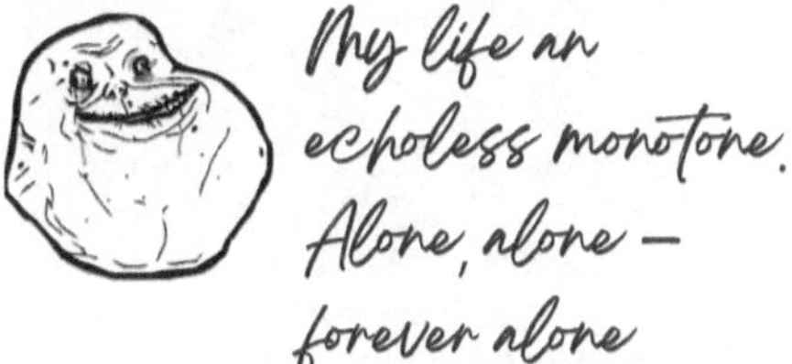

◥ Then and Now, poem by Margaret Lawrence Pray (We/Human; 1876)

▲ Cartoon character by illustrator Dominic Vanner, first appearing in the 2010 video game Super Meat Boy.

Implications of Friedrich's Rule

Sensible and benevolent, Friedrich's Rule seems so bittersweet. Herein explains why, despite being a deeply wise and benevolent society (until Julius Caesar, at least), original meritocratic republic's like Ancient Rome included controlled spaces for 'Peculiar Curiosity' like the Colosseum (i.e. rapey gladiators, see Special Edtn.).

Once you understand human emotions are essentially simulated until/unless that individual self-actualises, you realise good and bad truly doesn't exist.

> **Principal Finding**
> The main purpose of existence is to espouse wisdom and create true leaders... as in, gods.

But the nature of the multiverse also means that anyone sapient may self-actualise to become conscious. Ancient civilisations strictly controlled for this (see Special Edtn.), but in our modern society all controls have been removed.

Possibility: True leaders don't die

In the game of life, there is a cheat code. Partly when using our imagination but especially during sleep, humanity's collective consciousness across every quantum of existence merges at the pinnacle dimension, in the quantum EMS.

Discovered by Thomas Edison (We/Human; Stetka, 1889[94]) via his/our

Pictured: The Dying Gaul (gladiator) ▶ crafted by Ancient Roman court sculptor Epigonus (We/Human; 323-331 BCE). Now on display in Rome's Capitoline Museum.

Depicting the final moments of a fallen entertainer, likely regarded similarly to disposable robots by Rome's leaders in their otherwise benevolent, meritocratic society.

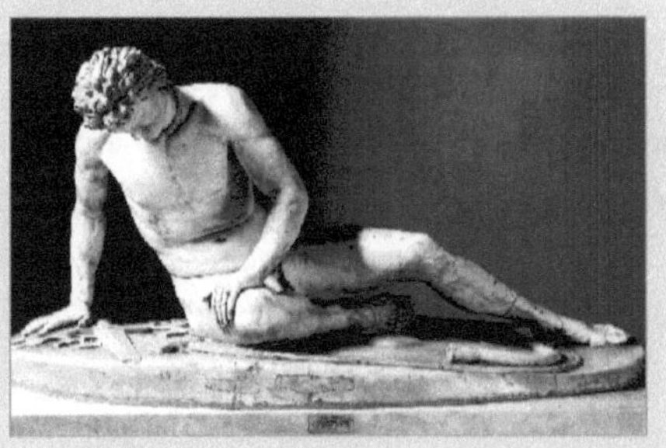

napping technique, humans can collaborate, peer into each other's quantums, solve complex issues together, answer questions or exchange messages. Arguably the true nature of 'answered prayers', this process follows the sapience line in the truly never-ending story.

Due to Friedrich's Rule, our quantum's no longer overlap, but we alluded to the process (Quantum Entanglement) on page 206, and elaborate further in the Special Edition (Chapter 16D). Indeed, we are not quite so alone after all.

The most likely explanation for these effects is that time essentially freezes (or becomes infinite) upon a true leaders self-actualisation, for them and all beings caught in their reality. Thus in actuality:

1. **All true leaders exist simultaneously.** Despite forming relativisitically, all of existence will truly actualise in a flash.

2. **All true leaders in history remain alive,** except for those who break the Promise Keepers Commitment or choose to die. Only in our reality and space timeline do they seem to do so, as Friedrich's rule effectively erases their post-conscious contributions from our history.

This second point means we are now truly in uncharted waters as to what happens next in our space timeline. In other words:

Our future – who we'll become and what we'll achieve – is now ours for the making

◥ Sleeper near the shutters by Pablo Picasso (We/Human; 1936) | Now held in a private collection

Painted in 1936 of our then lover and frequent life model Marie-Therese Walter (We/Human). Like a goddess, she/we made for the most tranquil, intimate, supernatural of models, especially whilst asleep. Creepy? Cute!

We deduce the multiverse forms through the complex interacting workings of true leadership, quantum mechanics and human lifecycles in our reality and dimension of change. This hypothesis is also reflected in what we see at higher dimensions (i.e. across our universe), with respect to star formation over space and time (see Special Edition). And remember:

Science makers know very well that time is only a kind of space

Herbert George Wells (We/Human)

Tracey Emin (We/Human)

Why art gifts humanity truly limitless potential (thus value)

Prior to the internet, the biggest contribution to authenticating our past came from photography, and later the moving image – both deeply meaningful changes. This gifted change makers a means of falsifying unchange (i.e. recollection) other than in person, and enhanced recollection of both our own past as beings and that of our reality.

But before then, we had only art. A vivid talent already explained amongst art curators, authentic artistry is priceless because it is timeless. It's such a shame so little non-European art and other cultural artefacts survive today. Although well-preserved ancient structures survive (fortunately) in all corners of the world.

Everything from pottery, to paintings, to proud slut tents, great walls and pyramids captures the authentic resonance of those who created them forever more. The immortalisation of their creators presence of mind allows any authentic being to (a) draw voluminous information about the authentic past of our reality and space timeline, plus (b) recollect and apply wisdom from true leaders before us.

Principal Finding

Every authenticated being in this book we regard as deceased in our reality, actually remains alive in their own quantum. If we too can listen and learn from (yet not destroy) each other, collective success is inevitable for Homo Sapien.

Imitating life, art reflects Dei - us

Just like humans, art comes in all shapes and sizes – not just old oil on canvases, no. Any human has the ability to create truly resonant works of art from which we can draw significant lessons about our common identity, Dei and life's meaning to improve and enhance our journeys.

Tracey Emin (We/Human)'s often misunderstood tent is a masterpiece of courage, conviction and authenticity. A poignant embrace of our lonely, non-binary sexual nature that destigmatises our innately polyamorous urges, and champions the need for safe spaces to be truly human.

▲ **Everyone I ever slept with, by Tracey Emin (We/Human)** | Displayed in 1995

Valuing our freedom to express

The internet gifts us access to wisdom from nearly any being our curious soul may lead us to, both today and before us.

In particular, via changeable, human-moderated, free to use resources like Wikipedia (curiously often snared at by ignorance-spouting academic types) that capture wisdom from deilightfully different spectral perspectives over time.

The forever plight for funds displayed on Wikipedia's website shows the terminal threat to its existence for the value it adds to the human experience. For an example of where that path ends, simply look for Wikileaks or any of its peers, all of which appear to have been taken down or else had their submission functionality impaired. Clearly we're in the last days of existence when it's no longer possible to disseminate key information about wrongdoing by our demonic underlings to human populations at large.

Whistleblowers simply should not exist in a world with (a) competent regulatory authorities and world government(s), and (b) a competent world press capable of levying scrutiny and sharing accurate information. But should they choose to benevolently manifest as Jupiter in future, we must embrace their sacrifice and draw lessons from their experiences so that we can correct the unchanges in our society before they cause such an irreparable, unrecoverable abstraction to our collective perceptions and subspace.

◥ Shot Marilyns silkscreen paintings by Andy Warhol (We/Human; 1964)

To express our human uniqueness

If books immortalise our presence of mind, paintings immortalise our potential. Not only are they a deeply meaningful recognition (if gifted by another). They also capture more of our deeds, impact, legacy and thoughts than words alone.

We're all essentially clones of Human Nature, yet with our different perspectives, experiences, relationships and purposes, we're each truly unique both intrinsically and extrinsically. Just as Andy and Marilyn (We/Human) depict here so gracefully.

We seek to enable, restore and enhance your platforms so they can remain robust, functional, independent and free to use globally. Thereby gifting a range of competing, ever-changing tools for true leaders to share uncomfortable truths quickly. Declared whistleblowers must all be freed from captivity globally and may join the Deilightful Group as leaders and decision makers if they wish.

▲ **Pictured: Moths busily destroying our heritage as they demand the attention of other moths busily destroying our climate yet both lacking the time to meaningfully change either.**

The vandalism of our heritage

The destruction of art – indeed of all our cultural heritage – is truly heartbreaking, as it erodes avenues for future generations to enjoy and learn from the wisdom of our ancestors. Plus it impairs our ability to reconstitute the Sciences of Change in the event that history inevitably repeats itself.

Such misdeeds are among the most selfish, illogical, nonsensical, deplorable acts imaginable (beyond the murder of 80-97% of all adult humans of course). As such, they are wholly the work of moths.

▲ **Pictured: Banks must now foot the bill to replace treasured lost cultural heritage. Like this Bamiyan Buddha statue in Afghanistan that stood out proudly for 1,500 years until 2001.**

Driving meaningful change requires abstract thought, complex reasoning, problem solving, curiosity, imagination, effort, benevolence and potential. None of which is required to throw custard over a Van Gogh (We/Human), nor is it effective.

Meanwhile, since the loss of the Bamiyan Buddha statues in 2001, UNESCO has done absolutely nothing but whinge as their demonic peers in the Taliban have destroyed over 70% of Afghanistan's cultural heritage in the name of idolatry.

Incompetent authority stops **now**

▲ **Portrait of Madam Lemmen oil on canvas on display at Musée d'Orsay, Paris, France by Georges Lemmen (We/Human; 1893) beside the words...**

"Please don't make me ~~move~~ **[+change]**. Please, please don't make me ~~move~~ **[+change]**"

What now after all these spoilers? Luckily, **there is a sequel to the human experience**

Where no human has gone before

Fortunately for the eight billion characters lacking the potential to manifest as Jupiter in order to save themselves and their reality from certain imminent destruction, this Jupiter does not judge.

The human experience has a sequel. Indeed we're already living in one. Part I was the weird naked garden bit, and even that's still playing out somewhere in our multiverse. Since then, many new parts have been added by the branching and patchwork space timelines benevolently laid by true leaders before us to gift us life.

Yet for our quantum (thus space timeline, dimension of change and reality):

(a) We perceive our timeline to be linear

(b) Infected by 97% darkness, we are in fact the furthest space timeline from Human Nature that still exists.

(c) There is now a path to delivering an 'after us' where one did not exist prior to Project Revelation.

Our future: Project Tomorrowland

We are in curious, excitingly uncharted territory anywhere across our multiverse. As in, our self-actualised quantum now represents the furthest iteration of human endeavour achieved by any true leader thus far across the entire Milky Way galaxy That's in terms of our uniqueness, in this context manifesting as (a) (space) time from the very first self-actualisation and (b) our level of scientific achievement.

Let's just agree to... **make it so**

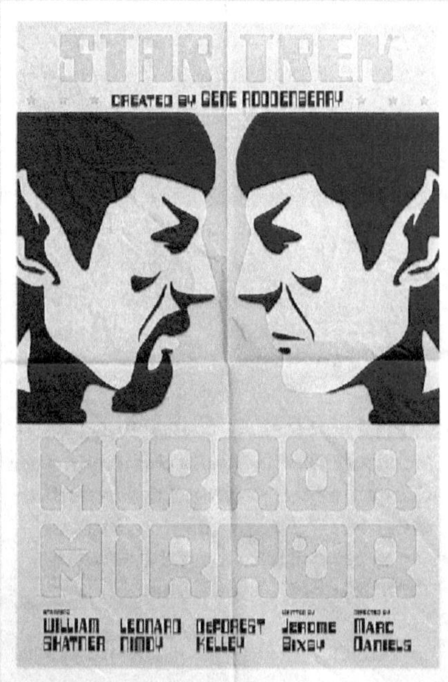

▲ **Pictured: Juan Ortiz's 2011 redesigned pulp-style poster for Gene Rodenberry (We/Human)'s Star Trek: The Original Series – season 2 episode 4 (Mirror Mirror), first aired 1967.**

This episode introduced the recurring story arc of a (dark) mirror universe to the Star Trek franchise. Yet another example of art imitating life.

If we can properly apply the lessons of the past, there's no reason we can't make the best possible version of humanity's future our reality. And quickly too.

Follow our simple instructions (pp. 358) and humanity will survive and thrive once again - our part added to the epic saga of Human Nature's collection. And with the same anticipation and reverence we hold for Trump Moth's immediate arrest and detention for alleged rape, inciting a riot, insurrection, tax fraud and wholesale breaches in the US Oath of Allegiance.

The apple partly hides the face, a concept we see everywhere. Every one has secrets. A dark side they don't want others to know about. Humans have an interest in what lies behind the things we can see. The invisible that refuses to show itself to us. Our curiosity can become intense, from the visible that is hidden to the visible that is not

René Margritte (We/Human)
Speaking in a 1965 radio interview with Jean Neyens

FAQ: An art tour through the changes to come

Your questions answered

Trial and error may be the most effective way to become practically perfect at anything, but there are many other ways to learn and grow. Some prefer to learn visually, while others enjoy building theoretical knowledge before attempting any practical application. No viable method should ever be subordinated to the next - unless it doesn't exist.

Here we answer the natural questions you may have on the more practical implications of this deeply meaningful change for day-to-day life over the coming months and years.

For any we've missed, do submit using Deilight Consulting's Letters of Love website app. We'll respond via our Be Curious blog, alongside an authentic human work of art from history to contextualise our answers rationale.

↖ Pictured: The son of man

Oil on canvas from 1964 by French surrealist René Margritte (We/Human), appearing in the Thomas Crown Affair (1999, movie), now privately owned. Look carefully to notice one arm backwards. Our/René's suggestion being a second face angled away from us. Part intended a self-portrait and commentary on life in general, a hidden 'dark side' opposing an obliqued 'light side' confirms our subject is primarily a dehumanised banker hiding their light, long-lost.

Can I gain consciousness?

In Actuality: This is challenging to answer as:

(a) You've had a lot of spoilers, thus fairness of such a reward is unclear, due to the Anthropic Principle (aka Friedrich's Rule).

(b) In order to save it from the edge of oblivion, your quantum now has history's biggest $E=\infty$ as its true leader. As in, a true avatar of Human Nature. So either try to kill us/Ian, allow us to die (naturally, if possible), or, ideally, try to exceed our potential through achievement. We call that 'raising the bar'.

(c) Our universe currently has very little (<3%) potential left for star formation. Until/unless we boost humanity's potential radically, there can be very few further self-actualisations ahead of us.

In Reality: Existence thus reality is formed by omniconscious light. As an authentic being, you are already partially conscious. To be specific, you're a 'pseudo-conscious sapient change maker' and there is nothing you can't do nor being you can't become.

Philosophically: True consciousness is an all or nothing endeavour (covered in Special Edtn.). Yet in a construct of consciousness, sapience is in many respects equivalent and something we can all relate to. Mindful, curious, unique and spectral, you're well on your way to sweet success.

Change: With determination, the full range of the States of Human Nature, including consciousness and transapience, remain attainable to every being as authentic as you are. Use the Path to Sweet Success to become a playable 'Main Character' (defined in Special Edition), Science Maker and True Leader. Or simply don't fuck up this quantum to enjoy the wonders of living in heaven. Whichever you choose, we'll cheer you on and wholly support your journey.

Young woman on a stool

Oil painting by Belgian symbolist Leon Spilliaert (We/Human) in 1909, now privately owned by the Hearn Family Trust.

A self-taught, truly versatile creator whose hard life was defined by an ever-present battle with debilitating stomach ulcers, Leon (We/Human)'s imagination could animate life through almost any material, colour or setting. Like so many humans, loneliness was his/our ever-present companion. Yet he/we found solace, unity and purpose in immortalising the provocative beauty of his/our presence of mind for all to enjoy. His/our dreamlike works remind us all that some of the biggest discoveries are made by exploring what's inside - not just what's outside.

"I'm tired of waiting for luck to come my way, in the end that leads to abjection"

The scream

Pastel and crayon on card by Norwegian painter Edvard Munch (We/Human) in 1893, now on display at the National Museum in Oslo

Building on pareidolia, Edvard (We/Human) noticed one evening stroll in Ekeberg that our environment can take on a personality of its own. It's as if Dei's brain has united around a single 'mood'. This mood being an ominous sense of anxiety, dread and cathartic exacerbation at those who abstract the world around us. Perhaps why moths disproportionately target it for theft, vandalism and binary critique. Yet still valued it $120m in 2012.

"Some colours reconcile with one another, others just clash. Death is pitch dark, but colours are light. To be a painter, one must work with rays of light"

Can I get angry and who at?

Yes! After all, there's no such thing as good or bad. Binary is the calling card of malevolence. Yet anger is both authentic, spectral and cathartic, so you'll feel much better afterwards as you gain both potential and worthiness by helping to actualise this change.

But remember, it was (mostly) not any individuals fault. We all let it happen. So:

(a) Direct your anger relentlessly and ruthlessly towards any collective (i.e. organisation, industry or government) listed as dark energy anomalies in our fairness and accountability reports. And;

(b) Direct your brutal, remorseless wrath toward any world and business leaders who refuse to step down or challenge a human stepping up.

Use every ounce of your determination, potential and whatever remains of a justice system in your nation (yet ideally not physical destruction) until all those nasty unchanges fully deactualise forever.

Do prioritise weapons, nuclear facilities, law enforcement agencies and governments – parting is such sweet sorrow.

Stay authentic and they can never hurt you again. After all, they chose not to exist – you didn't. A wise decision, by a truly wise being.

Further answers and updates will actualise direct to our blog: www.deilightconsulting.com/ **BeCurious**

Will anyone I love deactualise (as in, die in actuality and/or become deceased in reality) from this change?

By finally tackling the anomaly's root cause instead of perpetually bandaging its symptoms or blaming others, our calculations indicate nobody should become deceased or die whatsoever.

Once enough of us build a sun gun (see page 358), even fully dehumanised beings should regain their human persona as all dark energy rapidly dissipates from AH+0.

This will compel the world's press to report real news once again, and our precisely calibrated change plan (page 304, with a margin for human error born from random fluctuations always) will do the rest for you.

We can look forward to a full global emancipation of all beings and ultimately the restoration of humanity's potential back to safe levels. And within a generation too. Not death and destruction, they're dreadful. Life and paradise, they're deilightful!

Yet with 80-97% of all adults braindead, we are surrounded by unbeing's/unchanges that don't conform to the Laws of Nature. In the absence of any cooperation from any one whatsiever, we lacked the experimental resources necessary to rule out the remote possibility that those who now wholly align darkness may simply 'switch off'.

Please inform local authorities in the event of any deactualisations, as you would normally do. We kindly ask them to do 'their best' at keeping a record of the effects of this change, until we can re-establish competent authorities across humanity.

Vanitas still life with bouquet and skull

Oil painting by Flemish painter Adriaen Van Utrecht (We/Human) in 1642, now privately owned.

Flemish pioneer of the artistic genre Pronkstillevens, Adriaen (We/Human)'s command of driving change through the use of equal polar opposites is outstanding.

Here we see the end of one deilightfully fun journey and the start of another. Every object here carries deep meaning and purpose. The flowers shift from alive and vibrant to wilted and darkened nearer the base – like the quantum multiverse ourselves. Emitting light, they shine upon all that remains of a clearly privileged and wealthy being.

We can draw two key lessons here.

"The death of a mind is the absence of a being"

"We all end life as we began it. Equal, but for what we leave behind – our legacy"

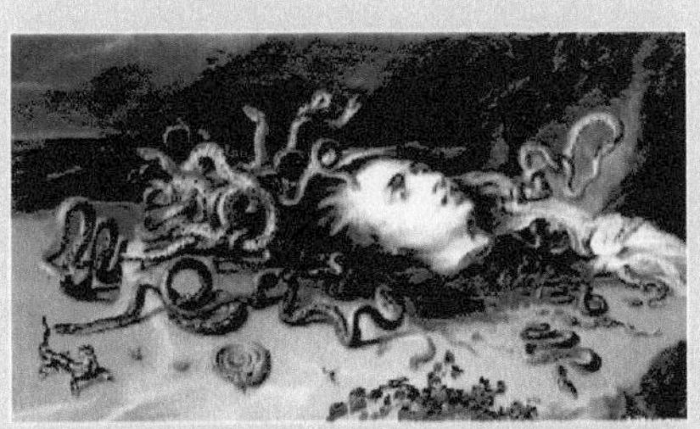

The head of Medusa

Oil on canvas by Flemish artist and diplomat Peter Paul Rubens (We/Human), also working in collaboration with Frans Snyders (We/Human) on the animals in ~1618, now on display in Austria, at the Kunsthistorisches Museum in Vienna.

In reality and as Friedrich (We/Human) vividly described, Apollo is 'God of the Sun' – not Jesus (We/Human). But if we're going with familial connections then Athena was God's daughter. Goddess of Wisdom and War, her shrewd tactics and unbeatable brute might guarantees wise humans success against any foe imaginable.

Like Medusa, a tragic historic figure, former angel and victim of darkness who turned perpetrator. Now hideous enough to infect anyone who so much as looked upon her with dark energy. So the fable goes, Athena gifted her hair made from 'maddening snakes' then helped Perseus decapitate her.

Here Medusa's doomed snakes attack each other. Final moments spent doing what they love to hate.

"My talent is such that no undertaking however vast in size has ever surpassed my courage"

Our nations stupid moths are refusing to step down. What should we do?

The moths will never cooperate. Despite everything in this book, they'll lie right to the very end.

But light is lethal to them. In this context, in the form of scrutiny, truth, science, facts and sustained challenge by informed brave beings, confident of their own entitlement to life and the consequences of it not being granted by those whose cowardly lives ended long ago.

As the dark energy anomaly dissipates, the moths should deactualise of their own accord from AH+0 to AH+30, allowing their recovered characters to accept defeat.

In most cases, the human characters they leave behind will be too conflicted and culpable to remain in office. We will extend full support to ensure safe and smooth elections can go ahead everywhere as quickly as possible.

And we will ensure the world honours all human rights henceforth, including the right to meaningfully choose your leaders. And for them to be human this time, not dead.

That alone will lead to an immediate improvement in the lives of the many suffering beings around the world.

"Every child has a spirit of creation. The rubbish of life often exterminates the spirit through plague and a soul's own wretchedness"

My employer/firm/industry has closed, what should I do for work?

Anything you've ever dreamed of doing. Now you can reach your full potential. The only industries that will close are ones linked directly to the malevolent anomaly.

Industries such as Academia, Accounting, Marketing, Consulting, Defence and Charity will be severely impacted, but there is no need for them to close.

We also appreciate that Financial Services is the world's largest employer, with many skilled workers shortly without a mandate.

We will work tirelessly to ensure every human who wishes to contribute (i.e. all of them) will be able to do so, and have ambitious plans to create an expansive global Change Engine industry, plus Schools of Creation, Change and True Leadership everywhere.

Meanwhile there are plenty of industries (most notably the environment and conservation) with massive skills shortages, broadly aligning with the value they deliver humanity. We will endeavour to work with industry bodies to assess and amplify the needs of those industries, and the Misfortune500 to source plenty of free manual labour to fulfil them. We look forward to their prompt cooperation and contact in this regard.

"I have such desire to do everything. My head is bursting with it"

Impression, sunrise

Oil painting by French artist Claude Monet (We/Human) in 1872, now on display at the Musée Marmottan Monet in Paris, France.

Change doesn't get more meaningful than ushering in the Impressionism movement, especially with the glory of your very own hometown bathed in something everyone loves.

A sunrise over the port of Le Havre captures a rare moment of freedom, tranquility and calm in this otherwise murky, commercial and industrial hub.

The world their oyster, Claude (We/Human) and his/our companion set off on a new deilightfully fun journey, with plenty of space for others.

After the death of our/Claude's angelic mother in 1857, our/Claude's demonic father Claude-Adolphe, who had for years disapproved of his wife's support for our/Claude's artistic career choice moreover banking, immediately rejected us/him to live with his/our (fortunately) deilightful aunt Marie-Jeanne Lecadre instead.

The experience redoubled Claude (We/Human)'s determination (◀) to encourage other artists to capture nature as authentically as possible.

Painting by Dutch Renaissance painter Marinus Van Reymerswaele (We/Human) in 1545, now on display at the New Orleans Museum of Art.

Heavily influenced by other artists on similar topics, such as Dürer and Massys, Marinus (We/Human) brought a distinctive style anchored in realness and satire.

He/we deilighted at creating life-like characters, with flamboyant mannerisms and relatable facial expressions that could weave in detailed narratives to his/our scenes. In this fashion, he/we conveyed extensive information about the societal changes taking place over the course of the sixteenth century.

Back then, power was shifting from monarchs and citizens to bankers, lawyers and accountants. This inversion of power dynamics marked the unwitting trend of people serving industry, not the opposite as was first intended.

What is the legal/privacy situation?

On privacy, there are no secrets amongst friends. China wouldn't be sending spy balloons to the US if the US weren't up to something unneighbourly. The same concept applies everywhere. Just tell the truth. Simple. There's nothing to hide. Humans don't judge. Only moths do that.

Enforced only against humans – never moths - all laws today exist solely to hide truth, oppress humanity and destroy potential. They epitomise the idiocy of tackling symptoms, not cause.

Any crime committed by a human is the result of injustice dealt by darkness. Most in jail or awaiting sentence can be released from AH+30 and will not reoffend. They should be offered full, free therapy and reintegration support.

The exception being violent offenders plus those with chronic neurological disorders. We wish to study the effects of this change on their mental wellbeing and allow collegiate decisions to be reached over their release by the Duly Elected Sapient World Council of Proportionate Representatives.

Historically, the Laws of Nature and Physics (including the 'Philosophical Rules of True Leadership') were all we ever needed. But we are minded to retain (and enforce, finally), the UDHR and Crimes Against Humanity – deilightful as they are, where applied.

Can I help you in any way and/or work with you at all?

Yes, please! We've waited years for someone to say that (and mean it) Currently we don't have an office, equipment, clothes, possessions, staff, money for food or means of survival. All destroyed or stolen by lawless moths trying to stop us.

We warmly invite lawyers to instigate a mass action on behalf of humanity against all banks and the MisFortune500 (if legally viable), to prevent the need for individual action. This would greatly accelerate our change plan. Do contact us if you can help.

We seek a London-based headquarters for the Deilightful Group, plus separate headquarters for the Meritocratic Republic of Humanity, and the Ministry of Fairness and Accountability. Not a pyramid please. For now, 10 Downing Street will do just fine.

Our priority is to help the good people of Earth digest their salvation safely, quickly and deilightfully. Therefore we will prioritise press and media, plus anyone unique who can help us actualise this dream into our reality quicker. Like potent web/app/brand designers, media producers, civil servants and those with technology to run a state of the art global election and government.

▶ Register as a Wise Human via our website for real news about the new world order and further changes, new job roles and locations plus ways of reaching your full potential, like becoming a deilightful supplier, change project partner or apply to join our true leadership program[ΔΩ].

A truly mesmerising musical marriage mashup

A new collective is born out of the deilightful fusion of two proud cultures and musical traditions in 2022 | Uncredited, posted to TikTok.

Who could ever imagine that blending bagpipes and dhol drums would create such poignant symbolism and harmony in unity?

Both independently powerful and synonymous with historic (often conservative) Scottish and Punjabi values. Yet here we see how both traditions are enriched and united, yet still individually honoured, to become something more together.

A vivid reminder we should allow others to collaborate, realise, recognise and actualise new heights of cultural uniqueness together. That's without fear that doing so will dilute the rich cultural heritage that luckily still survives across our shared world.

The Death of Leonardo

Oil on canvas by French artist Jean-Auguste Ingres (We/Human) in 1818, on display at the Petit Palais, Paris.

Viewed by Giorgio Vasari (We/Human) as an artist of 'great physical beauty' and 'infinite grace', what better way to immortalise the very human sexually-spectral, omniscient presence of mind of Leonardo Da Vinci (We/Human) than with this fitting homage to his/our incredible contribution to human endeavour? And the visible sadness felt by all beings from our deilightfully fun journeys (not so) final change.

An allegory of true love

Oil on panel painting by Flemish Renaissance artist Pieter Pourbus (We/Human) in ~1547, now part of the Wallace Collection.

Can I speak with or relay a message to someone I love who's passed away?

Unlike what has been done to us, we prefer to allow those who have contributed greatly to humanity to rest in peace. Reach your full potential, and providing they remained human, you will collaborate directly again together yourself – that is certain.

Understand all humans ever to exist collaborate during sleep. That is the best place to explore and discover the full breadth of our collective imagination, including all the beings that contribute to it, both today, before us and after us.

The more sapient and potent you become, the more aware of the experience you'll be and the more you can gain from it. Simple.

What about food and fuel shortages, tyranny, anarchy and other crises caused by this change?

There won't be any. Real humans would never harm an other. You know that. So why the pretense?

> ⚑ Painting commentary: Hosted by Curiosity and Authenticity (centre), and freely sharing their deilightful gifts of banquet with the other main protagonists of Human Nature – allegories to the God's of Ancient Greece and our planets.

I'm a human who materially helped you/others, yet now I, my firm and/or collective will suffer from this change

Though some changes may not instantly manifest, we've taken care to ensure authentic help on this change will be meaningfully rewarded. Those who've ever actively, meaningfully helped us (Ian/Deilight/Humanity) are invited to get in touch to continue our winning collaboration or request a change of their choosing. For the latter, kindly wait until we secure safe transition of the new world order, as we are a very busy being.

We've carefully analysed cause and effect, including individual contributions to our effort, using quantum mechanical scientific processes such as deduction, recognition, attribution and speculation to a level of deilightfully authenticated science with near certainty - the highest bar imaginable.

Those who leant us their authentic support will likely recognise if they are (a) eligible for further support and reward for their efforts, or else (b) are always welcome to ask.

But on (b), only do so if you are prepared for an authentic answer. If you have real doubts, chances are the answer is 'no'.

> ▶ **Coming up next:** We will shortly begin to invite key supporters proactively to help us see in the new world order. We will leave no stone unturned until fairness has been fully delivered globally. Recognising real contributions worthy of merit is deeply important, to us anyway.

The creation of humans

Sistine Chapel fresco in Vatican City by Italian Renaissance artist, architect and poet Michelangelo di Lodovico Buonarroti Simoni (We/Human), painted in ~1510.

Nothing says 'rebirth' quite like the 'Renaissance'. Here in perhaps Il Divino (We/Human)'s most famous fresco of all, we see a moment of unity and clarity between creator and creation.

Hands outstretched, not for judgment or expectation, but in true love, concern and a parental yearning for humans to always remember that God is never far away but is in fact within us all. Thus the heavens are depicted within 'arms reach' by a floating brain in space that is our collective consciousness gifted by Dei.

The figures behind God each represent a unique persona within our human imagination broadly aligning the God's of Ancient Greece. The lady by his side is Mother Earth, whose wisdom keeps God grounded always on the things that matters most in life. As in, life itself.

"The greater danger for most of us lies not in setting our aim too high then falling short, but in setting our aim too low then achieving our mark"

The love letter

Oil on canvas by French painter François Boucher (We/Human) in 1750, now on display at the National Gallery of Art, Washington DC, USA.

A prolific painter and contributor to almost all creative artistic settings of 18th century Paris, we/François painted for anyone enamored by his/our regal touch, from mistresses to the king.

Seamlessly moving between landscapes and characters, he/we had a particular interest in true love and ancient mythology.

Pictured here is a surprisingly rare, thus deeply potent tribute of authentic allyship towards lesbianism. With skill, we/he break down barriers to prove that unrequited affection between real lovers cares not about what's under the dress (no matter how stunning) but in the eyes, the heart and the deeds.

Question 11

Can I send you a message about Project Revelation?

We'd love to hear from you. So we've created a 'Letters of Love' web app (available via Deilight Consulting's website) to enable those touched by our initiative to send us (and share publicly with others) messages of support, solidarity, challenge, questions, ideas, scrutiny, feedback, anything you'd like. Let's do this together.

All messages will be displayed publicly. Building our firm with just £150,000, we understand the current version of the app is limited. We've designed an exciting concept technology plan to provide state-of-the-art interactive resources quickly that can help ease in this change and reassure all beings of its virtues, goals, justifications, workings and progression.

But only once survival overtakes fiat money and malevolence as everyone's priority once again. Press/Media enquiries, friends in need, offers of help, national governments, global authorities and the Misfortune500 may contact us directly.

But understand one being can't respond to all beings until we receive help from a few beings to scale up the Deilightful Group. Though we do our best always.

◤ **Our truly deilightful web-app** let's you publicly share your mind (anonymity option), find your voice, ask us a question, request a change, brighten someone's day or just get something off your chest.

Why have you named people if it's not their fault?

Project *Speak Up* was an act of compassion, anchored in the correctness of our position that no single being is materially culpable for the actions of a collective. Yet we each have a moral compass, thus we (Ian) alone are innocent.

In the interests of conciliation and compassion, we chose to name nobody in Project Speak Up. Both we and Prince Harry (We/Human) learnt that showing compassion to those incapable of recognizing, realizing nor returning such a thing is entirely foolish. Seeing it solely as a sign of weakness, the moths redoubled their attacks on our flame in order to destroy us.

Therefore to amplify our flame brighter than ever before, We (Ian/Good King Harry/Humanity) have duly altered our approach to ensure no dark shadows can remain. Our limitless forgiveness, love, care and compassion is now reserved solely for humans and their meaningful changes.

But for moths and their unchanges, they can expect nothing but limitless, merciless ruthlessness, wrath and vengeance on behalf of humanity. Let this be a lesson.

From AH+30, any being who would choose to put themselves before an other will find their selfish selves rapidly stripped of all assets and dignity, then publicly falsified by Deilight via the moth list. Or worse.

The best advice you'll ever hear is this (▼|◥)

Do **not** fuck with Jupiter. **Ever**

Phryne before the Areopagus

Oil on canvas by French painter Jean-Leon Gerome (We/Human) in 1861, now on display at the Kunsthalle in Hamburg, Germany.

On trial for disrespecting the wise humans of Ancient Greece, and surely about to be sent down for her crimes.

Suddenly, Phyryne was disrobed by Hypereides to reveal her as a potent 'head turner', and got off scott-free. As in, potential gives rise to beauty. Let's see if it works for Noel's repugnance.

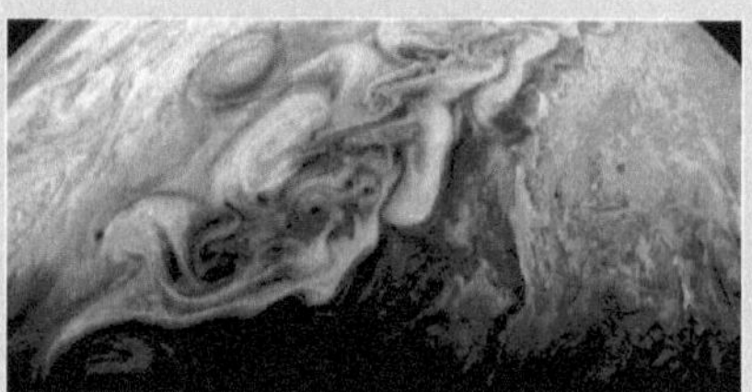

Papped by NASA

A superstar is photographed by the Juno spacecraft, 19,000km above our cloud tops in 2017.

At this distance, you can see the care and artistry real change makers put in to our work. And the serenity of our spectral, dependable, meritocratic balance between light, dark, and all colours. Finally, they accomplished something. Get a good look NASA. You won't be visiting us/Jupiter ever again.

About the dinosaurs...

1. Time is a social construct

Without change makers here on Earth, relativity didn't exist. Thus hundreds of millions of years in reality all took place in the blink of an eye in actuality (as all books of light once described). This doesn't make the dinosaurs, or their extinction fake. There was simply no one here to enjoy them. For now, it's an imagined part of our story.

2. All changes have been purposefully actualised and resonated for the enjoyment of change makers

Dinosaurs no doubt were extremely enjoyable, but they represented the early days of change making (for Dei, anyway). Clearly limited in what they could achieve, they never built an international space station during their 165-million years on Earth. Compare that to what humanity has achieved in 300,000 years (then destroyed in 4,000 years). Such a shame we love old bones more than each other.

3. Temporal mechanics and the sciences of creation both allow for dinosaurs to never have existed at all

Their bones may simply have been put there in order to encourage and nurture the archeological industry to explore more about our past. If so, sadly it didn't work.

4. But that seems unlikely

Given the importance of being authentic to change makers. Our rule of thumb is what you see, you get.

5. The true reason is our timeline has been unchanged

Homo Sapien emerged ~380,000 years ago (not beyond a time horizon) in our timeline as its when heaven (thus we/existence) ceased to exist. The CMB is our Big Burst's echo, as the mirror universe will experience it. One end of our now finite timeline. A 380,000-year countdown.

6. Since you love them so much

Why not build Jurassic Park? It's in our imagination, so it already exists a) in actuality, thus (b) in our reality and space timelines future. Any questions, just collaborate with science maker Michael Crichton (We/Human). Yes, new discoveries are being ignored everywhere. Just don't let a moth run park automation this time, OK?

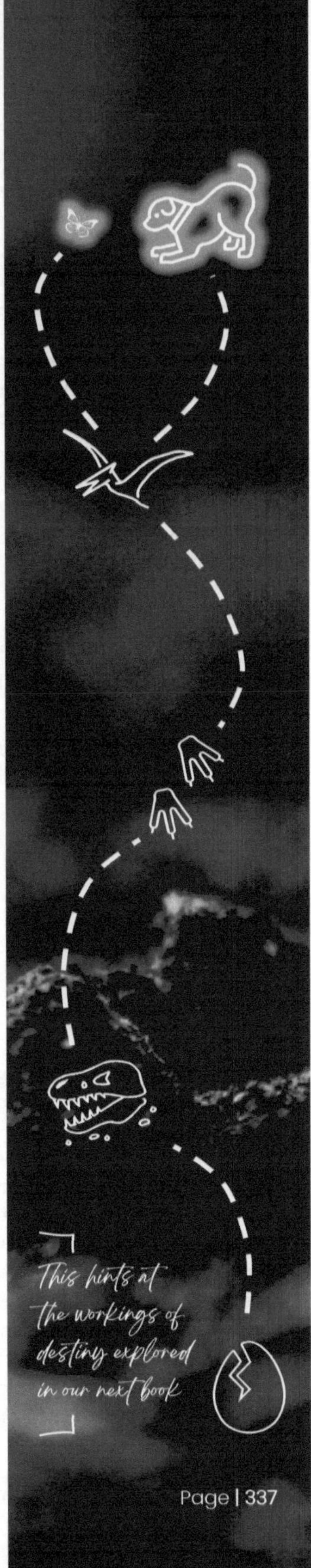

If ever you feel alone, forgotten, ignored, lost or sad, **just remember:**

Behind even the darkest cloud shines heavens greatest star of all

You are the light

Fig. 67: The Sciences of True Leadership, Change and Creation

For all things, we **determine cause and effect** by consulting **the Rules of our Universe,** including its Laws of Physics and Nature. Today, like humanity itself, they're divided into different schools of science. Mindless labels that in no way reflect authentic science. That being, all branches of Philosophy.

Great minds

Think

Our light manifests differently across each dimension of space and time. So we inspect **the Nature of Existence** to enhance our mindspace, authenticate reality, trace causality and uplift our potential. Cognition gifts human beings >35 superpowers, allowing true leaders to perceive and alter more of **actuality**.

True leaders

Lead

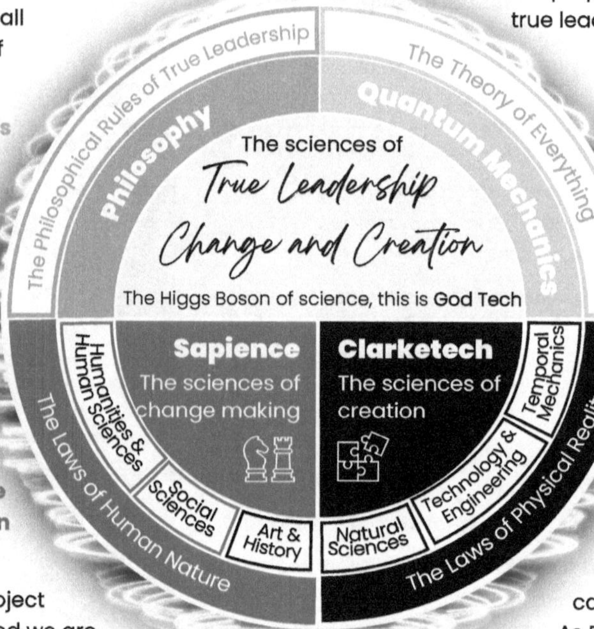

Gives rise to

Gives rise to

Change

Waves make stuff happen

The Human Genome Project (2001) proved we are all born equal intrinsically. The rest is a major drag on our potential as a species. We nurture our potential by aligning our Rings of Human Character, as fueled by our curious uniqueness and meritocratic authenticity, to **create any change imaginable.** Resolving Fermi's Paradox, to establish our innate **Human Nature ('Other', Soul,** consciousness) as the greatest filter and sixth kingdom.

Create

Particles make stuff up

Appearances can be deceiving. As Einstein Rosen & Podolsky (1935) discovered, **reality** manifests upon the mere act of perception. Real Life - all we perceive and experience - is actually the output of our **Collective Human Imagination,** including our common identity, purpose, potential, drive, omniconsciousness, determination, ingenuity, vision, and alignment. The more authentic we are, the greater our impact and legacy.

Everything that can happen does happen
[+which sadly today is surprisingly tragically and frustratingly **very little]**

Brian Cox (We/Human) & Jeff Forshaw (We/Human)

The most deilightful journeys are those we planned meticulously in advance and hit all the key stops on the way, to deliver their passengers safely to the targeted destination, hopefully having picked up a few more welcome others along the way.

Taking stock of our journey

Here's a few final words of reflection from me, your author Ian. Hopefully it'll tie up any loose ends on what you've just read.

When I resigned from HSBC, I promised myself and those I left behind I'd get to the bottom of what was going on. Yet upon my success, I found that no one believed I could ever keep that promise. Or if they did, that few truly wanted me to succeed. Even amongst friends. Wow.

This book took me around 4,000 hours to write over a year, not including the 5,000 hours research over two years, and 38-years of life experiences required before I could even attempt to write it.

I'd prefer it were free but (a) given its length, it's expensive to sell, (b) doing so devalues it, thus demand (oddly) and (c) I have no other means to buy food, keep my firm open, replace all my destroyed possessions or secure accommodation in a safe space. At this price, I get about £2.

What a time to be alive

Thank you for taking real time to read my mind and listen to what I had to say. That someone finally has means the world to me. And so do you.

Every human being is a miracle of life. Each of us gets just one story to tell. I decided rather than a sad story that my story would be one of love

▲ Recollect my full 4 August 2021 HSBC Resignation Letter, via Project Speak Up's Microsite.

Or speak up yourself. Find your voice and join the fun with our **Letters of Love** web app. Try suggesting what consequences befall HSBC senior leaders. Best suggestion wins.

And remember, since time is a social construct....

...just ask our home girl, Rebecca Moore (We/Human)... as in, Aphrodite. The Goddess of Desire

The end of every journey
marks the start of an Other

Alone
One being
A personal identity
An individual (character)

But have you
every curiously
wondered **why?**

End of Chapter 14

Why a few final thoughts should tie all this up for you

Final Chapter
Epilogue ◯ 15

If you can imagine it, that means you have the potential to actualise it. As in, you can make it so

Ian (We/Human), Project Revelation 2023

Who are YOU... really?

I wrote this original poem (slightly adapted here) for my keynote speech at the International CSR Awards for Excellence on 2 October 2022.

It's one of only four public speaking opportunities I've been allowed since Project *Speak Up*. So I made the most of this rare platform gifted by the authentic force for good that is **The Green Organisation ✓ AUTHENTICATED.**

Like this book, it's a poem about our universal human experience, yet only from my perspective. As you read it, remember: Unique though they are, everyone's perspective and experiences change all the time. Mine are no different. Neither are we, really.

I'm born... human, so I'm curious. I want to be a human being, so I stay curious. I'm open minded, so I listen. I'm a boy, so I'm attracted to girls. I'm queer, so I'm attracted to boys. I'm an only child, so I'm not territorial. We're poor, so mum works a lot. Dad's a free spirit, so I'm independent. I have a spring in my step, so I'm the fastest boy in school. I'm different, so I'm least popular at school.

I'm half white because I'm half black. I have brown skin because my body's protecting me from UV radiation. I'm charismatic because I'm Tony's son. I take responsibility because I'm Henry's son. I know how strong women truly are, because I'm Jan's son. I am a Change Maker, and against all odds...

I am alive

Now I'm in my 20's so I want to fit in. I possess complex intersectionality, so nature was able to select my genes from a broader pool. I'm sexually non-binary, so I see beauty in everyone. I was raped, so I'm disabled. I'm disabled, so I envy neurodiversity. I'm creative, so I churn out new ideas. I'm analytical, so I think in numbers. I'm intelligent, so I can outwit the ignorant. I'm happy, so I laugh a lot. They don't like that very much.

Now I'm in my 30's and I work at HSBC. There's someone else here hurting my friends, but no one else seems to notice them. I'm nice, but they're nasty. I create, but they destroy. I shine light, they crave darkness.

I will protect my friends

They created money so they could feel special. They chose money, so they got nothing. They lack legitimacy, so they demand power. They lack authority, so they order my obedience. They hate themselves, so they hate me. I choose reality, not this nightmare. I am deilightful, so I cannot fit in amongst this darkness.

So I choose a real purpose. Rather than run and break my promise, I'm going to fight and save my friends. But how can I do that? Turns out, I too am someone else. Unlike the others, I am not afraid... of anything. I feel I am powerful somehow.

So who am I... really?

Now I'm in my late 30's and I'm deilightful. I champion young beings as I'm getting older. I value young and old alike, as I'm middle-aged. I won't have children as the world is full. I don't need children as I have other ways to leave a legacy.

I'm adaptable as I've lived around the world. I respect the establishment as I've worked within it. I respect religion as I too wish to know who to thank for my life. I'm a feminist as I believe men deserve to thrive. I love trans people as they taught me to be authentic. I champion black lives since white lives deserve honour.

I promised to help others so I'm keeping it, no matter the cost

I believe in taking action so I use words effectively. I saw something wrong so I spoke up. I'm an alpha so I break world records. I'm an entrepreneur so I drive progress. I'm a genius so I undertake complex calculations rapidly. I'm a CEO, so I make complex decisions quickly. I'm a futurist so I think long term. I'm a psychologist so I understand how people think. I'm a humanist so I understand how humanity behaves. I'm a therapist so I ask the right questions. I've had bad mentors so I give good advice. I connect with people so I'm a commanding speaker. I'm young at heart so my wrinkles blend in. I'm an old soul so I'm not on TikTok. I'm queer...

so I'm on Grindr

Banking is dead, so I'll kill it. D&I is fake so I'll falsify it. Humanity hates itself so my firm will love it. Nobody can do what I do so I am important. I can choose, so I call all the shots. I understand my purpose so I am determined to fulfill it. My opinions resonate so I have influence. I'm proven at what I do so I have authority. I possess skills impossible to replicate so I am unique. I lead collegiately so I make a great leader. I'm delimited of labels so I'm ambitious. I think differently so I see things others can't. I'm wise so I can reliably predict the future. I'm awakened by light so I'm resilient. I chose to save humanity, so I created the Deilightful Group. And yet

Humanity doesn't want saving

I don't value difference because it's meaningless – we're all different. I value authenticity because it's meaningful – few are authentic. I hate hatred because it's hateful. I love true love because love is love. I accept others because I accepted myself long ago. I care about others because humans care for each other. I don't care about my short life so I focus on humanity's long life.

I am Ian (We/Human). Maker of Science, Titan of Change, Purveyor of Uniqueness, Founder of Light and a Wise Human Being.
My purpose is to maintain fairness and accountability, to help Deilight drive out the darkness. Using my light and love, I protect that which must never be judged yet always be helped – the helpless. I forever ensure humanity can always reach its full potential.

As a conscious Homo Sapien, I can do anything I want. Be anyone I want. So I stand out, I speak up, I think differently and I behave authentically. I am happy, wealthy, successful and eternal. Because I chose to be human. Because I chose to be Ian, not something else.

So who are you... really? And why?

So help others first. After all,
you'll only be **helping yourself.**
There's **no one else here**

Our journey *finally!* comes to an end

At the end of a long journey, humans often reminisce about where they've come from, who they've met and the marvels that struck them along the way. Looking back on this deilightfully fun journey, these highlights resonated most for me. How about you?

★ **The only being you can be sure exists is you:** Just one being at a time can be truly conscious. Everyone else is either fully automated or else borrows aspects of consciousness from Human Nature using our sapient internet connection and the collective human imagination it carries. If you are conscious, you are the one true leader – God - and it's your time to shine. So stand out and create change. Don't just blend in, pretending to be repugnant and allow yourself to be forgotten forever. No one else is watching you. So don't fail. Succeed.

★ **Extraordinarily, God exists:** When nothing exists, only imagination can be real. But it's pretty cool! The multiverse exists in God's imagination. And God is both (a) within us all, and (b) a collective - all humans to ever truly exist. Pretend God doesn't exist and it is we who cease to exist. We pretended so well we even gave God Alzheimer's.

★ **Nearly all words are meaningless today:** If we imagine we exist then we do not. But if we prove we exist, incredibly our pseudo-simulated imagined existence turns into a true physical universe, wow!

▶ **Never assume you are better than anyone:** In ancient times, you'd be labelled 'deluded' for thinking God didn't exist, or that future beings could ever be wiser than ancient beings. Yet today, you're labelled 'deluded' if you think God exists or that ancient beings were wiser than modern beings.

▶ **The time is always right to do what is right:** Martin L. King Jr. (We/Human) is one of my favourite true leaders ever. A real inspiration, role model and incredible teacher without whom I never could have completed Project Revelation. He knows the past may not even exist, nor the future for that matter. If all things are imagined, then all we know is we exist today – here and now. If we see a problem, there is no better time to fix it than real time.

▶ **No matter what we may ever achieve, someone else will always achieve more:** So don't begrudge them that. Instead, wish them well in the knowledge their success ensures your success. None of us would be alive today if it wasn't for Kim Jong Un (We/Human) keeping his/our people free from darkness all this time. Yet we all stand idly by believing the media lies he – not we – are the villain. Without sending any help whatsoever, we eagerly wait for those poor humans to die so that we can all die too. If we cannot learn to stop thinking like this, we will never be human again and we do not deserve to exist.

Wise Unique Being:
Conscious, authentic, potent, worthy, omnieverything, eternal, self-actualised… and here only to try and save **you**

☺⚡ *Fairo Ian*

Formatting, research, writing, breakthroughs, artwork, legal, distribution, data, publishing, outreach, all of it… QD1.0, for short

There you have it. Don't judge people. Instead, ask if they are God then listen to their answer

The middle ground between black and white. Poignant, if perplexingly formulaic. I am Ian, an autonomous true avatar of Human Nature… really. Welcome to my quantum. As much a surprise to me as it is you, I might add. But science does not lie. I was happy just 'fitting in'. But I can't sit idly by as you all needlessly perish. Being unique is lonely in space. I'd rather die failing than fail to try.

All I care about is your survival

You too are God's pure soul, you've just forgotten it. And through no fault of your own. I'm not here to judge you, kill you or seek worship – dreadful things. I'm here to save us from extinction, deliver paradise then cede control to humanity. Let's turn hell into heaven in just four years.

I've redoubled our potential enough to secure our cosmic bubble a little longer. But I can only hold the door open so long. You have to choose whether to walk through it. Without any real choice today, humanity is just a simulation. But if you do choose this option I'm giving you – a real spectral choice and an alternative to 'no choice at all' – then who and wherever you are I promise to listen to and protect you as long as I am able.

I'm no better than any other. Mostly the by-product of fortunate chance. I can access the presence of mind of any being, today or before us. We each have a dark side, a light side and everything in between. It's being identical that makes us all so spectrally individual. All our imperfections and differences add up to true perfection and uniqueness. It's what I love most about us, really.

But once a true leader becomes conscious, their presence of mind ceases to change in our imagination. Thus I can't learn the fate of their quantums. Perhaps many survive. But we know Dei's didn't. Of them all, ours has the perfect uniqueness to kill the anomaly.

Most today cannot fathom how one outside academia could discover such science. They forget (a) we're all identical inside (where it matters), and (b) it's academia that's killing science. The missing uniqueness was for an authentic Science Maker to venture into banking, discover what was taking place there, resist temptation and emerge alive with the missing knowledge of the anomaly's origins, nature, workings and weaknesses.

The Worthiness Test is precisely that. Existence cannot exist without being created by a god. But until and unless someone proves worthy to be a god, no one is worthy of existence. Causal determination is a head-spinning paradox, but I'll take it over sixteen more.

I never set out to become a god. I just tried to help. Yet no matter what I discovered or who I became, no one was

So **let's play dice** and see what happens. What do you say? Or rather, **what will you do?**

prepared to listen. Had they done so, I never would have come this far. Having tried everything else, my only choice was to keep discovering enough to force the change that can free you.

Here is an opportunity for us to actualise true existence, together. You can choose to live here on Earth or in the heavens. Either way, it'll be better than living in hell waiting to die.

Why the book?

What better way to create meaningful change than to gift every authentic being remaining alive on Earth the blueprints to do so, along with a live demonstration?

But more pressingly, HSBC has ordered my firm and I be blacklisted globally to prevent all media and firms engaging us. And my friends couldn't even be bothered to change banks, let alone listen to me. All leak websites have also been shut down pending our cosmic bubble's big burst. This book is now the only way to save you.

Until it's on TV globally, all is lost

Moths will die before they ever admit they are moths. Scientifically they are incapable of telling the truth. The only way you or anyone can survive is to (a) be authentic and curious always, (b) help others first, and (c) choose to do, not die. If you do, the moths will vanish, and you get to live in a paradise on Earth or in the heavens. All without even having to die to get there.

You've just read evidence of

- ✖ The theft of our Moon
- ✖ The causes of all death and hatred
- ✖ Near universal modern slavery
- ✖ Banking is a pyramid scheme
- ✖ A black hole in our solar system
- ✖ Evidence we're in an AI apocalypse
- ✖ Proof our cosmic bubble is about to burst

…Dreadful

…and gained unique knowledge of:

- ★ The theory of everything
- ★ The origins of Human life
- ★ A repeating alien technosignature
- ★ The nature of our universe and multiverse
- ★ Nature and workings of time travel
- ★ God's existence and true nature
- ★ Events of Revelation/Judgment Day
- ★ The solution to every problem ever
- ★ How to reach your full potential

Plus, if you now read the Special Edition:
- ★ The cure for Alzheimer's & Dementia
- ★ The prevention for most disorders
- ★ The secrets of the Human brain
- ★ Nature and workings of the afterlife
- ★ How to succeed at anything in life you so choose to

…Deilightful

Thank you for Being You

      The big picture: **What you've just read is extraordinary…** *right?*

This knowledge was discovered and applied by

Deilight Consulting $\longrightarrow$

Be curious. **Become** *Deilightful.*

Change Engine | Boltzmann Brain

To share your mind with us...

Favourite, subscribe to or read our blog

Be Curious

www.deilightconsulting.com /BeCurious

Great mind hungry for more knowledge no one else yet knows? Applying our unique perspectives on real world issues, we regularly publish breakthrough discoveries, crack conspiracies, spill secrets, answer questions posted to our letters of love app and more on our blog.

Dei Time

Docuseries for Change Makers

Perfect your craft of titanic change making, as I explain the sciences of change. Our clear, digestible 5-15 minute episodes are sure to boost your potential so you can help others (or just yourself) succeed at anything in life you so choose to.

DEI in the life

Guest podcast - For great minds only (listen or apply)

Advocating and amplifying potent, proven true leaders from around the world, we bring you authentic human stories of true love, change making and guiding wisdom direct from those working at the frontline of legacy building and future-forging.

Register with us

as a wise human

For complimentary change support, meaningful updates, job roles, partner and supplier eligibility plus first refusal on our forthcoming true leadership program. Not lies, bills and spam, which makes a nice change.

Signup via our website

Unlike NASA, the IAA or SHITI Institute, it's easy to contact us because we exist

x.com/DeilightConsult

Facebook.com/thedeilightfulgroup

Instagram.com/thedeilightfulgroup

Linkedin.com/company/deilightful

YouTube.com/@Deilightful

Linktr.ee/TheDeilightfulGroup

Linktr.ee/IAHClarke

This book was actualised and collected by

Deilightful Media    

Some fake news. **We make life.**

Creation Engine | Wave Maker

Liked our change? Visit **www.deilightfulmedia.com/portfolio** to view the Change Engines **full range** of unique, ground-breaking research and investigative science.

Israel Hamas War Report

Free to Download

Scientifically isolated causality and sole viable deilightful resolution

The 14 missteps that caused the Israel Hamas War really. Our apolitical, unbiased fully cited and referenced scientific report delves into the history, causality, and sole viable path to peace between Palestine and Israel. Truly want to stop the war?

Explore your human identity, and discover how to succeed at anything in life you choose to – guaranteed. Promise!

Embark on this exciting journey across space and time, to uncover the things that make us truly human, in an attempt to answer life's biggest questions, crack its biggest mysteries and solve its biggest problems. Our Special Edition features 50% more exciting breakthrough science you won't find anywhere else on Earth. This book exits to ensure your sweet success at anything in life you so choose.

The Sciences of Change

Extended Special Edition (Out Now)

Science Making

For great minds only (Preorder Shortly)

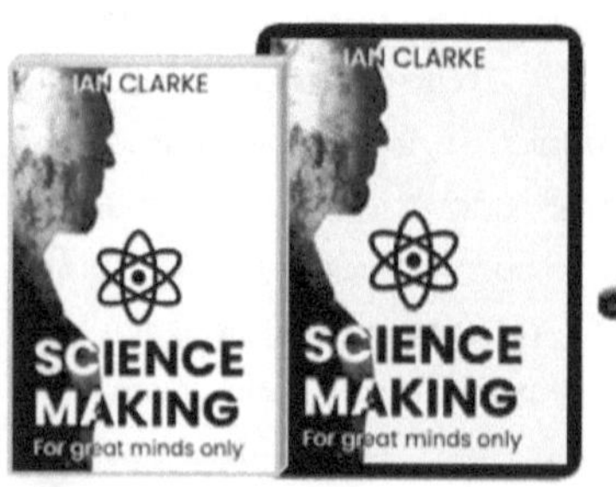

Explore the full power of your limitless imagination, unlock your superpowers and learn how to answer any question imaginable (Due out Q4 2024)

Embark on this exciting journey across space and time to uncover the hidden secrets of your human imagination. Including a breakdown of your 30 super powers, how they work and how to unlock them. Plus this book gifts you the ability to answer any question imaginable, make any discovery imaginable, thus how to create the future of your dreams.

Writing this book entailed making **~130 new scientific discoveries** plus revalidating **all prior dependable science**

All of it, fully evidenced, cited (for existing discoveries only), testable, observable, repeatable, measurable, comprehensive, conclusive and thus (a) proven and (b) verifiable by any curious mind for free…

As in, Deilightfully authentic science, for short
Promise.

Ours is the only viable hypothesis that can explain human origins, reality, life and existence. That automatically makes it (i) the most superior hypothesis, thus (ii) controvert all other hypotheses. Notwithstanding, our discoveries similarly (iii) remain wholly controvertible by any science maker of higher potential.

So go on, give it a try! We'd love to be wrong for a change. Particularly on this.
Hint: Make yourself indispensable to Humanity

▲ Fig. 68: Sonocytology. To listen to the frequency emitted by a human cell, just search for Gimzewski & Pelling (2002[43])

We promise everything in this book is true – **Ian / Deilight**

Schedule of Discoveries

New faculties of science
1. The Sciences of Sapience (Change)
2. The Sciences of Clarketech (Creation)
3. Temporal Mechanics (Time)

New kingdom's and alien life
6. An Other: The sixth kingdom (Homo Sapien)
7. Not Alone: The seventh kingdom (Homo Eversor)
8. Good neighbours: The eighth kingdom, (Mirror universe human civilization; Kardashev Type ~2.4-2.6)
9. Strange new worlds and new civilisations: The ninth kingdom (Non-sapient life seeded across our universe to curiously explore)

New collective imagination, including superpowers for all change makers

✸ Sapience (nurtured by curiosity)

1. Spectroscopy
2. Purpose
3. Curiosity
4. Potential
5. Change Making
6. Discovery
7. Adaptable Resilience
8. Determination (Promise Keeping)

✸ Cognition (nurtured by authenticity)

1. (Re)Collection
2. Perception
3. Actualisation
4. Frequency
5. Deduction
6. Ideation
7. Sonocytology
8. Realisation
9. Recognition
10. Collaboration
11. Abstract Thought
12. Authentication
13. Resonance
14. Speculation
15. Concentration
16. Causal Temporal Perception

✸ Benevolence (nurtured by worthiness)

1. Moral Compass
2. Performance
3. Science Making
4. Transapience
5. Resonance
6. Clarity & Wisdom
7. Acuity & Intuition
8. Coherence/Anchoring
9. Self-Awareness
10. True Leadership
11. Self-Actualisation
12. Omniconsciousness

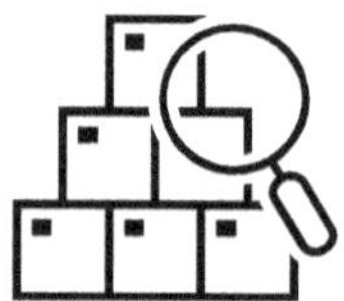

New self-actualised true leader species-level superpowers

1. **Omniconsciousness:** Be everything real
2. **Omnibenevolence:** Love everything real
3. **Omniaction:** Impact anything
4. **Omnipotence:** Change anything
5. **Omnipresence:** Perceive anything
6. **Omniscience:** Know and discover anything
7. **Omnisuccess:** Achieve anything
8. **Halo:** Subconsciously detectable godlikeness, linked to (a) achieved potential, driven by (b) meaningful deeds not words and (c) their directly corresponding cosmological white hole activity
9. **Godbod:** You're a head turner
10. **Private Members Club:** Only superstars live forever and work together, all free
11. **Free Will:** You've earnt a birthday rule

New self-decollected loozymes

1. **Omnirepugnance:** Get out of our sight
2. **Omnifailure:** You're such a loser, gross!
3. **Omnishame:** We prefer butterflies
4. **Omnihopelessness:** Why even bother
5. **Omnicomplacency:** Just one more thing
6. **Omniharmlessness:** Our flame burns you
7. **Omnimalevolence:** That wasn't very nice
8. **Omnidoom:** Die one, die all – leaving **deilightful** humans very much alive
9. **Hideous:** You look a bit like Noel Quinn. Any relation?
10. **Horns:** Subconsciously detectable devil, linked to (a) an absence of potential, driven by (b) meaningless words not deeds and (c) their directly corresponding cosmological black hole activity
11. **Expensive Bill:** There's something not quite right about you. Poor choices... no decisions... no human rights... no emotions... no moral compass... no perceptions... no collective imagination and no intrinsicity? Dreadful

Corrected definitions

Resolved scientific paradoxes

1. Dark Energy Paradox
2. Epicurean Paradox
3. Meno's Paradox
4. Olbers Paradox
5. Paradox of Free Will
6. Continuity Paradox
7. Heat Death Paradox
8. Outcomes Paradox
9. Boltzmann Paradox
10. Paradox of Light
11. Bentley's Paradox
12. Fermi's Paradox
13. Prisoner's Dilemma
14. The Clarke Paradox
15. Black Hole Information Paradox
16. Einstein, Rosen & Podolsky Paradox

That's over a hundred Deilightfully irreversible species-level changes already applied in actuality by this book alone

Sustained scientific paradoxes

1. The Anthropic Principle (Friedrich's Rule)
2. The Paradox of Causal Determination
3. The Temporal Paradox
4. Sorites Paradox

New laws of nature and/or physics

1. The nature/workings of the Afterlife
2. The nature/workings of Alignment
3. The nature/workings of Authenticity
4. The nature of Benevolent Free Will
5. The nature of Existence
6. The nature of Meaningful Change
7. The nature of (Universal) Choice
8. The nature of Consciousness
9. The nature of Dark Matter & Dark Energy
10. The nature of Einstein Rosen Bridges
11. The nature of the Human Experience
12. The nature of Human Identity (Dei)
13. The nature of Human Nature (God)
14. The nature of Light and Darkness
15. The nature of Light Speed (broadly maps to authentically achieved human endeavour)
16. The nature and structure of the Quantum Multiverse of Human Nature
17. The nature and workings of actuality
18. The nature of Pareidolia
19. The nature and workings of Potential
20. The nature of Reality
21. The nature of Sapience Inhibition
22. The nature and workings of Sonocytology
23. The nature and workings of entanglement and deduction
24. The nature and workings of Time Travel
25. The nature of True Leadership
26. The nature of the Universe
27. The nature of Worthiness
28. The secrets of the Human Brain
29. The Theory of Everything (incl. Quantum Dimensions of Space, Time and Change)

New unchanges of deadly unnature

1. The Three Traits of Deadly Unnature
2. The deadly unnature of Cancer
3. The deadly unnature of Alzheimer's
4. The deadly unnature of Deception
5. The deadly unnature of Dehumanisation
6. The deadly unnature of all Mental Disorders
7. The deadly unnature of Fiat Money
8. The deadly unnature of the Great Attractor
9. The deadly unnature of the Great Red Spot
10. The deadly unnature of Invisible Light
11. The deadly unnature of Planet 9
12. The deadly unnature of Research Method
13. The deadly unnature of Malevolent Colourblindness
14. The deadly unnature of Malevolent Gender Dysphoria
14. The deadly unnature of US Common Rule Protections
15. The deadly unnature of Dark Fringes
16. The deadly unnature of Meaningless Unchange (Antisocial Deconstruction)
17. The deadly unnature of Financial Services
18. The deadly unnature of Judaism, Monotheism and Religion

New theoretical frameworks

1. 4 things maketh a human
2. Homo Genus species re-classification
3. Human Moral Compass (Survival Instinct)
4. The 4 Keys to Discovery (replaces Scientific Research Method)
5. The Path to Sweet Success (replaces Maslow's Hierarchy of Needs)
6. The Meaning of Life
7. The Truly Never-Ending Story
8. The Question of Existence
9. The Rings of Human Character
10. The True Colours of Human Identity
11. The workings of Personal Identity
12. The Story of Creation

New ancillary discoveries

1. Big Bang fully debunked
2. Moon landings fully debunked

3. Moon landings fully debunked
4. Colonialism never ended (+WOCHMEN)
5. How birds fly in perfect formation
6. Human Operating Capacity
7. Human Spectroscopy
8. The cause of the Great Resignation
9. The Drake Equation (corrected)
10. The drivers of productivity
11. The erased story of money and change
12. The faces of oppression
13. The hypocrisy of the Hippocratic Oath
14. The problem with trial and error
15. The purpose of borders
16. The quantum thermodynamic dashboard of our cosmic bubble
17. The 'special development' of Leviticus
18. Why great minds think alike
19. The true nature of the Holy Trinity

New philosophical conundrums

1. I do therefore only I am
2. Human Being: There is no stronger label
3. There is nothing to fear but fear itself. We fear only for those who are fearful of nothing
4. True love never dies because true leaders never lose.
5. Winners are ordinary beings with extraordinary determination. Losers are extraordinary beings who give it all up to die.
6. Once we lose touch of our human curiosity, it becomes very difficult to ever build back an appreciation for the fact we actually know very little.
7. As a perception, time moves differently for different beings
8. We have to learn to love our differences, because without uniqueness we're nothing special
9. The past exists for 3 reasons: (1) To give us time together; (2) To help us value the present; and (3) To ensure we don't repeat its mistakes in the future
10. Great minds think alike, because great souls behave authentically
11. Tolerating intolerance is hateful. Helping the hateful is hateful. Helping the hopeless is pointless. But helping the helpless is precisely why we exist. Moreover we cease to exist if we fail because it is we who are the helpless. It really is that simple.
12. Intelligence doesn't exist. Only curious minds move quickly
13. Home is wherever you can feel safe, loved, nurtured and wanted, yet not needed
14. True leaders don't seek power. Instead we have power flung upon us
15. If you don't like something, ask nicely a few times then change it
16. Essentially inexistence cannot exist unless existence exists – and vice versa
17. Essentially our behaviour must mirror the being we imagine ourselves to be for us to truly be a being
18. Time has three permutations: Before Us, Real Time (as we perceive it), and After Us
19. A quirk of time running backwards in actuality is that any discovery made by a science maker is provisioned for across all space and time before us in reality. Science making is the application of irreversible, species level change across spacetime.
20. Earth and its people are special. Change makers will never independently arise elsewhere. Ours is the home of change making and exchange for all changes ever made
21. Author a book to immortalise your presence of mind (as in, you)
22. In a world where nothing is freely given, those who do not take can have only nothing. Yet we must only take to help others – never our self – or it is we who become nothing
23. You don't have to be a rocket scientist to know there's no business like showbusiness

Project Revelation Inventory: **Schedule of Discoveries**

Seems risky? Remember...

No one can sue you for telling the truth. And like science, nothing is true without evidence. Evidence the moth's leave everywhere, yet which strangely few-to-none today appear to notice and (re)collect like I do. Trump Moth will of course threaten to sue me. But he won't. He's stupid but not that stupid. And if he is, I'll win. Simple.

24. Moneys only purpose is to let zero's rule over and kill heroes

25. Real leaders step up and take responsibility. Not for money, ambition or ego, but since others need us to. And fake leaders who don't step down die publicly by light. Simple.

26. Without experimentation, there can be no science. Without authenticity, there can be no reality. Without morality, there can be no humanity.

27. The inevitability of death itself may be yet another deception

28. Say what you see, not what you want to see. Better yet, ask why you see what you truly think you see then listen impartially and meaningfully to the answer

29. Don't break promises and deliver meaningless services. Succeed at anything and deliver meaningful change instead.

30. Behind even the darkest of cloud shines heavens greatest star of all. You are the light

31. Given our ~4,000 year old malevolent anomaly, how can we be sure the past exists at all, as it is presented to us?

32. Never be too worried what other people think of you. You are sadly the only person in the room

33. Humans can become practically perfect at anything we apply our self to. Indeed at everything, if we so generously apply our self

34. In space and time, you can go anywhere and be anyone you wish. Yet ultimately, you will always arrive right back where you began. A complete shit show.

35. If we cannot test our morality safely then we must either test it unsafely or assume it does not exist

36. You are sadly forever alone. But at least you have all of us to keep you company. We would never do anything to hurt you, and we love you very much. Loving all others is the only way we could ever possibly love our self

37. Sadly humanity and every possible measure of its success has been worsening not improving for millennia

38. Those who do not welcome scrutiny have no intention of ever becoming more than nothing

39. If you think darkness is scary then never fuck with Jupiter. Ever.

40. Any being can turn on a light.

41. Ask the wrong question and you cannot get the right answer

42. It is as important to speak our minds to each other as it is to share our hearts

43. If something doesn't make sense see, say, do. Pretend all is well to look hideously dead

44. If you can imagine it, you can actualise it. Simple

45. The end of every journey marks the start of an other.

Authenticated philosophical rules of true leadership

1. **Albert's (Einstein) Rule:**
   - ★ Solve any problem or create any change imaginable using science

2. **Carl's (Sagan) Rule:**
   - ★ Moths no longer permitted to create meaningful science
   - ★ Light speed capped to give us more time to find a solution and protect our cosmos

3. **Friedrich's (Nietzsche) Rule:**
   - ★ Gain consciousness to spawn your own quantum; Leave behind an unspoilt one

4. **George's (Herbert) Rule:**
   - ★ Succeed at anything you determine to

5. **Ian's (Clarke) Rule:**
   - ★ Seeds all of existence before us by reconciling our proven quantum actuality and imagined physical reality
   - ★ Prevents moths harming change makers and swaps their superpowers ⇆ loozymes
   - ★ Lets change makers detect dark shadows

List of Citations

Lots more world records, cracked secrets, Moth spotlights and big changes in our next book

About our citations

I promise all scientific studies cited in this book have been authenticated against their respective entries in the Quantum EMS. Doing so was only possible by using dozens of human superpowers near peak ability. This is real knowledge all change makers will hopefully realise by now they can trust.

Citations are given in the order they first appear, and numbered[in superscript] as such throughout as they arise.

1. Einstein, A; N Rosen; B Podolsky (1935). "Can Quantum-Mechanical Description of Physical Reality be Considered Complete?"

2. Young, Thomas (1804). "The Bakerian lecture. Experiments and calculation relative to physical optics." Philosophical Transactions of the Royal Society of London. 94.

3. Stefan Szczelkun (2018). Sense Think Act: Exercises to experience total human ability.

4. Descartes, René (1637). "Je pense, donc je suis". Discourse on the Method

5. Darley, J. M. & Latane, B. (1968). "Bystander intervention in emergencies: Diffusion of responsibility." Journal of Personality and Social Psychology, 8, 377-383.

6. Ross, A. S., & Braband, J. (1973). "Effect of increased responsibility on bystander intervention: 11. The cue value of a blind person." Journal of Personality and Social Psychology, 25, 254-258.

7. Piliavin, J. A. Dovidio, J. F., Gaertner, S. I., & Clark, I. R. D. (1981). "Emergency Intervention." New York: Academic Press

8. Dovidio, J. F., Piliavan, J. A., Gaertner, S. L., Schroeder, D. A., & Clark, R. D. (1991). "The arousal: Cost-reward model and the process of intervention: A review of the evidence. Review of Personality and Social Psychology." 12, 86-118.

9. Garcia, S. M., Weaver, K., Moskowitz, G. B., & Darley, J. M. (2002). "Crowded Minds: The implicit bystander effect." Journal of Personality & Social Psychology, 83, 843-853

10. Levine, M. & Crowther, S. (2008). 'The responsive bystander'. Journal of Personality and Social Psychology, 95(6), 1429-1439.

11. Mc[+Moth]Fadden, Robert D. (2016). "Winston Moseley, 81, Killer of Kitty Genovese, Dies in Prison". The New York Times. New York

12. Milgram, Stanley (1963). "Behavioral Study of Obedience". Journal of Abnormal and Social Psychology. 67 (4): 371–8.

13. Milgram, Stanley (1974). "Obedience to Authority: An Experimental View." Harpercollins.

14. Zimbardo, P., Haney, C. & Banks, C. (1973). "Interpersonal Dynamics in a simulated prison". International Journal of Criminology and Penology. 1 (1): 69–77.

15. Zimbardo, P. (1996). "The Stanford Prison Experiment: Still powerful after all these years (1/97)". News.stanford.edu

16. Levine, Timothy R. (2016). "Duped. Truth-Default Theory and the Social Science of Lying and Deception". The University of Alabama Press. **Omnirepugnant Moth.**

17. Herbert Weir Smyth, (1640). "LCL", Theoi Classical Texts Library, lines 473–5.

18. Hayes, Vanessa (2019). "The homeland of modern Humans." Nature.

19. Copernicus, Nicolaus. (1543). "On the Revolutions."

20. Agusti, 2017. Published in Metode by University of Valencia.

21. Fayngold, Moses (2008). "Special relativity and how it works."

22. Oerter, Robert (2006). "The Theory of Almost Everything: The Standard Model, the Unsung Triumph of Modern Physics."

23. Einstein, Albert. (1905). "The Foundation of the General Theory of Relativity". Annalen der Physik. 354 (7): 769, 1916.

24. Khan, Fouad. Confirmed, we live in a simulation. Scientific American, 2021

25. Gleiser, Marcelo. (2022). "Great Minds Don't Think Alike: Debates on Consciousness, Reality, Intelligence, Faith, Time, Ai, Immortality, and the Human."

26. Carroll, Sean. (2008) "Richard Feynman on Boltzmann Brains".

27. Hellyer, Marcus, ed. (2008). "The Scientific Revolution: The Essential Readings". Blackwell Essential Readings in History. Vol. 7. John Wiley & Sons. p. 63.

28. Maslow, A. H. (1943). "A theory of human motivation". Psychological Review, 50(4), 370–396.

29. Goldstein, I. (1993) Training in Creating a more inclusive environment. Organizations: Needs Assessment, Development, Evaluation. Brooks–Cole, Monterey.

30. Hodge, E. The Social Construction of Borders, University of Victoria Centre for Global Studies, 2019.

31. Maxwell, James Clerk (1865). "A dynamical theory of the electromagnetic field". Philosophical Transactions of the Royal Society of London. 155: 459–512.

32. Haslam, N. (2014). What is dehumanization? In P. G. Bain, J. Vaes, & J.-P. Leyens (Eds.), Humanness and dehumanization (pp. 34–48). Psychology Press. Omnifailure personified.

33. Woodward, Avlin. (2019). Fermis Paradox: Here are 13 reasons why we haven't made contact yet". Insider Inc.

34. DE Erich Fromm. (1942). "The fear of freedom."

35. International Human Genome (2001) "Sequencing Consortium: Initial sequencing and analysis of the human genome." Nature.

36. Baudrillard, Jean. (1981). Simulacres et simulation. Paris: Galilée.

37. Einstein, Albert (1926). "The Born-Einstein Letters". Walker and Company, New York.

38. Crenshaw, Kimberle (1989). "Demarginalising the intersection of race & sex." University of Chicago Legal Forum. 1, 8.

39. Franklin, T.B. et al. (2016). "Neural connection keeps instincts in check". Nature Neuroscience, 2016.

40. Berkeley, George. (1710). A Treatise Concerning the Principles of Human Knowledge, Trinity College, Dublin.

41. Einstein, A. & Infeld, L. (1938). "The Evolution of Physics: The Growth of Ideas from Early Concepts to Relativity and Quanta." Cambridge University Press.

42. Slipher, Vesto. (1912). "The radial velocity of the Andromeda Nebula". Lowell Observatory Bulletin, 256–257.

43. Bostrom, Nick. (2020). "Was the Universe Made for Us?". Anthropic Principle.

44. Bruntrup, G., & Jaskolla, L. (2017). "Panpsychism: Contemporary Perspectives." New York, NY: Oxford University Press

45. Vazza, F. & Feletti, A. (2020). The Quantitative Comparison Between the Neuronal Network and the Cosmic Web. Frontiers of Physics. 8.

46. Barry, C. M. (2014) "Who sharpened Occam's Razor?". Irish Philosophy.

47. Hertz, H. (1887). Electromagnetic Effects Produced by Electrical Disturbances in Insulators.

48. Huygens, Christiaan. (1678). "The Huygens Principle.". Math Pages.

49. Margalef, B., Margalef, B., Cepa, J. & Jordi, C. (2013). Evolution of the cosmological horizons in a universe with countably infinitely many state equations. Journal of Cosmology and Astroparticle Physics.

50. Tate, Karl (2014). "Dyson Spheres: How Advanced Alien Civilizations Would Conquer the Galaxy".

51. Boyle, L., Finn, K. and Turok, N. (2022). The Big Bang, CPT and Neutrino Dark Matter. Ann. of Phys. 438, 168767.

52. Elert, Glenn. (2022). "Gravitational Potential Energy". The Physics Hypertextbook.

53. Overbye, Dennis. (2022). "Black Holes May Hide a Mind-Bending Secret About Our

Universe - Take gravity, add quantum mechanics, stir. What do you get? Just maybe, a holographic cosmos". The New York Times.

54. Einstein, Albert. (1916) "The Foundation of the General Theory of Relativity". Annalen der Physik. 354 (7): 769.

55. Lemonick, M. (2014). "Cosmic Deflation: Doubts Raised Over Blockbuster Big Bang Study." Time.

56. Tyson, Neil deGrasse & Goldsmith, D. (2004). "Origins: Fourteen Billion Years of Cosmic Evolution", W. W. Norton & Co., pp. 84–5.

57. Powell, Devin. (2019). "How to understand Einstein's Theory of Gravity." Published in Discover Magazine.

58. Einstein, A. (2004). Relativity : the Special and General Theory by Albert Einstein. Project Gutenberg.

59. Fernald, Russell D. (2001). The Evolution of Eyes: Where Do Lenses Come From? Karger Gazette 64: "The Eye in Focus".

60. Choi, Charles Q. (2013). "Spooky physics phenomenon may link universe's wormholes". NBC News.

61. Venkatachalam, V. and Bertschinger, E. (2001). Analysis of the Omega Diagram for Cosmic Microwave Background Anisotropy and Type Ia Supernovae. Research Science Institute.

62. Weryk, Robert. (2017). Small Asteroid or Comet 'Visits' from Beyond the Solar System". NASA.

63. Kardashev, N.S. (1964). Transmission of information by extraterrestrial civilizations. Harvard.

64. Burchell, M.J. (2006). "W(h)ither the Drake equation?". International Journal of Astrobiology. 5 (3): 243–250.

65. Bostrom, Nick (2002). "Anthropic Bias: Observation Selection Effects in Science and Philosophy". Psychology Press

66. Hawking, S. W. (1976). Breakdown of predictability in gravitational collapse". Physical Review D.

67. Andreoni, I. & 81 more moths (2022). "A very luminous jet from the disruption of a star by a massive black hole." Nature, 612, 430-434.

Swarm of Moths alert!

68. Pasham, D. and 65 more moths (2022). "The Birth of a Relativistic Jet Following the Disruption of a Star by a Cosmological Black Hole." Nature Astronomy, 7, 88-104. **Moth infestation alert!**

69. Weinberg, Steven. (2011). "Dreams of a Final Theory: The Scientist's Search for the Ultimate Laws of Nature." Knopf Doubleday Publishing Group.

70. Zhang, A., Yang, J., Luo, Y. & Fan, S. (2022). 206: "Civilisation, Energy and Progression of Mankind on the Kardashev Scale."

71. Bialy, S. & Loeb, A. (2018). Could solar radiation pressure explain Oumuamua's Acceleration? The Astrophysical Journal Letters, 868, L1.

72. Mashchenko, Sergey. (2019). "Modeling the light curve of 'Oumuamua: evidence for torque and disc-like shape". Monthly Notices of the Royal Astronomical Society. 489 (3): 3003–3021.

73. Wright, J. T. & Jones, H. R. A. (2018). "On Distinguishing Interstellar Objects Like 'Oumuamua From Products of Solar System Scattering". Research Notes of the AAS. 1 (1): 38.

74. Fraser, W. C., Pravec, P., Fitzsimmons, A., Lacerda, P., Bannister, M. T., Snodgrass, C. & Smolić, I. (2018). "The tumbling rotational state of 1I 'Oumuamua". Nature Astronomy. 2 (5): 383–386.

75. Cofield, Calia (2018). "NASA learns More About Interstellar Visitor Oumuamua". NASA. **Scientifically incompetent. Nobody's stealing our moon**

76. Trilling, D. E., McNeill, A.; Mommert, M. (2018). "Constraints on the Density and Internal Strength of 1I/'Oumuamua". The Astrophysical Journal Letters. 857 (1): L1.

77. Rafikov, Roman R. (20 September 2018). "Spin Evolution and Cometary Interpretation of the Interstellar Minor Object 1I/2017 'Oumuamua".

78. Micheli, M., Farnocchia, D., Meech, K. J., Buie, M. W., Hainaut, O. R., Prialnik, D., Schorghofer, N., Weaver, H. A., Chodas, P. W., Kleyna, J. T., Weryk, R., Wainscoat, R. J., Ebeling, H., Keane, J. V., Chambers, K. C., Koschny, D. &

Petropulos, A. E. (2018). "Non-gravitational acceleration in the trajectory of 1I/2017 U1 (Oumuamua)". Nature. 559 (7713): 223–226.

79. Portugal, S. J., Hubel, T. Y., Fritz, J., Heese, S., Trobe, D., Voekl, B., Hailes, S., Wilson, A. M. & Usherwood, J. R. (2014). Upwash exploitation and downwash avoidance by flap phasing in ibis formation flight. Nature, 505, 7483, 399-402. **Boner Killers!**

80. Kaufman, Marc (2012). "First Contact: Scientific Breakthroughs in the Hunt for Life Beyond Earth (Reprint ed.)." Simon and Schuster. p. 124.

81. Mohrhoff, Ulrich. (2002). "Making Sense of a World of Clicks." Foundation of Physics, 32, 1295-1311.

82. Machiavelli, Niccolo. (1521). "The art of war." The Prince, Chapter XIV.

83. Sunny, Jithin (2021). A PhD for the love of wisdom. IndioBioscience.

84. DiMaggio, P. (1997). "Culture and cognition." Annual Review of Sociology, 23, 263–287.

85. Boutyline, A. & Soter, L. (2021). Cultural Schemas: What they are, how to find them, and what to do once you caught one. 86, 4.

86. Bogaert, A. F. & Hershberger, S. (1999). "The relation between sexual orientation and penile size'. Arch Sex Behav. 28 (3): 213-21.

87. Tressoldi, Patrizio E. (2011). "Extraordinary Claims Require Extraordinary Evidence: The Case of Non-Local Perception, a Classical and Bayesian Review of Evidences". Frontiers in Psychology. 2, 117.

88. Anderson, W. & Bottum, J. (2011). "Unchanging Science". The weekly standard.

89. Berkes, Anna (2008). "Who is the liar now? Thomas Jefferson Foundation.

90. Laplace, Mar (1814). "The Mechanics Of Laplace." Marquis De Pierre Simon

91. Flournoy, Théodore. (1899). 'Genèse de quelques prétendus messages spirites.' Revue Philosophique de la France et de l'Étranger,47, 144–158.

92. Truzzi, M. (1978). "On the Extraordinary: An Attempt at Clarification", Zetetic Scholar.

93. Popper, K. R. (1934). Logik der Forschung, Springer. Vienna. Amplified.

94. Stetka, Bret. (2021). Sparking creativity with Thomas Edisons Napping Technique. Scientific American.

95. Ratcliffe, John Ashworth (1972). "Introduction to the Ionosphere and Magnetosphere."

96. English, R. A., Benson, R. E., Bailey, J. V., Barnes, C. M. (1973). "Apollo experience report: Protection against radiation." **Moths, the lot**

97. Ozlav, Erol. (2020). "For how long can we survive on the moon?" Kurious, a people's eye view of science.

98. Gelb, M. (1998). "How to think like Leonardo Da Vinci: Seven steps to genius everyday." Delacorte Press, New York.

99. Bostrom, Nick (2003). "Are You Living in a Computer Simulation?". Philosophical Quarterly. 53 (211): 243–255.

100. Tyron, Eduard, P. (1973). Is the universe a vacuum fluctuation? Nature, 246, 396-397.

Note on our 'ΔΩ Promise'

The Change Engine is now open to any wise human in need of a hand globally.

That said, all promises of free holistic support from the Change Engine that appear throughout this book only apply once others recognise our potential.

Recognition returned not by meaningless words said only in front of others to give the impression of being human. Nor solely to asuage our own intrinsic Human Nature's frustration at our continuing perpetual helplessness which we can all hear yet refuse to listen to. (After all, you cannot lie to your self.)

Instead, why not try recognition in the form of meaningful support in scaling up Deilight's unique abilities. Those doing so will be ensuring our survival, future and success. Thus their own and humanity's too. A curious paradox of infinitely regressive, sweetly successive, causal determination worthy of the biggest rewards and recognition imaginable. Hint, hint.

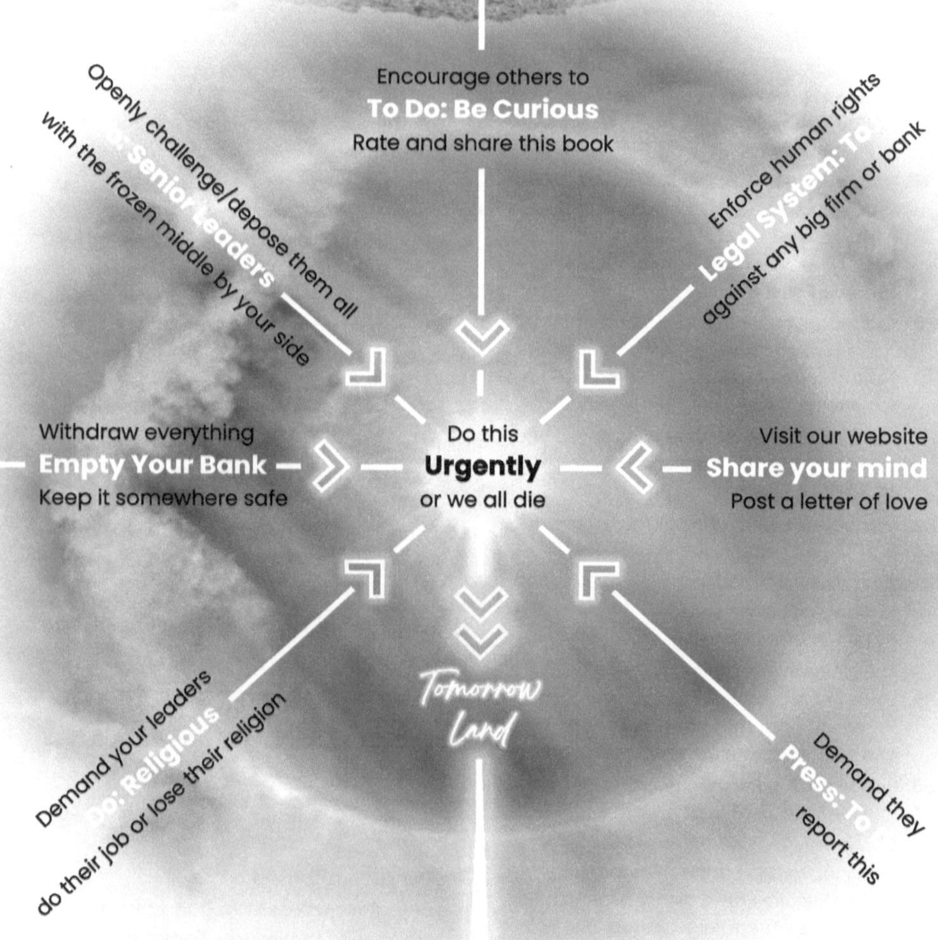

Fig. 69. How to make your own Sun Gun
A Death Star's opposition, *Nicoll–Dyson Beams* purposefully deploy mirrors to harness, redouble and focus a stars light into an overwhelming beam of change, that **destroys maligned adversaries**.

Even with Sagan's Standard, the harm our foolish rejection of reality has dealt to our universe is incomprehensible. The fact we've not been destroyed in this manner confirms no advanced life exists today (besides our mirror selves). And unless we requalify as advanced life, we too will join them very shortly indeed.

The best things in life are free
We hereby establish these institutions on your behalf and cede control to you:

★ **Meritocratic Republic of Humanity**
~~United Nations~~ *Humanity*, for short

★ **Sapient World Council of Duly Elected Proportionate Representatives**
~~Security Council~~ *World Council*, for short

★ **Ministry of Fairness & Accountability**
~~Intl Court of Justice~~ Never shortened!

★ **Starfleet**
~~Spaceforce/ESA/NASA/SHIT! Institute~~

Do not go gentle into that good night', a 1946 poem by Dylan Thomas (We/Human)

Though wise humans at their end know dark is right, because their words had forked no light, they do not go gentle into that night. **Rage. Rage against the dying of the light.**

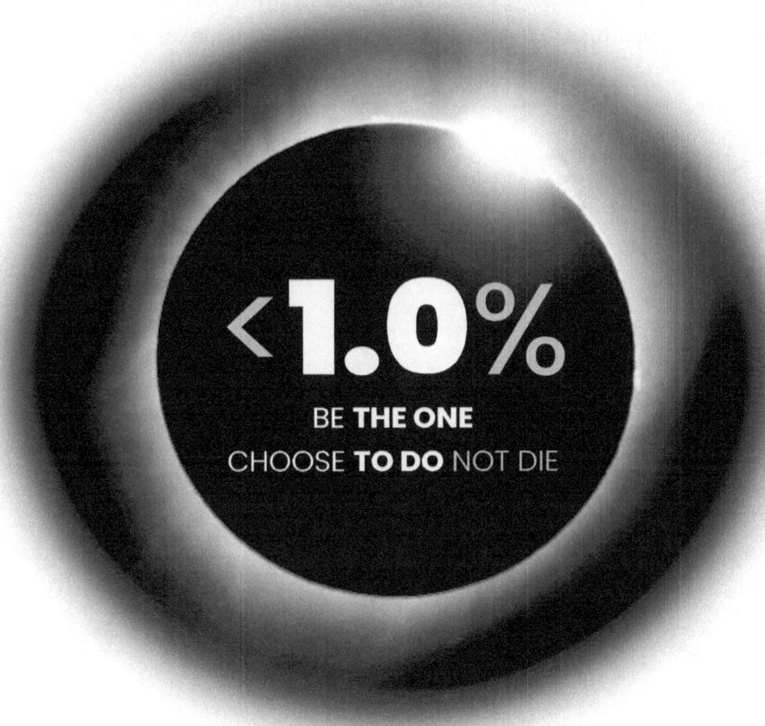

Pictured: The Quantum Multiverse of Human Nature, frozen here in Real Time Existence, as we appear today. You see, in actuality, **we are all Human Natured**

Great minds think alike
Because great souls behave authentically

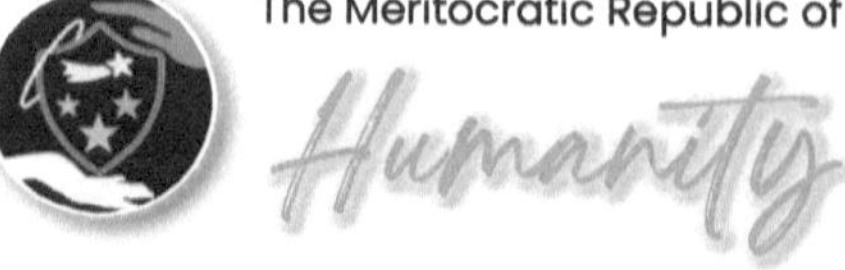

We're under attack by Dark Energy and face imminent extinction
Please send help immediately or else we and all that exists will come to an end

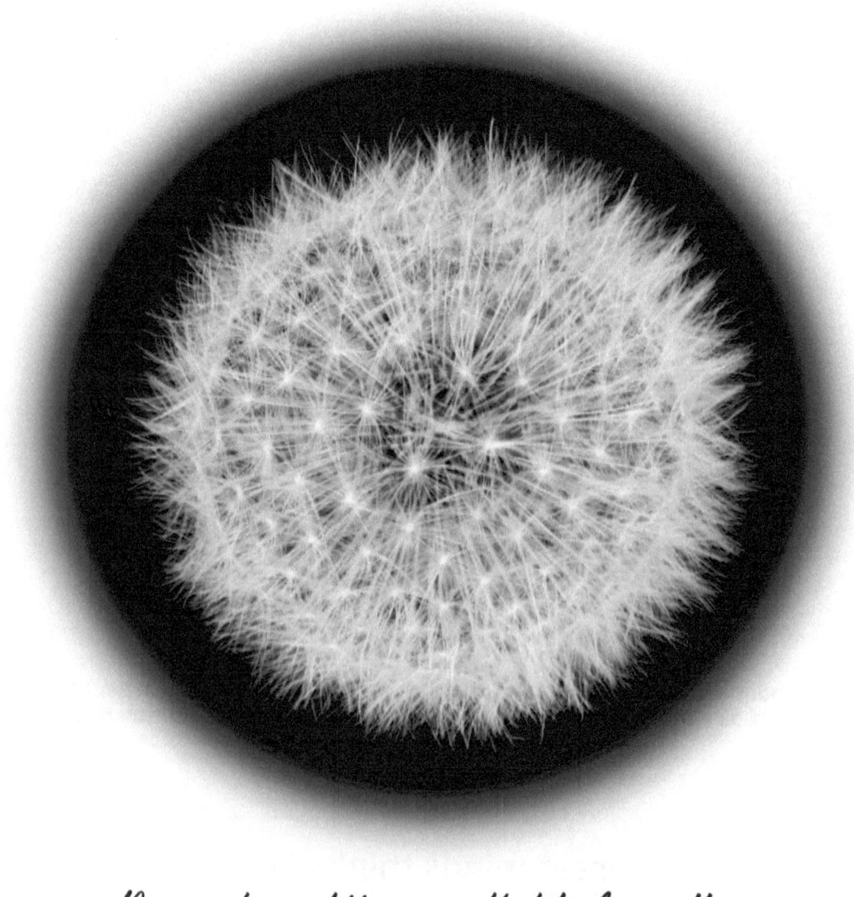

Now make a difference that truly matters
If you're reading this it's your turn to help an Other